ELECTRONIC ENGINEERING SYSTEMS SERIES

Series Editor
J. K. FIDLER
University of York

KNOWLEDGE-BASED SYSTEMS FOR ENGINEERS AND SCIENTISTS
Adrian A. Hopgood, The Open University

OPTIMAL AND ADAPTIVE SIGNAL PROCESSING
Peter M. Clarkson, Illinois Institute of Technology

CIRCUIT SIMULATION METHODS AND ALGORITHMS
Jan Ogrodzki, Warsaw University of Technology

OPTICAL ENGINEERING
John Watson, University of Aberdeen

PRINCIPLES AND TECHNIQUES OF ELECTROMAGNETIC COMPATIBILITY
Christos Christopoulos, University of Nottingham

ACTIVE RC AND SWITCHED-CAPACITOR FILTER DESIGN
T. Deliyannis and I. Haritantis
University of Patras

DESIGN AUTOMATION OF INTEGRATED CIRCUITS
K.G. Nichols, University of Southampton

OPTIMAL

and

ADAPTIVE SIGNAL

PROCESSING

PETER M. CLARKSON

Associate Professor
Electrical and Computer Engineering Department
Illinois Institute of Technology
Chicago, Illinois

CRC Press

Boca Raton   Ann Arbor   London   Tokyo

Library of Congress Cataloging-in-Publication Data

Catolog record is available from the Library of Congress.

© 1993 by CRC Press, Inc.

International Standard Book Number 0-8493-8609-8 (hardcover)
International Standard Book Number 0-8493-8627-6 (softcover)
Printed in the United States 1 2 3 4 5 6 7 8 9 0
Printed on acid-free paper

Preface

This book was born from a short course on "Applications of Adaptive Filtering to Underwater Problems" taught by Professor Joe Hammond and myself. Between 1985 and 1988, the course ran successfully at the Institute of Sound and Vibration Research in Southampton, and at other locations on both sides of the Atlantic Ocean. Later, at the Illinois Institute of Technology I became involved in teaching graduate courses in digital signal processing. The emphasis on underwater topics diminished, but the ideas of optimal and adaptive processing remained central to these courses. From the time Joe and I first planned our short course, we were aware that there was no text available that covered all, or even most of the material we had in mind. There were certainly books on adaptive signal processing, and some of these were excellent, but there were none which gave the subject material, and more importantly the balance of theory, application, and example that we were looking for. As a result we wrote a set of notes for the course. Later, teaching at IIT, I was faced with a similar problem, and so starting with the original short course material, I created a set of course notes for a two-semester graduate level digital signal processing sequence. Several semesters and many iterations later this text has resulted.

In creating the sequence I tried to cover many of the most important topics in signal processing, using as central themes the ideas of *optimality* and *adaptation*. Being aimed at a graduate audience, the book is intended to pick-up where a first course in digital signal processing leaves off. I have strived for readability and continuity throughout, and the book should be easily accessible to graduate engineering students. It is assumed that the student has been exposed to the fundamentals of sampling, discrete Fourier transforms, z transforms, and digital filter design. Some exposure to random signals is also helpful although the essentials are reviewed in Chapter 2. The subject matter proper begins in Chapter 3. The ideas of optimality in parameter estimation are introduced, and related to techniques for the estimation of system models, recovery of signals from noisy measurements, and system identification. Maximum Likelihood, Maximum Aposteriori, and Minimum Mean-Squared Error (MMSE) methods are reviewed, and linear MMSE estimation is considered in detail. Applications of linear MMSE estimation to prediction and to inverse filter design are used as illustrations.

The coverage of adaptation is intended to introduce the reader to a broad range of issues in adaptive processing. We begin in Chapter 4 with an intuitive derivation of the simple Least Mean Squares (LMS) algorithm, followed by an in-depth review of the properties of that algorithm. The mathematics underlying the analysis of adaptive filter designs can be approached at various levels. Many adaptive algorithms have a simple structure, but the analysis is sufficiently complex as to have defied complete solution for 30 years. Some of the most recent research results rely on very sophisticated mathematical techniques. Progress in the analysis of such algorithms has been made by restricting consideration to relatively simple idealized system inputs and by making a number of simplifying assumptions. These simplifications produce an analysis that is less precise, but still gives meaningful predictions of system performance. The presentation in this text largely follows this approach, and consequently mathematically there should be little here that is unfamiliar to most graduate engineering students.

Before considering more advanced algorithmic issues, we introduce the reader to some of the applications of adaptive filters. We examine Adaptive Noise Cancellation (ANC) in its various forms, spectral line enhancement, time-delay estimation, echo cancellation, and Adaptive Differential Pulse Code Modulation (ADPCM) systems, all using the LMS algorithm. Having thus motivated the use of adaptive filters we proceed in Chapter 6 to examine alternative algorithms, and implementations for adaptive filters. These include simple ad-hoc variants of the LMS algorithm, but also include coverage of the more powerful Least-Squares family. Discussion of relatively recent topics such as nonlinear adaptive filters is also included. Alternative forms such as IIR filters and lattice structures are discussed, as are frequency domain implementations of transversal filters.

In the last segment of the text (Chapters 7 and 8), we expand our ideas about optimal and adaptive systems to the spatial and frequency domains, devoting a Chapter each to the related problems of spectral estimation and array processing. In both cases we review elementary concepts and classical methods before proceeding to develop optimal processors, constrained optimal processors and adaptive implementations.

Examples and case studies are given throughout the text, and these are supplemented by a very diverse mix of problems. Some of the questions cover revision of fundamental digital signal processing material, some give numerical examples on topics developed in the text, some deal with proofs of results given in the text. Finally, some questions deal with related subject material drawn from recent research papers.

It would be remiss of me not to thank the people who have in some way contributed to this text. At the Institute of Sound and Vibration Research I would like to thank all the faculty, staff and students of the signal processing group. In particular, Joe Hammond, now Director of the Institute, who introduced me to signal processing. Joe was a superb advisor and an inspiration to all his graduate students. I must also thank Rob Harrison and John Mourjopoulos, my fellow members of Data Anal — an institution that was definitely ahead of its time. The principles we learned in formulating and conducting the MFOR contract series have served me well, and were diligently applied in the preparation of this manuscript. At IIT I must thank department chairmen Henry Stark, and Joe LoCicero who hired me, both of who have helped me a great deal. I also appreciate the support shown to me by IIT President Lew Collens and Provost Darsh Wasan.

In the preparation of the manuscript I am indebted to Don Chmielewski who worked on many of the block diagrams, and to Paul Hybert who helped me in the seemingly endless struggle with the word processor. My own graduate students, past and present, provided the inspiration for many of the problems and numerical examples in the manuscript. Of these, I must especially thank Miroslav Dokic who directly contributed to several of the simulation studies described. I would also like to thank Navin Sullivan of CRC, and his anonymous reviewers, who made numerous useful suggestions. Finally, all would be lost without Linda, who put up with the whole thing.

Peter Clarkson
Chicago, Illinois

The Author

Peter Clarkson was born in Altrincham, England in 1958. He obtained the B.Sc., with first class honors, from Portsmouth Polytechnic, England, in 1980, and the Ph.D. degree from the University of Southampton, England, in 1984. From 1980 to 1987 he was associated with the Institute of Sound and Vibration Research at the University of Southampton; from 1980-1984 as a research assistant, from 1984-1985 as a post-doctoral fellow of the U.K. Science and Engineering Research Council, and from 1985-1987 as a research lecturer funded by the U.K. Admiralty Research Establishment. Since 1987 he has been with the Department of Electrical and Computer Engineering at the Illinois Institute of Technology, Chicago, where he is currently Associate Chairman and an Associate Professor.

This book is respectfully dedicated
to Dr. Demetrius G. Vakaleris

Contents

Contents

chapter one

Introduction

The primary objective of this text is to develop the ideas of *optimality* and *adaptation* in signal processing. As a definition, we may say that optimal signal processing is concerned with the design, analysis, and implementation of processing systems that extract information from sampled data in a manner that is 'best' or optimal in some sense. Defining optimality, formulating criteria for optimality, and developing signal processing algorithms corresponding to the various optimality criteria are some of the major themes of this text. We shall define optimal strategies for digital filter design, parameter estimation, spectral analysis, and for spatial filters (optimal arrays). These operations encompass the full range of digital signal processing activities.

The tools of optimal processing are the techniques of statistical signal analysis. In Chapter 2 we review these ideas, focusing on sampled signals, or time-series, and in Chapter 3 we develop the techniques of estimation theory for random signals. Estimation of system parameters and signal characteristics plays an important role in statistical signal analysis. We shall see that estimation is an imperfect process. Associated with the estimation of any parameter there will be errors. Our objective is to define procedures for the estimation of these parameters that are in some sense 'optimal'. Defining such procedures involves assigning a 'cost' to the estimation errors and then minimizing this cost. We discuss several powerful methods for optimal estimation: The methods of Maximum Likelihood (ML), Maximum A Posteriori (MAP) and Minimum Mean-Squared Error (MMSE) estimation. Some of these techniques may be familiar from the study of statistics and probability. We shall review these ideas from that perspective, but also try to set the techniques into a signal processing framework. Of the basic estimation strategies, the method of greatest practical value (largely because it has least reliance on prior knowledge) is the MMSE or least-squares approach. In MMSE estimation, the 'cost' of the procedure is the square of the error between the estimate and the desired parameter; we minimize the average of that squared error. This approach accounts for the majority of all

optimal processing methods, and is the single most important
theme of this book. In Chapter 3, we take this MMSE philosophy,
apply it to digital filter design, and illustrate its power through two
important, and as it turns out closely related problems; linear
prediction, and inverse filter design.

Prediction represents a basic problem in time-series analysis:
How do we use current and previous samples of a time-series to
predict the future behavior of that signal? In linear prediction we
use a linear combination of those previous values, and in MMSE
linear prediction we choose the coefficients of that weighting so as
to minimize the mean-squared prediction error. Linear prediction
is one of the most important topics in signal processing with
applications as diverse as speech processing and exploration
seismology. The subject occupies this central role in signal
processing theory essentially because in successfully predicting a
signal *we identify the mechanism underlying the generation of that
signal.*

In inverse filtering, the objective is to construct a filter that can
unravel (or deconvolve), the effects of a convolution of two signals.
We shall see that when dealing with realizable Finite Impulse
Response (FIR) filters, it is generally not possible to precisely
achieve this inversion. Instead, we may construct an inversion
that approximately deconvolves the two signals. The method of
MMSE provides a convenient and powerful framework for
optimizing the design of this inverse. The application also
demonstrates the fact that the method is equally applicable to
both deterministic and random signals; for deterministic signals, we
simply minimize the sum of squared errors in the design.

In Chapter 4, we turn our attention to our second major
objective; the study of adaptation in signal processing. Adaptive
signal processing is concerned with the design, analysis and
implementation of systems whose structure changes in response to
the incoming data. That is, adaptive processing deals with the class
of *data adaptive* techniques. There are two basic factors that
motivate adaptive processing. Firstly, we often need to analyze
data whose properties are unknown *a priori*. It is difficult to
construct a sensible processing strategy under these conditions. An
adaptive processing scheme iterates towards the required
processing strategy using each sample of data as it is measured.
The second motivating factor for adaptive schemes is the existence
of systems whose properties change with time. For such signals,
adaptive processing and specifically adaptive filtering, provides a

method whereby one may track the changes in the data and thus maintain a strategy which is consistent with the processing aims. Essentially, an adaptive filter is one whose structure can be adjusted in response to changing signal properties.

In Chapter 4, our introduction to adaptive processing is based on a single adaptive filtering algorithm known as the Least Mean Squares or LMS. This adaptive filter is a natural development of MMSE methods, and can be viewed as a procedure that iteratively adapts the structure of the filter towards the optimal MMSE solution, and tracks changes in the optimal solution. We can think of the MMSE cost function as a quadratic surface defined as a function of the values of the filter coefficients. Such a quadratic surface has as many dimensions as filter coefficients, but has a single minimum value corresponding to the MMSE solution for the filter. In LMS, the filter coefficients are iteratively updated, searching the surface for the minimum value, descending towards the minimum by taking steps proportional to the instantaneous gradient of the surface.

The LMS algorithm has a remarkably simple structure. However, the properties of adaptive filters, including LMS, are difficult to characterize analytically. In conventional filter design we consider stability of the system in terms of conditions for 'Bounded Input-Bounded Output' (BIBO). This BIBO stability is also important in adaptive filter design. However, in addition to this consideration, we must also consider the stability of the iterative update of the impulse response coefficients, and its convergence, or non-convergence to the MMSE solution. Adaptive filters driven by random data are themselves random. Even if an adaptive filter converges towards the MMSE solution, the filter coefficients will typically undergo fluctuations about that MMSE solution. In analyzing adaptive filters driven by random inputs we look for convergence *on average*. However, we must also consider the perturbations of the filter response coefficients about the average, and the impact that such fluctuations have on the overall performance of the system. In particular, the perturbations of the adaptive filter coefficients impact the mean-squared error achieved by the system, causing this to increase above the MMSE for the fixed optimal solution. In Chapter 5 we illustrate some of the applications for adaptive filtering by considering in detail the use of the LMS adaptive filter for noise cancellation, estimation of an unknown time-delay between two signals, enhancement of a noisy spectrum, cancellation of unwanted echos in a telecommunication system, and the use of adaptive filters for coding signals for efficient transmission.

In fact, the tremendous range of filter structures and adaptation laws leads to a bewildering range of algorithms for adaptive filtering. For much of this text we focus on the simple LMS. Not until Chapter 6, after we have discussed the FIR transversal implementation of the LMS algorithm in considerable detail, and examined a number of applications, do we venture to explore other possibilities. LMS is simple both in concept and implementation. Practically, the algorithm is known to be well-behaved in a broad range of signal environments. On the other hand, convergence is driven by the gradient of the quadratic mean-squared error surface. This produces non-uniformity in the convergence due to disparity in the magnitude of the gradient component corresponding to different axes of the mean-squared error surface. We consider alternative algorithms derived by modifying the LMS to achieve some specified objective; normalized algorithms for data with fluctuating power levels, algorithms designed to improve convergence rate, algorithms that can adapt successfully in the presence of extraneous noise, and computationally simplified algorithms. We also consider the more powerful Gauss-Newton or Least-Squares algorithms. These algorithms have been applied to signal processing problems more recently than LMS. They have the potential to provide more uniform and ultimately faster convergence. Against this, we shall see that Least-Squares algorithms require more computational power and are more prone to instability problems than LMS. Adaptive implementations of digital lattice structures also have the potential to reduce non-uniform convergence behavior, and thereby increase overall convergence speed. We examine gradient descent lattice algorithms which represent a good compromise having faster convergence than LMS, and computation midway between LMS and Least-Squares algorithms, with reasonably robust performance.[1]

We also illustrate Infinite Impulse Response (IIR) adaptive filters. In some applications, an FIR adaptive filter may require many coefficients to give acceptable performance. By incorporating poles into the filter, IIR adaptive filters have the

---

1. There are also lattice implementations of Least-Squares algorithms, though these are not considered here.

potential to give comparable performance with far fewer coefficients. Against this, stability and convergence are harder to ensure for the IIR form.

Frequency domain implementations of the transversal FIR filter are another useful adaptive form. In the frequency domain, the simple LMS algorithm can be implemented with reduced computation (by exploiting the speed of computation possible with the Fast Fourier Transform (FFT) algorithm), and with improved convergence properties. Finally, at the end of Chapter 6 we return to some of the applications described earlier, illustrating the use of the more powerful algorithms and structures developed.

In Chapter 7 we turn our attention to our third major objective; to expand our consideration of optimal and adaptive techniques into the frequency and spatial domains. Although these two subjects are addressed separately in Chapters 7 and 8, they are in fact closely related. Both are concerned with detection, estimation and classification of signals. In spectral analysis, these processing objectives are met by exploiting the frequency domain properties of the signals. Array signal processing is concerned with the application of signal processing techniques to the outputs from a spatially distributed array of sensors. Such arrays receive signals generated by propagating waves. In array processing, estimation and detection problems are solved using the mechanism of spatial discrimination. Thus, the techniques and the optimal forms for spectral estimators have close parallels in array processing with a duality between the spatial and frequency domains.

In spectral estimation, we review the classical (non-parametric) estimators before introducing the model based (parametric) methods corresponding to MMSE and ML optimizations. In array processing, we review the elements of delay-and-sum beamforming and control of the array response. We develop optimal solutions based on maximizing the output Signal-to-Noise Ratio (SNR), and on minimizing the mean-squared error. We also illustrate the relationships between these optimal forms and the corresponding spectral estimators. Finally, adaptive solutions for these optimal array processors are developed.

It is important to recognize at the outset that the optimal and adaptive processors that we shall consider have not become the cure-all processing strategy that some early researchers envisaged. They *have* proved very valuable in numerous applications in areas as diverse as telecommunications systems, sonar and radar systems, audio and acoustics, and biomedical engineering. As we shall see, these applications cover a diverse range of functions from areas of digital signal processing including spectral analysis, array processing, signal enhancement (and noise cancellation), equalization and control. Moreover, as low-cost real-time processors continue to become more powerful, so the list of applications continues to grow.

chapter two

Random Signal Analysis

2.1 Discrete Random Signals

2.1.1 Definitions and Notation

Many of the fundamental tools of digital signal processing; the rules of convolution, z and Fourier transforms, and digital filter designs, are usually first encountered in the framework of deterministic signal analysis. A deterministic signal may be defined as one which is exactly specified or predictable according to some mathematical model or rule. Many practically arising signals cannot be reasonably described as deterministic, however, having elements or components which defy precise mathematical analysis. Of course, it is tempting to pose questions as to whether such signals are truly random, or whether some hidden physical property or simply a more accurate measurement procedure would show this to be merely a case of apparent randomness. From our point of view these distinctions are irrelevant in the sense that given the data and the available information the signal exhibits randomness, apparent or otherwise. Once one accepts this point it becomes clear that most practically occurring signals are in some part random. In digital signal processing our concern is with sampled random signals. These sampled or discrete random signals are typically characterized as being infinite in length and as not possessing finite energy. Consequently, the summations of the z and Fourier transforms do not converge and we may say that the transforms do not exist. Such signals are characterized in probabilistic terms and the analysis is conducted using the tools of statistical analysis.

The results we present in this Chapter comprise a brief review of those aspects of random signal analysis which are important for the optimal and adaptive processing which follows. We shall be very selective in our coverage and the reader who seeks a broader perspective should consult one of the many excellent texts on this material. Of these may cite in particular, Papoulis [1],

Davenport [2] and Stark and Woods [3] for coverage of fundamental statistical concepts and random processes. Detailed discussion of time-series analysis, correlations and spectra may be found in the classic works of Box and Jenkins [4], Jenkins and Watts [5] and Priestley [6], and more recently in the text by Wei [7]. The material of this Chapter is drawn largely from these and other similar texts.

We define a **discrete random signal**, {**x(n)**}, as a sequence of indexed random variables assuming values:

$$\mathbf{x(0),x(1),...,x(i),...}$$

This sequence is discrete with respect to the sampling index n, which may represent time or some other physical variable. Such an ordered sequence is also known as a **time-series** and this term is generally used whether or not time is the independent variable [7]. We shall deal exclusively with sequences that are equally spaced with respect to the physical variable. The random variable associated with each sample is not specified by this definition and may be real or complex, and be of continuous, discrete or mixed types. We shall be principally concerned with real random signals and, except where stated, this form should be assumed. We further assume that each sample is represented with infinite precision so that the sampled random signal may be obtained from a continuous waveform via the mechanism of **ideal sampling**. In principle, each random sample may be drawn from a continuous, discrete or mixed density. We shall restrict our attention to continuous random variables. Thus we assume that the amplitude of each sample may take any of a continuous range of values. Unless stated otherwise we assume that the range of these values will be the entire real axis.

A particular sample $x(n)$ is characterized by a **cumulative density function (cdf)**

$$F_{\mathbf{x}}(x(n)) = p(\mathbf{x}(n) \leq x(n)) ,\qquad (2.1.1)$$

and by a **probability density function (pdf)**

$$f_{\mathbf{x}}(x(n)) = \frac{\partial F_{\mathbf{x}}(x(n))}{\partial x(n)} .\qquad (2.1.2)$$

Generally, where confusion is unlikely to occur we shall ignore the subscript and write simply

$$f_x(x(n))=f(x(n)) \qquad ; \quad F_x(x(n))=F(x(n)) \ .$$

Note that we pay lip service to the practice in the statistical literature of using bold face characters for random variables and italic for particular values of random variables. Once we have completed the definitions, however, in line with the usual signal processing practice, the bold typeface variables will cease and we shall refer to the random signal exclusively as $\{x(n)\}$ or simply as $x(n)$.

There is no particular reason to expect the probability functions associated with sample indices n_1 and n_2 to be the same and in general each sample will be associated with its own distinct probability density function. We can relate the random samples at different indices through definitions for bivariate and multivariate distributions. The bivariate cumulative density function is

$$F(x(n_1),x(n_2))=p(\mathbf{x}(n_1)\leq x(n_1),\mathbf{x}(n_2)\leq x(n_2)) \ , \qquad (2.1.3)$$

and the corresponding joint density function is

$$f(x(n_1),x(n_2))=\frac{\partial^2 F(x(n_1),x(n_2))}{\partial x(n_1)\partial x(n_2)} \ . \qquad (2.1.4)$$

More generally, the k-th order multivariate distribution is defined by

$$F(x(n_1),...,x(n_k))=p(\mathbf{x}(n_1)\leq x(n_1),...,\mathbf{x}(n_k)\leq x(n_k)) \ , \qquad (2.1.5)$$

and the joint density is

$$f(x(n_1),...,x(n_k))=\frac{\partial^k F(x(n_1),...,x(n_k))}{\partial x(n_1),...,\partial x(n_k)} \ . \qquad (2.1.6)$$

A complete probabilistic description of the random sequence $\{x(n)\}$ requires knowledge of $F(x(n_1),...,x(n_k))$ for all k and all $n_1,n_2,...,n_k$. More realistically, we usually have only a single sequence available; the result of a single experiment. Each experimental result represents a single **realization** from the infinite set, or **ensemble**, of possible results (see Figure 2.1.1). This is another useful way to think of the distinction between deterministic and random signals. A deterministic signal is **repeatable**. Each experiment produces the same result. By contrast the random signal produces a different result for each experiment. Each result is just a single member of the ensemble.

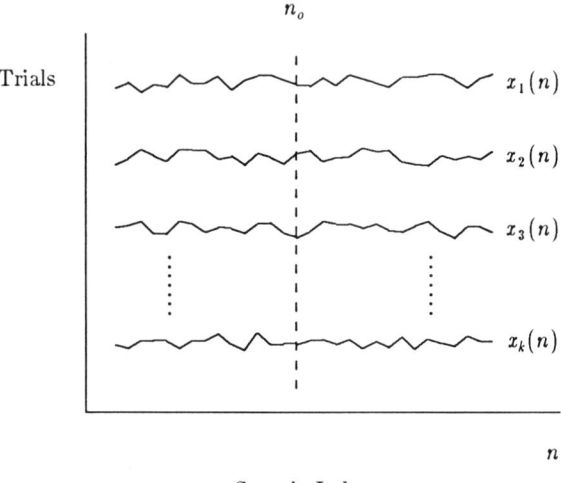

Figure 2.1.1 Realizations of a random signal $x(n)$. $x_1(n)$, $x_2(n)$,... are sample realizations. Ensemble averaging (for index n_0) is performed by averaging $x_1(n_0)$, $x_2(n_0)$....

2.1.2 Averages

By its very nature, the value of a random signal cannot be exactly predicted. We typically characterize such random signals in terms of **averages**. Since our definition of discrete random signals is in terms of random variables, the averages we require are obtained as extensions of the usual statistical definitions:[1]

a) Mean Value

The simplest average is the mean or expected value denoted μ_n. Formally, the expected value of random variable $x(n)$, for a

---

1. In line with earlier comments, all of the definitions given will assume continuous distributions for the sample amplitudes.

probability density $f(x(n))$, is

$$\mu_n = E\{x(n)\} = \int\limits_{-\infty}^{\infty} x(n)f(x(n))\,dx(n) \ . \tag{2.1.7}$$

Expectation can also be interpreted as an average value determined by repeating an experiment many times (the so-called 'relative frequency' interpretation). That is

$$\mu_n = \lim_{N\to\infty} \left[\frac{1}{N} \sum_{i=1}^{N} x_i(n) \right] , \tag{2.1.8}$$

where $x_i(n)$ is the result of the ith experiment at sample index n. The most important point is that the expected value is found by averaging *over the ensemble not over the sample index* (see Figure 2.1.1). Note that, in general

$$\mu_m \neq \mu_n \qquad ; \ m \neq n \ ,$$

since the distributions at m and n generally differ. Three important properties of expectation should be noted:

i) The expectation operator is linear. That is

$$E\{ax(m)+by(n)\} = aE\{x(m)\}+bE\{y(n)\} \ , \tag{2.1.9}$$

where a,b are arbitrary constant values and $x(m),y(n)$ are random variables taken at arbitrary sample indices m and n.

ii) The expectation of a product of two random variables is generally *not* equal to the product of the expectations

$$E\{x(m)y(n)\} \neq E\{x(m)\}E\{y(n)\} \ . \tag{2.1.10}$$

The exception to this rule occurs when random variables x and y are independent, see Section 2.1.3.

iii) The expectation of a function $y = g(x)$ of a random variable x is given by

$$E\{y\} = \int\limits_{-\infty}^{\infty} g(x)f(x)\,dx \ . \tag{2.1.11}$$

Thus the expected value of a function of a random variable can be obtained without explicitly determining the density of the function y.

b) Correlation and Covariance

In addition to first order moments it is possible to define an entire set of higher moments of the discrete random signal $x(n)$. The most important of these are the second moments:

Correlation — The correlation of a random signal $x(n)$ is defined by

$$R(m,n)=E\{x(m)x(n)\} \ . \tag{2.1.12}$$

This may be written

$$R(m,n)= \int\limits_{-\infty}^{\infty} \int\limits_{-\infty}^{\infty} x(m)x(n)f(x(m),x(n))dx(m)dx(n) \ . \tag{2.1.13}$$

A relative frequency interpretation of this is given by

$$R(m,n)= \lim_{N\to\infty} \{\frac{1}{N}\sum_{i=1}^{N} x_i(m)x_i(n)\} \ , \tag{2.1.14}$$

where once again the average is taken across the ensemble. The correlation is a measure of the degree of association or dependence between x at index n_1 and at index n_2. In particular

$$R(n,n)=E\{x^2(n)\} \ , \tag{2.1.15}$$

is the average power of $x(n)$. The correlation is often called the **autocorrelation**, reflecting the fact that $R(m,n)$ measures the correlation of signal $x(n)$ with itself.

Covariance — The covariance of a random signal is defined by:

$$C(m,n)=E\{(x(m)-\mu_m)(x(n)-\mu_n)\} \ , \tag{2.1.16}$$

or

$$C(m,n)= \int\limits_{-\infty}^{\infty} \int\limits_{-\infty}^{\infty} (x(m)-\mu_m)(x(n)-\mu_n)f(x(m),x(n))\,dx(m)\,dx(n) \ .$$

Equation (2.1.16) may be expanded to give

$$C(m,n)=E\{x(m)x(n)\}-\mu_m\mu_n \ . \tag{2.1.17}$$

In particular

$$C(n,n)=\sigma_n^2=E\{(x(n)-\mu_n)^2)\} \ , \tag{2.1.18}$$

is called the **variance**. Comparing equations (2.1.12) and (2.1.17) we see that if $\mu_n=\mu_m=0$, then the correlation and covariance are identical. The use of such a zero-mean assumption is widespread and is justified on the grounds that a non zero-mean, or d.c. value, can easily be removed by highpass filtering.

Covariance and Correlation Matrices – The $(L\times L)$ matrix

$$C= \begin{bmatrix} C(0,0) & C(0,1) & . & . & C(0,L-1) \\ C(0,1) & C(1,1) & . & . & . \\ . & & . & . & . \\ . & & . & . & . \\ C(0,L-1) & & . & . & C(L-1,L-1) \end{bmatrix} , \tag{2.1.19}$$

with elements $C(i,j)$ defined by equation (2.1.16), is called the **covariance matrix** of order L. The matrix is symmetric because $C(m,n)=C(n,m)$, which follows directly from (2.1.16). Similarly,

$$R= \begin{bmatrix} R(0,0) & R(0,1) & . & . & R(0,L-1) \\ R(0,1) & R(1,1) & . & . & . \\ . & & . & . & . \\ . & & . & . & . \\ R(0,L-1) & & . & . & R(L-1,L-1) \end{bmatrix} , \tag{2.1.20}$$

with elements $R(m,n)$ defined by equation (2.1.12), is called the **correlation matrix** of order L. These matrices assume considerable importance in optimal signal processing. In view of the relation between $C(m,n)$ and $R(m,n)$, we may write

$$C = R - \underline{\mu}\, \underline{\mu}^t \; , \qquad\qquad (2.1.21)$$

where $\quad \underline{\mu} = E\{[x(0), x(1), ..., x(L-1)]^t\} = [\mu_0, \mu_1, ..., \mu_{L-1}]^t \quad$ Once again, if $\underline{\mu} = 0$ then the correlation and covariance matrices are identical. If we define a vector of random variables $\underline{x} = [x(0), x(1), ..., x(L-1)]^t$, then R may be expressed as

$$R = E\{\underline{x}\,\underline{x}^t\} \; ,$$

and C as

$$\begin{aligned} C = Cov\{\underline{x}\} &= E\{\underline{x}\,\underline{x}^t\} - \underline{\mu}\,\underline{\mu}^t \\ &= E\{(\underline{x}-\underline{\mu})((\underline{x}-\underline{\mu})^t\} \; . \end{aligned}$$

c) Higher Moments

Moments of arbitrary order can be defined with respect to the random signal $x(n)$. For example, third order moments can be defined about the mean as in

$$C^{(3)}(l,m,n) = E\{(x(l)-\mu_l)(x(m)-\mu_m)(x(n)-\mu_n)\} \; , \qquad (2.1.22)$$

or about the origin

$$R^{(3)}(l,m,n) = E\{x(l)x(m)x(n)\} \; . \qquad\qquad (2.1.23)$$

The generalizations to N-th order moments are by direct extension:

$$C^{(N)}(l_1, l_2, ..., l_N) = E\{(x(l_1)-\mu_{l_1})(x(l_2)-\mu_{l_2})...(x(l_N)-\mu_{l_N})\} \; , \qquad (2.1.24)$$

and

$$R^{(N)}(l_1, l_2, ..., l_N) = E\{x(l_1)x(l_2)...x(l_N)\}. \qquad\qquad (2.1.25)$$

Such moments have been used far less in signal processing applications than the first and second order moments defined above, though this is changing as interest in nonlinear systems increases (see Nikias and Raghuveer [8] for a recent review).

2.1.3 Independence, Correlation and Orthogonality

Samples $x(m)$, $x(n)$ are said to be **independent** random variables if

$$f(x(m),x(n))=f(x(m))f(x(n)) .$$

It follows immediately that for independent random variables

$$E\{x(m)x(n)\}=E\{x(m)\}E\{x(n)\} . \qquad (2.1.26)$$

Inspection of equation (2.1.17) shows that if $x(m)$, $x(n)$ are independent then

$$C(m,n)=0 . \qquad (2.1.27)$$

If equation (2.1.27) holds, the random variables are said to be **uncorrelated**.[2] Thus we see that independent samples are also uncorrelated. The converse is not generally true, though an important exception is the case of $x(m)$, $x(n)$ Gaussian, for which uncorrelated samples *are* independent.

Samples $x(m)$, $x(n)$ are said to be **orthogonal** if

$$E\{x(m)x(n)\}=0 . \qquad (2.1.28)$$

Consideration of (2.1.17) and (2.1.27) leads to the conclusion that $x(m)$, $x(n)$ are orthogonal if they are uncorrelated and if either of μ_m or μ_n is zero. Since we routinely employ the zero-mean assumption for $x(n)$, we often refer to the condition (2.1.28) as indicating that the samples are uncorrelated.

2.1.4 Stationarity

A discrete random signal is said to be **strictly** stationary or **strongly** stationary if its k-th order distribution function

---

2. The fact that the term 'uncorrelated' relates to the covariance sequence rather than the correlation is a potential source of confusion. Fortunately, most of the time we assume that signals encountered are zero-mean. Under these conditions, covariance reduces to correlation. On the few occasions where the zero-mean assumption is not employed, we shall be careful to define correlation properties with care.

$$F(x(n_1), x(n_2), ..., x(n_k)) \; ,$$

is shift-invariant for any set $n_1, n_2, ..., n_k$ and for any k. That is if, for all k

$$F(x(n_1), x(n_2), ..., x(n_k)) = F(x(n_1+n_0), x(n_2+n_0), ..., x(n_k+n_0)) \; ,$$

$$(2.1.29)$$

where n_0 is an arbitrary shift. Thus, stationarity is the invariance of all statistical properties of the sequence $x(n)$ to location (sample index). In practice, this definition is difficult to use. A weaker but more useful definition of stationarity can be obtained by insisting not that the distribution function be shift-invariant, but that the *moments of the sequence* be shift-invariant. In particular, a discrete random signal is said to be **weakly** or **wide-sense** stationary if all its first and second order moments are finite and independent of sample index. Wide-sense stationarity is widely used in practice and will be adopted throughout this text. Note that, although strict stationarity is generally a much stronger condition, a strictly stationary sequence may not be weakly stationary as defined here because it may not have finite moments.

For weak stationarity we require that the mean value be independent of location so that

$$E\{x(m)\} = E\{x(n)\} = \mu. \qquad (2.1.30)$$

Similarly the correlation becomes[3]

$$R(m,n) = E\{x(m)x(n)\} = r(n-m) = r(i) \; , \qquad (2.1.31)$$

where $i = n - m$ is called the **correlation lag**. Hence the correlation depends only on the lag i, not on the particular value of the sample index. We may therefore write the correlation for stationary $x(n)$ as

$$r(i) = E\{x(n)x(n+i)\} \; . \qquad (2.1.32)$$

---

3. We denote the correlation elements for stationary signals by the lower case symbol r. This helps to distinguish between correlation elements and matrices, and between time domain correlations and the frequency domain spectra defined in Section 2.2.

Three important properties of the autocorrelation sequence $r(i)$ should be noted:

i) The autocorrelation is an even sequence

$$r(i)=r(-i) , \qquad (2.1.33)$$

and hence is symmetric about the origin. Because of this symmetry, plots of the autocorrelation sequence sometimes show only non-negative lags.

ii) The mean-square value is greater than or equal the magnitude of the correlation for any other lag

$$E\{x^2(n)\}=r(0)\geq |r(i)| . \qquad (2.1.34)$$

iii) Substituting (2.1.31) into (2.1.20) we see that for stationary signals the correlation matrix becomes

$$R=\begin{bmatrix} r(0) & r(1) & . & . & r(L-1) \\ r(1) & r(0) & . & . & . \\ . & & . & . & . \\ . & & & . & r(1) \\ r(L-1) & r(L-2) & . & . & r(0) \end{bmatrix} . \qquad (2.1.35)$$

Observe that the elements on each diagonal are identical. Such a matrix is called **Toeplitz**. The $(L\times L)$ correlation matrix is a symmetric Toeplitz matrix, and has a total of only L distinct elements. The correlation matrix is a **positive semi-definite** matrix which means that

$$\underline{a}^t R \underline{a} \geq 0 \qquad \text{for any } (L\times 1) \text{ vector } \underline{a}.$$

This may be expanded as

$$\sum_{i=1}^{L}\sum_{j=1}^{L} a_i a_j r(|i-j|) , \qquad (2.1.36)$$

where $\underline{a}=[a_1,a_2,...,a_L]^t$. Note, however, that the matrix does not have the stronger **positive definite** property:

$$\underline{a}^t R \underline{a} > 0 \ ,$$

for any non-zero $\underline{a}$. The proof of the positive semi-definite property is given in Appendix 3A. As we shall see, this property has important practical implications.

2.1.5 Some Important Random Signals

a) Purely Random Signals (White Noise)

A zero-mean, stationary discrete random signal which satisfies

$$f\left(x(0), x(1), ...\right) = f\left(x(0)\right) f\left(x(1)\right)... \tag{2.1.37}$$

and where the marginal density $f(x(i))$ is the same for each i, is called a **purely random sequence**. In words, equation (2.1.37) says that the elements of the sequence $x(n)$ are **independent, identically distributed (iid)**. As we have seen, samples that are independent are also uncorrelated, so that for a zero-mean iid sequence

$$r(n-m) = E\{x(m)x(n)\} = \sigma_w^2 \delta(n-m) \ , \tag{2.1.38}$$

where $\sigma_w^2 = E\{x^2(n)\}$ is the variance of the signal, and $\delta(i)$ is the Kronecker delta. Signals with such a correlation structure are referred to as **white sequences** or **white noise**. White noise is a convenient and simple idealization which, because of the independence of the samples and resulting structure of its correlation sequence, is often useful for analytic purposes. If the reader is unfamiliar with the term white noise the reason for this name will soon become apparent.

Notes:

i) Since independent sequences are necessarily uncorrelated, all iid sequences are white but, since uncorrelated does not necessarily imply independent, the converse is not necessarily true. In fact, some authors (see, for example, Priestley [6]), use the term purely random to describe sequences of uncorrelated random variables. The definition used here for purely random – that of independent random variables – includes uncorrelated sequences as a subset. To avoid any possibility of confusion we shall assume that all white sequences encountered are purely random − that is generated by independent random variables. This device allows the use of the terms iid, purely random, and white interchangeably, though the reader should keep the above discussion in mind when dealing with such sequences.

ii) For white sequences the correlation matrix of equation (2.1.35) takes the diagonal form

$$R = \sigma_w^2 I \, , \qquad (2.1.39)$$

where I is the $(L \times L)$ identity matrix.

iii) The density $f(x(n))$ is unspecified by the definition (2.1.37), so that it is equally possible to generate white noise, using say Gaussian or Uniform densities.

iv) An iid sequence is always strictly stationary, and is also weakly stationary except for sequences with unbounded moments (as, for example, with an iid sequence with Cauchy distribution [7]).

b) First Order Markov Signals

The white noise signal is characterized by the independence of each sample from any of the preceding samples. The next least complicated random signal is one for which the conditional density satisfies

$$f(x(n)/x(n-1), x(n-2), ..., x(0)) = f(x(n)/x(n-1)) \, , \qquad (2.1.40)$$

where $f(x/y)$ denotes the density of the random variable x conditioned on the observed value y. According to the definition (2.1.40), the random signal at time n is statistically completely determined by the density conditioned on the previous sample alone. Such a signal is called **first-order Markov**.

c) Gaussian Random Signals

A Gaussian random signal is one for which any set of L samples has a joint Gaussian density. Thus, the random sample $x(i)$ is distributed as

$$f\left(x(i)\right)=\frac{1}{(2\pi)^{1/2}\,\sigma_i}\,e^{\frac{-1}{2\sigma_i^2}\left(x(i)-\mu_i\right)^2}\quad,$$

denoted $N(\mu_i,\sigma_i)$, and the joint density of L samples $x(n_0),x(n_1),...,x(n_{L-1})$ is

$$f\left(x(n_0),x(n_1),...,x(n_{L-1})\right)=f(\underline{x})=\frac{1}{(2\pi)^{L/2}\,\Delta^{1/2}}\,e^{-\frac{1}{2}(\underline{x}-\underline{\mu})^t C^{-1}(\underline{x}-\underline{\mu})}\quad,$$

$$(2.1.41)$$

where $\underline{x}=[x(n_0),x(n_1),...,x(n_{L-1})]^t$, $\underline{\mu}=[\mu_{n_0},\mu_{n_1},...,\mu_{n_{L-1}}]^t$ and Δ is the determinant of C, where as in (2.1.19), C is the covariance matrix for the elements of $\underline{x}$. A simplification occurs when $x(n)$ is zero-mean, stationary, then we have

$$f(\underline{x})=\frac{1}{(2\pi)^{L/2}\,\Delta^{1/2}}\,e^{-\frac{1}{2}\underline{x}^t R^{-1}\underline{x}}\quad,\qquad(2.1.42)$$

C being equal to the stationary correlation matrix R of equation (2.1.35) for the zero-mean case. A further simplification follows when $x(n)$ is zero-mean Gaussian iid, that is Gaussian white noise. In this case R is the diagonal matrix of equation (2.1.39) and hence

$$R^{-1}=\frac{1}{\sigma_w^2}I\ .\qquad(2.1.43)$$

The determinant is given by

$$|R|=\Delta=\sigma_w^{2L}\ .\qquad(2.1.44)$$

The density $f\left(\underline{x}\right)$ may then be written

$$f(\underline{x})=\frac{1}{(2\pi)^{L/2}\,\sigma_w^L}\,e^{-\frac{1}{2\sigma_w^2}\sum\limits_{i=0}^{n_{L-1}}x^2(i)}\ .\qquad(2.1.45)$$

The form of the exponent can be verified simply by expanding the vector representation of equation (2.1.42). The Gaussian random signal enjoys several key properties [6]:

i) Recall that in general independent sequences are uncorrelated but that the converse is not necessarily true. Gaussian random variables provide the exception to that rule because for such random variables uncorrelated implies independent and consequently Gaussian white noise signals are necessarily generated from iid samples (see also Problem 2.31).

ii) Any linear operation (including linear filtering) applied to a Gaussian signal produces a result (output) which is also Gaussian.

iii) All higher moments of a Gaussian distribution can be expressed in terms of the first and second moments of the distribution only. Hence, a Gaussian distribution is completely characterized by its first two moments. It follows from this that for Gaussian random signals the definitions of strict and weak stationarity are equivalent.

2.1.6 Complex Random Signals

A complex random signal $\{x(n)\}$ is described by

$$x(n) = x_R(n) + jx_I(n) , \qquad (2.1.46)$$

where $x_R(n)$ and $x_I(n)$ are real random signals. The definitions of moments given above for real signals may be naturally extended as:

Mean Value

$$\mu_n = E\{x(n)\} = E\{x_R(n)\} + jE\{x_I(n)\} . \qquad (2.1.47)$$

Covariance

$$C(m,n) = E\{(x(m) - \mu_m)^*(x(n) - \mu_n)\} , \qquad (2.1.48)$$

where the superscript * denotes the complex conjugate.

Correlation

$$R(m,n) = E\{x^*(m)x(n)\} , \qquad (2.1.49)$$

or, for stationary $x(n)$

$$r(i)=E\{x^*(n)x(n+i)\} ,\qquad\qquad (2.1.50)$$

and the average power is

$$r(0)=E\{|x(n)|^2\} .\qquad\qquad (2.1.51)$$

2.2 Spectral Representations of Discrete Random Signals

A zero-mean stationary sequence $\{x(n)\}$ does not have finite duration or finite energy. As such, Fourier and z transforms of this signal do not converge. On the other hand, the correlation sequence is deterministic and usually has finite energy (though not always, a simple counterexample is provided by the sinusoidal process of Problem 2.3). When this condition holds, the envelope of the correlation decays as the lag number increases, and the correlation is absolutely summable. Hence the z transform of the sequence

$$R(z)= \sum_{i=-\infty}^{\infty} r(i)z^{-i} ,\qquad\qquad (2.2.1)$$

has some region of convergence for z [9]. Note that since

$$r(i)=r(-i) ,$$

then

$$R(z)=R(z^{-1}) .\qquad\qquad (2.2.2)$$

2.2.1 The Power Spectrum

The transform obtained from $R(z)$ by setting $|z|=1$ is

$$R(e^{j\omega})=R(z) \qquad ; z=re^{j\omega} \text{ with } r=1 ,$$

or

$$R(e^{j\omega})= \sum_{i=-\infty}^{\infty} r(i)e^{-j\omega i} .\qquad\qquad (2.2.3)$$

$R(e^{j\omega})$ is called the **power spectrum** or **power spectral density** of the random signal $x(n)$. This definition is equivalent to the Fourier transform of the sequence $r(i)$. The inverse transform is given by

$$r(i) = \frac{1}{2\pi} \int_{-\pi}^{\pi} R(e^{j\omega}) e^{j\omega i} d\omega \; . \tag{2.2.4}$$

The particular result for lag $i=0$ gives the signal power

$$E\{x^2(n)\} = r(0) = \frac{1}{2\pi} \int_{-\pi}^{\pi} R(e^{j\omega}) d\omega \; .$$

$R(e^{j\omega})$ represents the distribution of the signal power over frequency. As a consequence of the real and even nature of $r(i)$, $R(e^{j\omega})$ is a real, even and non-negative function of ω (see Problem 2.5). Also, since $R(e^{j\omega})$ is obtained as the transform of the discrete sequence $r(i)$, the power spectrum is periodic in frequency with

$$R(e^{j(\omega + 2k\pi)}) = R(e^{j\omega}) \qquad ; \; k=1,2,... \tag{2.2.5}$$

As usual with Fourier transforms of discrete signals, because of the symmetry and periodicity of $R(e^{j\omega})$, plots of the spectrum typically cover the range $0-\pi$ only.

An important special case is $x(n)$ white, then from equation (2.1.38)

$$r(i) = \sigma_w^2 \delta(i) \; ,$$

and transforming using equation (2.2.3) gives

$$R(e^{j\omega}) = \sigma_w^2 \; . \tag{2.2.6}$$

Thus the power spectral density of white noise is constant, indicating that the power of a purely random signal is equally distributed over all frequencies (see Figure 2.2.1). This, by the way, explains the name white noise − the equal distribution of power at all frequencies is analogous to the spectral content of white light.

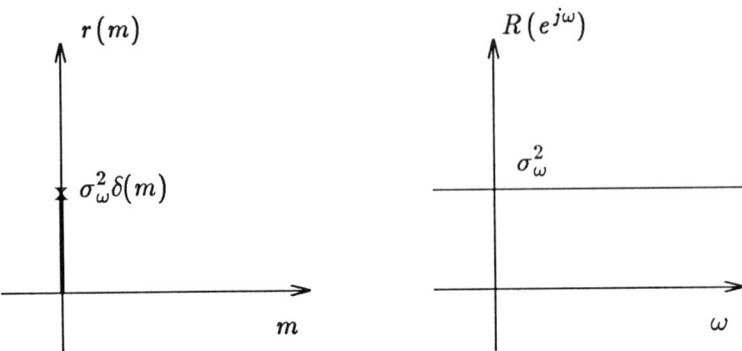

Figure 2.2.1 Correlation $r(m)$ and power spectrum $R(e^{j\omega})$ for a white noise signal.

2.2.2 The Cross-Correlation and Cross-Spectrum

For two random sequences $x(n)$, $y(n)$ we can extend the ideas of autocorrelation and power spectrum to form analogous definitions of cross-correlation and cross-spectrum. The **cross-correlation** is defined by

$$R_{xy}(m,n)=E\{x(m)y(n)\} , \qquad (2.2.7)$$

where the subscripts have been introduced for clarity. Henceforth, where any confusion may arise we shall write the autocorrelation of $x(n)$ as R_{xx}. For random signals $x(n)$, $y(n)$, strict stationarity requires that the statistical properties of the signals at sample index n be identical to those of $x(n+n_0)$, $y(n+n_0)$ for any n_0. As with stationarity of a single random process, wide-sense stationarity requires only shift-invariance of moments up to order two. **Joint stationarity** of sequences $x(n)$ and $y(n)$ requires that $x(n)$ and $y(n)$ are each wide-sense stationary, *and* that the cross-correlation between $x(n)$ and $y(n)$ depend only on the correlation lag. Thus for $x(n)$, $y(n)$ jointly stationary

$$r_{xy}(i)=E\{x(n)y(n+i)\} . \qquad (2.2.8)$$

Two simple properties of $r_{xy}(i)$ should be noted:

i)

$$r_{xy}(i)=r_{yx}(-i) . \qquad (2.2.9)$$

ii)

$$|r_{xy}(i)| \le (r_{xx}(0)r_{yy}(0))^{1/2} . \qquad (2.2.10)$$

The **cross-spectrum** $R_{xy}(e^{j\omega})$ is defined by

$$R_{xy}(e^{j\omega}) = \sum_{i=-\infty}^{\infty} r_{xy}(i)e^{-j\omega i} . \qquad (2.2.11)$$

$r_{xy}(i)$ is not an even sequence as is $r_{xx}(i)$, consequently $R_{xy}(e^{j\omega})$ is complex. In view of equation (2.2.9), however

$$R_{xy}(e^{j\omega}) = \sum_{i=-\infty}^{\infty} r_{xy}(i)e^{-j\omega i},$$

$$= \sum_{i=-\infty}^{\infty} r_{yx}(-i)e^{-j\omega i} ,$$

or, changing the index of summation

$$R_{xy}(e^{j\omega}) = \sum_{l=-\infty}^{\infty} r_{yx}(l)e^{j\omega l} = \left[\sum_{l=-\infty}^{\infty} r_{yx}(l)e^{-j\omega l} \right]^* .$$

Hence

$$R_{xy}(e^{j\omega}) = R_{yx}^*(e^{j\omega}) , \qquad (2.2.12a)$$

or

$$R_{xy}^*(e^{j\omega}) = R_{yx}(e^{j\omega}) . \qquad (2.2.12b)$$

2.3 Response of Linear Systems to Random Signals

Consider a zero-mean, stationary sequence $x(n)$ as input to a linear shift-invariant system with impulse response function $h(n)$ (see Figure 2.3.1). The output $y(n)$ is represented in terms of the convolution summation

$$y(n) = h(n) \, ^* x(n) = \sum_{k=-\infty}^{\infty} h(n-k)x(k) . \qquad (2.3.1)$$

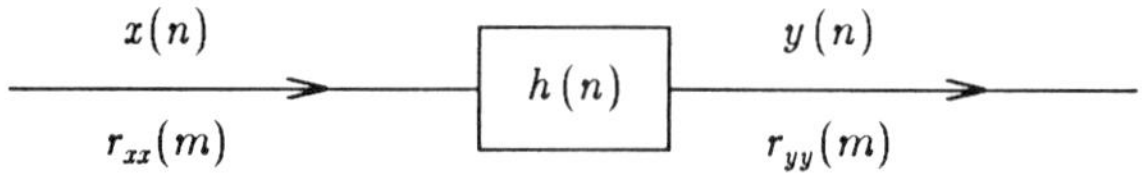

Figure 2.3.1 A zero-mean stationary signal $x(n)$ is input to a stable linear shift-invariant system $h(n)$ producing a zero-mean stationary output $y(n)$.

First, note that the expected value of the output is

$$E\{y(n)\}= \sum_{k=-\infty}^{\infty} h(n-k)E\{x(k)\}=0 \ .$$

Thus we see that the zero-mean input signal $x(n)$ produces a zero-mean output $y(n)$. From the definition (2.1.31), the correlation of the output signal may be written

$$R_{yy}(n,n+m)=E\{y(n)y(n+m)\} \ , \tag{2.3.2}$$

where we use the upper case form for R_{yy} and show dependence on two arguments because stationarity of $y(n)$ has not, at this stage, been established. Substituting for $y(n)$, equation (2.3.2) becomes

$$R_{yy}(n,n+m)=E\{ \sum_{k_1=-\infty}^{\infty} \sum_{k_2=-\infty}^{\infty} h(k_1)h(k_2)x(n-k_1)x(n+m-k_2)\} \ .$$

Taking the expectation operation inside the summation, and using the fact that $x(n)$ is stationary we may write

$$R_{yy}(n,n+m)= \sum_{k_1=-\infty}^{\infty} \sum_{k_2=-\infty}^{\infty} h(k_1)h(k_2)r_{xx}(m+k_1-k_2).$$

Note that the right hand side of this equation has no dependence on n and hence

$$R_{yy}(n,n+m)= \sum_{k_1=-\infty}^{\infty} h(k_1) \sum_{k_2=-\infty}^{\infty} h(k_2)r_{xx}(m+k_1-k_2)=r_{yy}(m) \ .$$

Thus we have seen that the output signal $y(n)$ is zero-mean and has correlation that depends on lag only. Hence $y(n)$ satisfies the conditions for weak stationarity.

Now let us try to simplify the expression relating input and output correlations. Let $k=k_2-k_1$ so that $k_2=k+k_1$. We have

$$r_{yy}(m)= \sum_{k=-\infty}^{\infty} r_{xx}(m-k) \sum_{k_1=-\infty}^{\infty} h(k_1)h(k+k_1) \ . \qquad (2.3.3)$$

From this, $r_{yy}(m)$ can be written as the convolution of $r_{xx}(m)$ with a sequence $g(m)$

$$r_{yy}(m)= \sum_{k=-\infty}^{\infty} r_{xx}(m-k)g(k)=r_{xx}(m)\ ^*g(m) \ , \qquad (2.3.4)$$

where

$$g(m)= \sum_{k_1=-\infty}^{\infty} h(k_1)h(m+k_1) \ . \qquad (2.3.5)$$

$g(m)$ is called the **sample autocorrelation** of h. From (2.3.4), the z transform of $r_{yy}(m)$ is given by

$$R_{yy}(z)=G(z)R_{xx}(z) \ . \qquad (2.3.6)$$

From equation (2.3.5), $g(m)$ is itself the convolution

$$g(m)=h(m)\ ^*h(-m) \ , \qquad (2.3.7)$$

so that

$$G(z)=H(z)H(z^{-1}) \ , \qquad (2.3.8)$$

and hence

$$R_{yy}(z)=H(z)H(z^{-1})R_{xx}(z) \ . \qquad (2.3.9)$$

Finally, taking $|z|=1$ gives the power spectrum

$$R_{yy}(e^{j\omega})=|H(e^{j\omega})|^2 R_{xx}(e^{j\omega}) \ . \qquad (2.3.10)$$

In words, the power spectrum of the system output is the product of the input spectrum and the modulus squared frequency response of the system $h(n)$. This fundamental result will be used extensively later.

Another important result relates the cross-correlation between input and output to the autocorrelation of the input as:

$$r_{xy}(m) = \sum_{k=-\infty}^{\infty} h(k) r_{xx}(m-k) \ , \tag{2.3.11}$$

and in the frequency domain

$$R_{xy}(e^{j\omega}) = H(e^{j\omega}) R_{xx}(e^{j\omega}) \ . \tag{2.3.12}$$

The derivation of (2.3.11) and (2.3.12) follows a similar approach to that for (2.3.4) and (2.3.10), and is considered in the Problems section (see Problem 2.9).

Special Case — White Noise Input

As a simple example consider an input $x(n) = w(n)$ where $w(n)$ is a zero-mean white sequence with variance σ_w^2. From (2.1.38) the correlation of $x(n)$ is

$$r_{xx}(m) = \sigma_w^2 \delta(m) \ . \tag{2.3.13}$$

Fourier transforming $r_{xx}(m)$ yields the power spectrum as in (2.2.6)

$$R_{xx}(e^{j\omega}) = \sigma_w^2 \ . \tag{2.2.6}$$

Applying $x(n)$ as input to a linear shift-invariant system, $h(n)$, yields output correlation $r_{yy}(m)$ as given by equation (2.3.4). Using (2.3.4) and (2.3.5) gives

$$r_{yy}(m) = g(m)\sigma_w^2 = \left[\sum_{k=-\infty}^{\infty} h(k) h(m+k) \right] \sigma_w^2 \ . \tag{2.3.14}$$

In particular, the output power is given by

$$\sigma_y^2 = r_{yy}(0) = \left[\sum_{k=-\infty}^{\infty} h^2(k) \right] \sigma_w^2 \ . \tag{2.3.15}$$

Also, from equation (2.3.10) the output power spectrum is given by

$$R_{yy}(e^{j\omega}) = |H(e^{j\omega})|^2 \sigma_w^2 \ , \tag{2.3.16}$$

and from (2.3.12) the cross spectrum is

$$R_{xy}(e^{j\omega}) = H(e^{j\omega})\sigma_w^2 \ . \tag{2.3.17}$$

These relations illustrate how a white noise signal is effectively 'colored' by the system, $h(n)$. Note also that equations (2.3.16) and (2.3.17) can be exploited in **system identification** and render white noise excitation as a possible competitor of impulse testing for such problems.

2.4 Random Signal Models

2.4.1 The Linear Process

One of the most useful random signal models is the representation of a random signal as the output of a white noise excited linear shift-invariant filter. For stationary signals this model is entirely general as the following result establishes:

Result − Wold's Theorem

Any stationary sequence $x(n)$ which has no deterministic (that is, completely predictable) components can be represented as the output of a stable, causal, shift-invariant linear filter, $h(n)$, with a white noise input, $w(n)$. That is

$$x(n) = \sum_{i=0}^{\infty} h(i)w(n-i) \ . \tag{2.4.1}$$

$x(n)$ is called a **linear process**.

This is a simplified form of a theorem due to Wold which dates back to 1938 [11], and is one of the most fundamental results in random signal analysis. The theorem says simply that any non-deterministic stationary random sequence may be generated by passing white noise through a shift-invariant linear filter. In spectral terms we see that the white noise input corresponds to an input/output spectral relation of the form of equation (2.3.16).

2.4.2 Modelling the Linear Process

For practical reasons, models of $H(e^{j\omega})$ are usually restricted to having a finite number of parameters. A fairly general model is obtained by using a system with transform $H(z)$ having the rational form[4]

$$H(z) = \frac{b_0 + b_1 z^{-1} + \ldots + b_N z^{-N}}{1 - a_1 z^{-1} - \ldots - a_M z^{-M}} \, , \qquad (2.4.2)$$

where the order of both numerator and denominator is finite and where the poles of $H(z)$, that is the roots $z = p_i$ for $i = 1, 2, \ldots, M$, of

$$1 - a_1 z^{-1} - \ldots - a_M z^{-M} = 0 \, ,$$

satisfy $|p_i| < 1$. Processes generated using the general pole-zero form of equation (2.4.2) in (2.3.16) are called **Autoregressive Moving Average (ARMA)**. Two special cases of this model are:

1. Autoregressive (AR) models — For which $N = 0$ in (2.4.2), giving an all-pole transfer function

$$H(z) = \frac{1}{1 - a_1 z^{-1} - \ldots - a_M z^{-M}} \, , \qquad (2.4.3)$$

where for simplicity we have set the numerator gain (b_0 in (2.4.2)) to unity.[5] When driven with a white input, the use of this all-pole model gives rise to an **Autoregressive (AR)** process of order M denoted AR(M).

2. Moving Average (MA) models — For which $M = 0$ in (2.4.2) giving an all-zero transfer function

$$H(z) = b_0 + b_1 z^{-1} + \ldots + b_N z^{-N} \, . \qquad (2.4.4)$$

When driven with a white input, the resulting output is a **Moving**

---

4. The use of negative signs for the denominator polynomial coefficients is merely a matter of convenience designed to produce positive terms in the difference equation representation developed below.

5. Note that this represents no loss of generality since we can control the gain through the variance of the input sequence.

Average (MA) process of order N denoted MA(N).

These models describe the way in which the white noise spectrum is shaped or colored. Taking

$$H(e^{j\omega})=[H(z)]_{|z|=1},$$

the coloration is explicitly indicated by equation (2.3.16). The corresponding spectra are thus given by:

ARMA Spectrum

$$R_{xx}(e^{j\omega})=\frac{|b_0+b_1e^{-j\omega}+...+b_Ne^{-j\omega N}|^2}{|1-a_1e^{-j\omega}-...-a_Me^{-j\omega M}|^2}\sigma_w^2. \qquad (2.4.5)$$

AR Spectrum

$$R_{xx}(e^{j\omega})=\frac{\sigma_w^2}{|1-a_1e^{-j\omega}-...-a_Me^{-j\omega M}|^2}. \qquad (2.4.6)$$

MA Spectrum

$$R_{xx}(e^{j\omega})=|b_0+b_1e^{-j\omega}+...+b_Ne^{-j\omega N}|^2\sigma_w^2. \qquad (2.4.7)$$

In each case the spectrum is a rational function of $e^{-j\omega}$.

Models of this form are used very widely in digital signal processing in areas such as data compression, signal synthesis, signal enhancement and spectral analysis among others. Many of these applications will be illustrated in later Chapters.

The general ARMA process is more complicated than either the MA or AR, having a generating model containing both poles and zeros. The reason for admitting this increased complexity is that the number of terms required to obtain an adequate representation using either an AR or an MA process may be very large. An ARMA model may obtain the same degree of fidelity with a greatly reduced parameter set compared to either of the simpler models. Such **parsimonious modelling** is desirable for both computational and analytic reasons.

2.4.3 Analysis/Synthesis

We can think of the ARMA, AR and MA models defined above in two basic roles:

1. *Process synthesis* — In which we use a white signal $w(n)$ as input to the system $H(z)$ to synthesize a signal $x(n)$.[6] Thus we create a single realization of the process by using an input of white noise to the generating difference equation (see Figure 2.4.1a). For the ARMA case, from equation (2.4.2) the difference equation is

$$x(n)=a_1x(n-1)+a_2x(n-2)+...+a_Mx(n-M)+$$
$$+b_0w(n)+b_1w(n-1)+...+b_Nw(n-N) \ . \qquad (2.4.8)$$

For the Autoregressive (AR) case, from equation (2.4.3) the difference equation is

$$x(n)=a_1x(n-1)+a_2x(n-2)+...+a_Mx(n-M)+w(n) \ , \qquad (2.4.9)$$

and for the Moving Average (MA) process, from (2.4.4) we have

$$x(n)=b_0w(n)+b_1w(n-1)+...+b_Nw(n-N) \ . \qquad (2.4.10)$$

In fact, by generating different iid excitation sequences, we may synthesize an infinite number of different realizations corresponding to any parameter set $\{a_1,...,a_M,b_0,...,b_N\}$. Being generated by the same model, and the same parameters, each of these synthesized sequences will have identical statistical properties.

An important difference between the AR and MA models is the duration of the influence of a particular input sample. For an MA process, the effect of any particular input sample $w(n_0)$, say, has a finite duration in the output signal (specifically $N+1$ samples). By contrast, the recursive nature of the AR process gives each input the potential to impact *all* future values of $x(n)$.

---

6. It is a common mistake to refer to any output signal generated by models of the form specified by equations (2.4.2), (2.4.3) and (2.4.4) as ARMA, AR and MA, respectively. In fact, these names are reserved for signals generated by filtering *white noise* using such models.

2. *Process analysis* – In which the sequence $x(n)$ and the parameters of the AR, MA or ARMA process which gave rise to $x(n)$ are known, and $x(n)$ is filtered using the inverse system:

$$G(z) = \frac{1}{H(z)} , \qquad (2.4.11)$$

to obtain $w(n)$ (see Figure 2.4.1b). The system $G(z)$ is referred to as the **analysis** filter. Note that the fact that $H(z)$ is stable and causal does not guarantee that $G(z)$ will be stable and causal. As we shall see in the next section, however, it turns out that a stable, causal analysis filter which recovers the white spectrum can always be found.

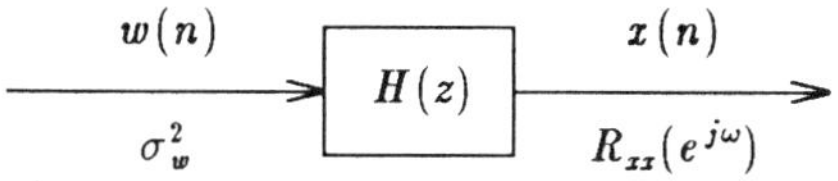

Figure 2.4.1a Process synthesis – A white noise signal $w(n)$ is input to a system $H(z)$ defined according to equations (2.4.2), (2.4.3) or (2.4.4), producing an ARMA, AR or MA process $x(n)$.

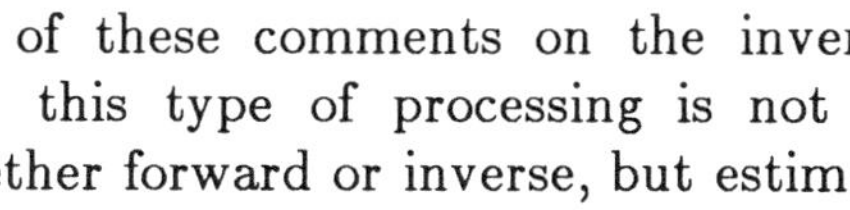

Figure 2.4.1b Process analysis – Sequence $x(n)$ is input to inverse system $G(z) = 1/H(z)$, producing a white output signal $w(n)$.

In spite of these comments on the inverse system, the real difficulty in this type of processing is not the filtering of the process, whether forward or inverse, but estimating the parameters of the model. We discuss parameter estimation in detail in Chapter 3.

Example – AR(1) process

The sequence generated by passing a stationary white noise sequence through a system

$$H(z) = \frac{1}{1 - a_1 z^{-1}} , \qquad (2.4.12)$$

is a first order Autoregressive signal. We impose the constraint $|a_1|<1$ so that the pole of $H(z)$ at $z=a_1$ lies inside $|z|=1$, and hence the causal system $H(z)$ is stable. In difference equation form we have

$$x(n)=a_1x(n-1)+w(n) \ . \tag{2.4.13}$$

Now, assuming zero initial conditions ($x(n)=0$ for $n<0$), we have

$$x(0)=w(0) \ ,$$
$$x(1)=a_1x(0)+w(1) \ ,$$
$$x(2)=a_1x(1)+w(2) \ .$$

This generating model gives rise to a first order Markov sequence (see Section 2.1). To see this, note that due to the recursive nature of the model, $x(n)$ influences all future samples, but only directly influences $x(n+1)$. Essentially all the information about $x(n-1),x(n-2),...,x(0)$ is contained in $x(n)$ and hence

$$f\left(x(n+1)/x(n),...,x(0)\right)=f\left(x(n+1)/x(n)\right) \ ,$$

which it will be recalled (see Section 2.1), is the defining condition for the first order Markov sequence.

2.4.4 *Stationarity and the Stability of the Generating System*

General Result: The random sequence $x(n)$ generated using a model of the form (2.4.1) is stationary if and only if the generating system $H(z)$ is stable and then only after transient effects have decayed [10].

Some confusion may arise between this result and that expressed by equation (2.3.10), where stationarity of the system output was demonstrated for a general $H(z)$. The distinction lies in the two-sided nature of $H(z)$ considered there. In particular, for the infinite two-sided system, no initial condition effects arise. By reviewing the analysis presented in Section 2.3 with limits other than $-\infty$ to ∞ the reader will appreciate the difficulties that such initial conditions pose. This fact has greater importance for AR and ARMA processes than for MA processes. In the latter case the transfer function, which is given by (2.4.4), is guaranteed to be stable provided only that the constants $b_0 b_1,...,b_N$ are finite. For AR and ARMA processes the requirement is that all M poles z_p of $H(z)$ satisfy $|z_p| < 1$.

Example – AR(1) Process

We can illustrate this general principle using the example of an AR(1) sequence discussed above. For the difference equation (2.4.13), the correlation function may be written

$$R_{xx}(n, n+m) = E\{x(n)x(n+m)\} ,$$

where for now we assume that $m \geq 0$. The double argument for the correlation is necessary at this stage because stationarity has not been established. Using repeated substitution, the difference equation (2.4.13) can be expanded as

$$x(n) = a_1(a_1 x(n-2)+w(n-1))+w(n)$$

$$= a_1^2(a_1 x(n-3)+w(n-2))+a_1 w(n-1)+w(n) ,$$

and so on. Finally, we may write

$$x(n) = w(n)+a_1 w(n-1)+...+a_1^n w(0) . \qquad (2.4.10)$$

Hence the correlation is given by

$$R_{xx}(n, n+m) = E\{x(n)x(n+m)\} E\{[w(n)+a_1 w(n-1)+...+a_1^n w(0)] \times$$

$$\times [w(n+m)+a_1 w(n+m-1)+...+a_1^m w(n)+...+a_1^{m+n} w(0)]\} .$$

Now,

$$E\{w(i)w(k)\} = \sigma_w^2 \delta(k-i) .$$

Hence

$$R_{xx}(n, n+m) = (a_1^m + a_1^{m+2}+...+a_1^{m+2n})\sigma_w^2 ,$$

$$R_{xx}(n, n+m) = a_1^m \left[\frac{1-a_1^{2(n+1)}}{1-a_1^2} \right] \sigma_w^2 . \qquad (2.4.14)$$

We see that for $|a_1|<1$, as $n\to\infty$

$$R_{xx}(n,n+m)\to\left(\frac{a_1^m}{1-a_1^2}\right)\sigma_w^2=r_{xx}(m) \quad ; m\geq0 .$$

Finally, in view of the even nature of the correlation sequence

$$r_{xx}(m)=\sigma_w^2\left(\frac{a_1^{|m|}}{1-a_1^2}\right) \quad ; -\infty<m<\infty . \qquad (2.4.15)$$

Thus R_{xx} loses its dependence on n only if $|a_1|<1$, which from (2.4.12) corresponds to the stability condition for the AR model. Even then the result only holds for large n, that is after initial condition effects have decayed. The practical consequences of this are clear: If it is intended to generate a stationary random sequence by passing white noise through a generating filter, care should be taken to reject initial values and thus to reduce nonstationarity effects.

It is more usual when calculating correlations and spectra to assume stationarity from the outset. This leads to a simpler calculation for $r_{xx}(m)$. Let us return to the first-order AR example:

$$x(n)=a_1x(n-1)+w(n) . \qquad (2.4.13)$$

Multiplying both sides by $x(n-m)$ (where we take $m>0$) and taking expectations gives:

$$E\{x(n)x(n-m)\}=a_1E\{x(n-1)x(n-m)\}+E\{w(n)x(n-m)\}$$

$$; m>0 . \qquad (2.4.16)$$

The second term of the right hand side is zero because $x(n-m)$ depends only on $w(n-m),w(n-m-1),...,w(0)$. Hence, assuming stationarity

$$r_{xx}(m)=a_1r_{xx}(m-1) .$$

Or, using repeated substitution

$$r_{xx}(m)=a_1^mr_{xx}(0) . \qquad (2.4.17)$$

This simple relation needs only an initial condition to be complete. This is provided by squaring both sides of (2.4.13) and taking expectations

$$r_{xx}(0)=E\{x^2(n)\}=a_1^2 E\{x^2(n-1)\}+\sigma_w^2+2a_1 E\{w(n)x(n-1)\} \; .$$

Again, since $E\{w(n)x(n-1)\}=0$, then

$$r_{xx}(0)=a_1^2 r_{xx}(0)+\sigma_w^2 \; ,$$

$$r_{xx}(0)=\frac{\sigma_w^2}{1-a_1^2} \; . \qquad (2.4.18)$$

Combining (2.4.17) and (2.4.18) and in view of the symmetry of $r_{xx}(m)$, the result of equation (2.4.15) follows.

Note that care must be taken when applying the procedure of equation (2.4.16). In principle $r_{xx}(m)=E\{x(n)x(n+m)\}$ and the same result could have been obtained by multiplying (2.4.13) by $x(n+m)$ and taking expectations. In this case, however,

$$E\{w(n)x(n+m)\}\neq 0 \; ,$$

because $x(n+m)$ is indirectly dependent on $w(n)$ due to the recursive nature of (2.4.13), and the complexity of the calculation is increased considerably.

The correlation sequence of (2.4.15) is depicted in Figure 2.4.2a. The corresponding power spectrum is easily calculated by applying the definition of (2.2.3) to the correlation sequence (2.4.15)

$$R_{xx}(e^{j\omega})=\frac{\sigma_w^2}{1-a_1^2}\sum_{m=-\infty}^{\infty}a_1^{|m|}e^{-j\omega m} \; ,$$

$$=\frac{\sigma_w^2}{1-a_1^2}\left(\sum_{m=0}^{\infty}a_1^m e^{-j\omega m}+\sum_{m=-\infty}^{-1}a_1^{-m}e^{-j\omega m}\right) \; ,$$

$$=\frac{\sigma_w^2}{1-a_1^2}\left[\frac{1}{1-a_1 e^{-j\omega}}+\frac{1}{1-a_1 e^{j\omega}}-1\right] \; .$$

After some simplification this gives

$$R_{xx}(e^{j\omega})=\frac{\sigma_w^2}{1-2a_1\cos\omega+a_1^2} \; . \qquad (2.4.19)$$

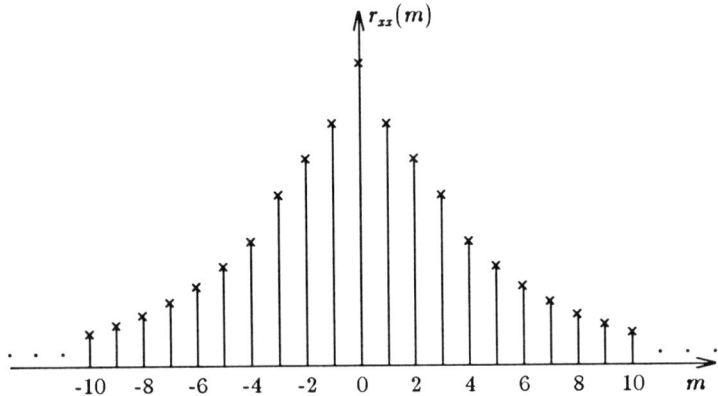

Figure 2.4.2a Correlation sequence $r_{xx}(m)$ for the first order AR process with generating model given by equation (2.4.13). In this example $a_1 = 0.8$.

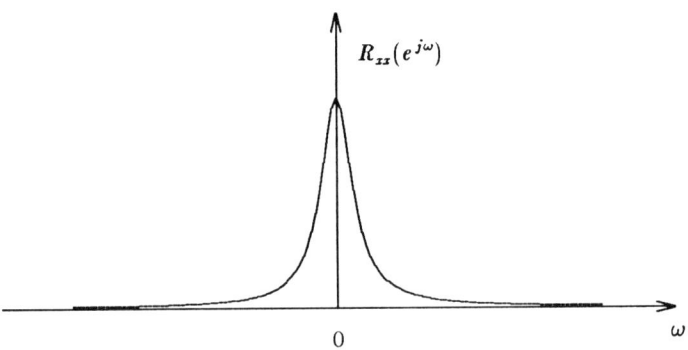

Figure 2.4.2b Power spectrum $R_{xx}(e^{j\omega})$ for the first order AR process with generating model given by equation (2.4.13). In this example $a_1 = 0.8$. Note that only one period of the spectrum is shown.

This spectrum is displayed in Figure 2.4.2b.

The extension of this example to an AR(2) process is considered in Appendix 2A. For the general M-th order AR process, explicit determination of the autocorrelation is generally difficult. We may at least obtain a recursive relation for the correlation coefficients of the AR process by multiplying both sides of the generating model (2.4.9) by $x(n-m)$ and taking expectations:

$$r_{xx}(m) = E\{x(n)x(n-m)\}$$

$$= a_1 E\{x(n-1)x(n-m)\} + \dots$$

$$\dots + a_M E\{x(n-M)x(n-m)\} + E\{w(n)x(n-m)\} \ .$$

For $m=1,2,\dots$ the last term of this equation disappears because $x(n-m)$ has no dependence on $w(n)$. Hence we obtain

$$r_{xx}(m) = a_1 r_{xx}(m-1) + a_2 r_{xx}(m-2) + \dots + a_M r_{xx}(m-M)$$

$$; \ m=1,2,\dots \qquad (2.4.20)$$

Thus, we have a purely recursive difference equation for $r(m)$. Also, using standard results from the theory of constant coefficient difference equations [12] it follows from equation (2.4.20) that provided the roots of the AR generating system are distinct, then the solution of (2.4.20) for $r_{xx}(m)$ consists entirely of a sum of exponential terms and of damped sinusoids, with the exponentials arising from real roots, and the sinusoidal component arising from complex conjugate roots in the generating system [4].

2.5 The Power Spectrum and the Phase of the Generating System

2.5.1 Ambiguity and the Power Spectrum

As we indicated in Section 2.2, the power spectrum of a stationary sequence $x(n)$ is real and even, and hence contains no information about the phase of the sequence. This can also be seen from equation (2.3.16) which shows that the power spectral density is derived solely from the magnitude squared of the generating system $H(z)$ and the variance of the input white noise σ_w^2

$$R_{xx}(e^{j\omega}) = |H(e^{j\omega})|^2 \sigma_w^2 \ . \qquad (2.3.16)$$

From this equation we see that replacing $H(e^{j\omega})$ by any other system $H'(e^{j\omega})$ which satisfies

$$|H'(e^{j\omega})|^2 = |H(e^{j\omega})|^2 , \qquad (2.5.1)$$

gives rise to the same spectrum $R_{xx}(e^{j\omega})$. Thus, we may conclude that the generating model associated with spectrum $R_{xx}(e^{j\omega})$ is not unique. Since any equivalent model has the same amplitude spectrum, this ambiguity is associated with the *phase* of the generating system.

Let us consider this phenomenon in the z domain As an example, consider a sequence $x(n)$ generated using a system $h(n)$ with transform $H(z)$ given by

$$H(z) = 1 + h(1)z^{-1} + ... + h(N)z^{-N} . \qquad (2.5.2)$$

In the z domain, the spectrum corresponding to this model has the form

$$R_{xx}(z) = H(z)H(z^{-1})\sigma_w^2 . \qquad (2.5.3)$$

$H(z)$ can be factored as

$$H(z) = (1 - z_1 z^{-1})(1 - z_2 z^{-1})...(1 - z_N z^{-1}) , \qquad (2.5.4)$$

and has zeros located at $z = z_i$ for $i = 1, 2, ..., N$. Since $x(n)$ is real, the zeros of $H(z)$ are either real or occur in conjugate pairs. Now, $H(z^{-1})$ is given by

$$H(z^{-1}) = (1 - z_1 z)(1 - z_2 z) \cdots (1 - z_N z) , \qquad (2.5.5)$$

and thus has zeros at $z = 1/z_i$ for $i = 1, 2, ..., N$. Substituting (2.5.4) and (2.5.5) into (2.5.3) we see that $R_{xx}(z)$ has zeros at $z = z_i$ *and* at the reciprocal locations (see Figure 2.5.1). Now consider another generating sequence $H'(z)$, say, with

$$H'(z) = (z^{-1} - z_1)(z^{-1} - z_2)...(z^{-1} - z_N) . \qquad (2.5.6)$$

This sequence has zeros at $1/z_1, 1/z_2, ..., 1/z_N$. Therefore $H'(z^{-1})$ has zeros at $z_1, z_2, ..., z_N$ and the spectrum generated using $H'(z)$ takes the form

$$R'_{xx}(z) = H'(z)H'(z^{-1})\sigma_w^2 . \qquad (2.5.7)$$

This spectrum has an identical zero structure to that generated

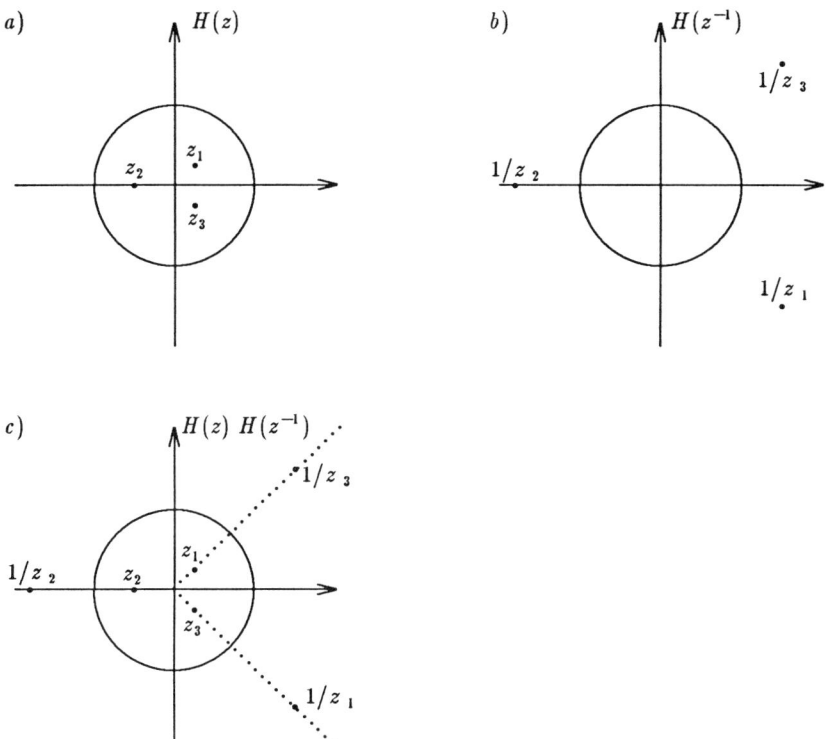

Figure 2.5.1 Pole-zero positions and the power spectrum. In this example $H(z)$ has three zeros z_1, z_2, $z_3 (=z_1^*)$.

using $H(z)$, even though the actual generating sequences $H(z)$ and $H'(z)$ were quite different. Moreover, we recognize that equation (2.5.6) can be written

$$H'(z)=z^{-N}(1-z_1z)(1-z_2z)...(1-z_Nz) .$$

Hence, from equation (2.5.5) we have

$$H'(z) =z^{-N}H(z^{-1}) . \qquad (2.5.8)$$

Substituting (2.5.8) into (2.5.7) gives

$$R'_{zz}(z)=z^{-N}H(z^{-1})z^NH(z)\sigma_w^2=H(z)H(z^{-1})\sigma_w^2=R_{zz}(z) .$$

Hence the spectrum generated using $h'(n)$ is identical to that obtained via $h(n)$. The generating sequences with transforms $H(z)$ and $H'(z)$ may be said to be **spectrally equivalent**.

Note that $H(z)$ and $H'(z)$ are just two among many possible sequences which give rise to the same spectrum. For this particular pair, equation (2.5.8) indicates that $h(n)$ is related to $h'(n)$ simply by *time-reversal and delay*. Other spectrally equivalent sequences are related via more complex operations. To pursue this further it is helpful to examine the phase structure of the generating system, and in particular to introduce the ideas of minimum and maximum phase.

2.5.2 Minimum and Maximum Phase Sequences

Consider a causal $(N+1)$ point sequence with transform $H(z)$

$$H(z)=h(0)+h(1)z^{-1}+...+h(N)z^{-N} , \qquad (2.5.9)$$

where the introduction of the coefficient $h(0)$ makes the model (2.5.9) slightly more general than in the example of (2.5.2). Equation (2.5.9) can be factored as

$$H(z)=K(1-z_1 z^{-1})(1-z_2 z^{-1})...(1-z_N z^{-1}) , \qquad (2.5.10)$$

where $z_1, z_2, ..., z_N$ are zeros of $H(z)$ and $K=h(0)$ is a constant. Then:

i) $H(z)$ is called **Minimum Phase** if the zeros z_i satisfy $|z_i|<1, \; ; i=1,2,...,N.$

ii) $H(z)$ is called **Maximum Phase** if the zeros z_i satisfy $|z_i|>1, \; ; i=1,2,...,N.$

iii) $H(z)$ is called **Mixed Phase** if neither i) nor ii) is satisfied.[7]

The concept of minimum phase is of particular importance in the study of inverse filtering (deconvolution) which is discussed in the

---

7. We conveniently ignore the possibility of zeros on the unit circle $|z|=1$, which are a limiting case and are strictly non-minimum phase. Oppenheim and Schafer [9] report that many of the properties of minimum phase systems hold for such cases.

next Chapter. Minimum phase can be defined more formally via consideration of the Hilbert transform, as in Oppenheim and Schafer [9] for example, though that interpretation is not particularly helpful here.

Two factorizations of sequences will be useful later, one into minimum/maximum phase and the other into minimum phase/all-pass. In particular, given a general sequence of the form (2.5.10)

$$H(z) = K(1 - z_1 z^{-1})(1 - z_2 z^{-1})...(1 - z_N z^{-1}) , \qquad (2.5.10)$$

where

$$|z_i| < 1 \quad ; i = 1, ..., N_1$$

$$|z_i| > 1 \quad ; i = N_1 + 1, ..., N$$

where $N_1 \leq N$, and where the ordering of the zeros is purely for notational convenience. Then $H(z)$ may be factored into:

a) Minimum/Maximum Phase

$$H(z) = K H_{\min}(z) H_{\max}(z) , \qquad (2.5.11)$$

where

$$H_{\min}(z) = (1 - z_1 z^{-1})(1 - z_2 z^{-1})...(1 - z_{N_1} z^{-1}) , \qquad (2.5.12)$$

$$H_{\max}(z) = (1 - z_{N_1+1} z^{-1})(1 - z_{N_1+2} z^{-1})...(1 - z_N z^{-1}) . \qquad (2.5.13)$$

b) Equivalent Minimum Phase /All-Pass

$$H(z) = K H_{eqmin}(z) H_{ap}(z) , \qquad (2.5.14)$$

where

$$H_{eqmin}(z) = (1 - z_1 z^{-1})...(1 - z_{N_1} z^{-1})(z^{-1} - z_{N_1+1})...(z^{-1} - z_N) , \qquad (2.5.15)$$

$$H_{ap}(z) = \frac{(1 - z_{N_1+1} z^{-1})...(1 - z_N z^{-1})}{(z^{-1} - z_{N_1+1})...(z^{-1} - z_N)} . \qquad (2.5.16)$$

$H_{ap}(z)$ is an **all-pass** transfer function, that is it satisfies

$$| H_{ap}(z)| = constant .$$ (2.5.17)

Thus, the modulus response of $H_{ap}(e^{j\omega})$ is independent of frequency and the influence of such a system is restricted purely to phase distortion (see Problem 2.21). We see that $H_{eqmin}(z)$ contains all the zeros of $H(z)$ which lie inside $|z|=1$ together with the reciprocals of those which lie outside (see Figure 2.5.2). $H_{ap}(z)$ contains all the zeros which lie outside $|z|=1$, and poles at the reciprocal locations.

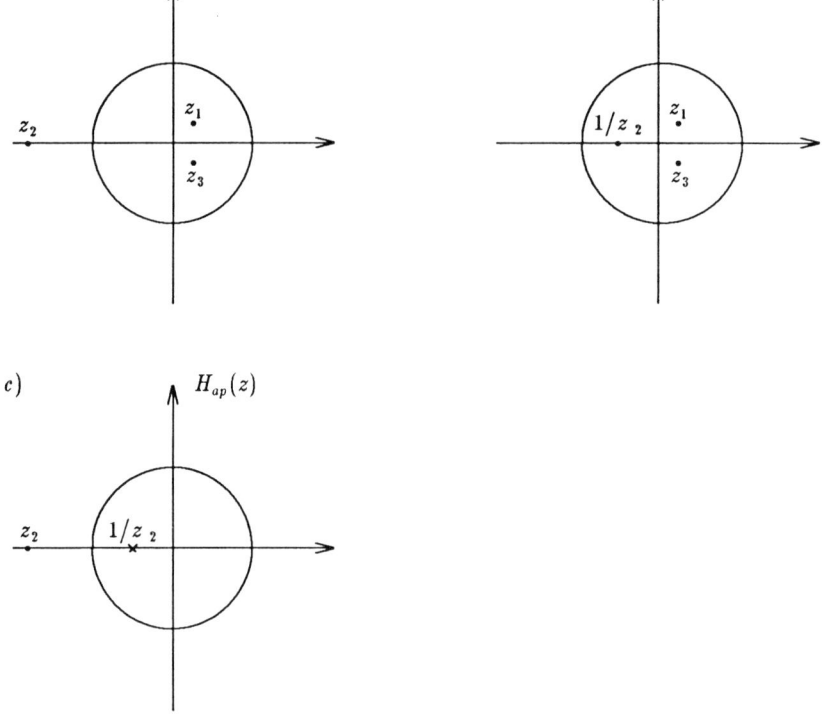

Figure 2.5.2 Factorization of $H(z)$ into equivalent minimum phase and all-pass. In the example shown, $H(z)$ has 3 zeros; two lie inside $|z|=1$ (z_1 and z_3), and one lies outside (z_2).

Let us now return to the question of the power spectrum. If $R_{xx}(z)$ is generated via a finite impulse response system $H(z)$ with N zeros $z=z_i$, then any system $H'(z)$ constructed using any combination of N zeros obtained by choosing one zero from each of the N zero pairs with locations $z=z_i$ and $z=1/z_i$ has the same spectrum $R_{xx}(z)$. For an $(N+1)$ point sequence this gives a total of 2^N distinct sequences which have the same spectrum. Of these, one is minimum phase, one is maximum phase and the rest are mixed phase. Note that not all the spectrally equivalent sequences will have real coefficients, this will only occur if the zeros are selected in complex conjugate pairs. All of the 2^N spectrally equivalent sequences are related by

$$H_j(z) = H_i(z) H_{ap}(z) , \qquad (2.5.18)$$

where $H_i(z)$, $H_j(z)$ are the transforms of any two spectrally equivalent sequences, and where $H_{ap}(z)$ is an all-pass transfer function with the form of (2.5.16). As a simple example consider the transform

$$H_1(z) = (1 - az^{-1})(1 - bz^{-1}) , \qquad (2.5.19)$$

where a, b are real. This generating model gives rise to the same spectrum as do

$$H_2(z) = (z^{-1} - a)(1 - bz^{-1}) , \qquad (2.5.20)$$

$$H_3(z) = (1 - az^{-1})(z^{-1} - b) , \qquad (2.5.21)$$

$$H_4(z) = (z^{-1} - a)(z^{-1} - b) , \qquad (2.5.22)$$

which may easily be verified by simply forming

$$R_{xx}(z) = H_i(z) H_i(z^{-1}) \sigma_w^2 , \qquad (2.5.23)$$

for each case.

The spectrally equivalent sequences may also be compared through the use of **partial energy**. Parseval's relation for a discrete signal of length $(N+1)$ with impulse response $\{h(0), h(1), ..., h(N)\}$ has the form (see, for example, [9])

$$\sum_{n=0}^{N} |h(n)|^2 = \frac{1}{2\pi} \int_{-\pi}^{\pi} |H(e^{j\omega})|^2 d\omega . \qquad (2.5.24)$$

Hence the energy of a signal is dependent only on the amplitude spectrum, $|H(e^{j\omega})|^2$. From this it may be concluded that all of the 2^N sequences giving rise to the same spectrum $R_{xx}(e^{j\omega})$ have the same total energy. The partial energy is defined by

$$P_l = \sum_{n=0}^{l} |h(n)|^2 \quad ; l < N . \qquad (2.5.25)$$

P_l will generally differ for each spectrally equivalent sequence for any given $l < N$.

One interesting property of the minimum phase member of the spectrally equivalent set is the following:

Result: The minimum phase sequence has the highest partial energy of any of the 2^N spectrally equivalent sequences [13].

Rather than proving this result we will give a simple example. Consider the three point sequence

$$x_1(0)=1, \ x_1(1)=1.3, \ x_1(2)=0.4 , \qquad (2.5.26)$$

with z transform

$$X_1(z)=(1+1.3z^{-1}+0.4z^{-2})=(1+0.8z^{-1})(1+0.5z^{-1}) .$$

We see that $X_1(z)$ has zeros at $z_1=-0.8$, $z_2=-0.5$. Hence, from the definition $X_1(z)$ is minimum phase. The remaining spectrally equivalent sequences are given by

$i)$ $\qquad X_2(z)=(z^{-1}+0.8)(1+0.5z^{-1})$

$\qquad\qquad x_2(0)=0.8, \ x_2(1)=1.4, \ x_2(2)=0.5 , \qquad (2.5.27)$

which is mixed-phase with zeros at $z_1=-0.5$, $z_2=-1/0.8=-1.25$.

$ii)$ $\qquad X_3(z)=(1+0.8z^{-1})(z^{-1}+0.5)$

$\qquad\qquad x_3(0)=0.5, \ x_3(1)=1.4, \ x_3(2)=0.8 , \qquad (2.5.28)$

which is mixed-phase with zeros at $z_1=-0.8$, $z_2=-1/0.5=-2.0$.

$iii)$ $\qquad X_4(z)=(z^{-1}+0.8)(z^{-1}+0.5)$

$$x_4(0)=0.4, \quad x_4(1)=1.3, \quad x_4(2)=1.0 \ , \tag{2.5.29}$$

which is maximum phase with zeros at $z_1=-1/0.8=-1.25$ and $z_2=-1/0.5=-2.0$.

Computing partial energies for the sequences is easily accomplished using the formula (2.5.25) for each of $x_1(n)$ through $x_4(n)$. For example, for $x_1(n)$ we have

$$P_0=(1)^2=1 \ ,$$

$$P_1=(1)^2+(1.3)^2=2.69 \ ,$$

$$P_2=(1)^2+(1.3)^2+(0.4)^2=2.85 \ .$$

The partial energies for the remaining sequences may be computed similarly and are tabulated in Table 2.1. Note that the table confirms the assertions made above — the partial energy for the minimum phase system $x_1(n)$ is highest for any $l<2$. Also, as expected the total energy P_2 is identical for all of the spectrally equivalent systems (see Table 2.1).

l	x_1	x_2	x_3	x_4
0	1	0.64	0.25	0.16
1	2.69	2.60	2.21	1.85
2	2.85	2.85	2.85	2.85

Table 2.1 Partial energy values for spectrally equivalent sequences derived from 1,1.3,0.4.

Note that, the term minimum phase is more correctly expressed as **minimum phase lag** and refers to the property of such systems of having the minimum delay at any frequency among all the spectrally equivalent sequences [13]. The term maximum phase can be defined similarly.

Pole/Zero Form

The extension of the definitions of minimum and maximum phase to include both poles and zeros is straightforward. Consider a stable two-sided system

$$H(z)=\frac{b_0+b_1z^{-1}+...+b_Nz^{-N}}{1-a_1z^{-1}-...-a_Mz^{-M}} \ , \qquad (2.5.30)$$

which can be factored as

$$H(z)=K\frac{(1-z_1z^{-1})...(1-z_Nz^{-1})}{(1-p_1z^{-1})...(1-p_Mz^{-1})} \ , \qquad (2.5.31)$$

where the $z=z_i$'s are the zeros and the $z=p_i$'s are the poles of $H(z)$ with

$$|z_i|<1 \quad ; \ i=1,2,...,N_1$$

$$|z_i|>1 \quad ; \ i=N_1+1,N_1+2,...,N$$

and where

$$|p_i|<1 \quad ; \ i=1,2,...,M_1$$

$$|p_i|>1 \quad ; \ i=M_1+1,M_1+2,...,M \ .$$

In general, $H(z)$ is minimum phase if it is a stable system with all its zeros and poles inside $|z|=1$, that is if $M_1=M$, and $N_1=N$. In this case $h(n)$ is also a causal (right-sided) sequence. $H(z)$ is maximum phase if it is a stable sequence with all its poles and zeros outside $|z|=1$. For stability, with all the poles outside $|z|=1$, $h(n)$ must be a purely acausal (left-sided) sequence. The extension of factorizations into minimum/maximum phase and equivalent minimum phase/all pass for sequences containing poles and zeros is straightforward by extension and is left as an exercise.

2.6 Estimation of Moments

2.6.1 General Remarks

In previous sections, the discussion of correlations and spectra has been entirely based on formulas derived directly from the definitions of Section 2.2. As we have seen, spectra are obtained by means of formal integrals using the probability densities of the signals, or by ensemble averaging. In a practical world, however, we usually do not know *a priori* the density associated with any sample $x(n)$. Similarly, in place of an ensemble of sequences, usually only a single experimental result, one member of the ensemble, is available. Moreover, this is likely to be finite in length. In summary then we usually have only a portion, of length M say, of one realization from the random sequence $x(n)$. Under these conditions it is necessary to replace the true expected values with estimates derived from whatever information is available.

Before discussing the estimation of the moments of random signals in detail, a few general comments on estimation are in order. A procedure for estimating a deterministic parameter θ, say, derived from M points of a random data sequence $x(n)$, produces an estimate that is itself a random variable $\hat{\theta} = T(x(n))$, say. The random variable $\hat{\theta}$ is referred to as an **estimator**. Any particular value taken by the random variable $\hat{\theta}$ is an **estimate**.[8]

The quality of an estimator is assessed through certain properties [6]:[9]

a) Bias

The bias $B(\hat{\theta})$ of an estimation procedure is defined by

$$B(\hat{\theta}) = E\{\hat{\theta} - \theta\} = E\{\hat{\theta}\} - \theta . \tag{2.6.1}$$

Clearly, if $B(\hat{\theta}) = 0$ then

$$E\{\hat{\theta}\} = \theta ,$$

and the estimation procedure is **unbiased**. As usual, the expectation operation denotes ensemble averaging and empirically we may say that a procedure is unbiased if, over a large number of repetitions, 'on average' it produces the right answer. This is certainly a desirable property, though an estimator may still be

---

8. Since we are using identical notation for random variables and their realizations some confusion could arise. This can be simply resolved by thinking of the estimator as the estimation procedure and the estimates as the values that it produces.

9. For simplicity we mainly discuss these measures in the context of a scalar parameter estimation problem. Note that similar measures may be defined for the general problem of estimating L parameters $\theta(0), \theta(1), ..., \theta(L-1)$ (see, for example, Van Trees [14]).

useful provided it satisfies the weaker condition of being
asymptotically unbiased in the sense that

$$\lim_{M \to \infty} \{E\{\hat{\theta} - \theta\}\} \to 0 \ . \tag{2.6.2}$$

b) Efficiency

The efficiency of an estimation procedure is determined by the
variance associated with $\hat{\theta}$. The lower the variance, the more
efficient the estimator. Given two estimators producing estimates
$\hat{\theta}_1, \hat{\theta}_2$ the relative efficiency, expressed as a percentage, is defined
by

$$R_e = relative\ efficiency = \left[\frac{Var\{\hat{\theta}_1\}}{Var\{\hat{\theta}_2\}} \times 100\% \right] \ . \tag{2.6.3}$$

Obviously a lower variance is desirable, and $R_e < 100\%$ would imply
an advantage for $\hat{\theta}_1$.

c) Mean-Squared Error

If an estimation procedure is biased then the variance, and hence
the relative efficiency, is not a good measure of quality. For
example, $\hat{\theta}_1$ may be unbiased but have a high variance, whereas $\hat{\theta}_2$
may be biased but have low variance. Under these conditions it is
unclear which procedure is the better estimator. A measure which
embraces both bias and variance is the **mean-squared error**. For
an estimator $\hat{\theta}$, the mean-squared error is defined as:

$$mse(\hat{\theta}) = E\{(\hat{\theta} - \theta)^2\} \ . \tag{2.6.4}$$

The variance of $\hat{\theta}$ is

$$Var(\hat{\theta}) = E\{(\hat{\theta} - \bar{\theta})^2\} \ ,$$

where

$$\bar{\theta} = E\{\hat{\theta}\} \ .$$

From equation (2.6.1) the bias is

$$B(\hat{\theta}) = \bar{\theta} - \theta \ .$$

Using these relations it is easy to show (see Problem 2.23) that

$$mse(\hat{\theta}) = Var(\hat{\theta}) + B^2(\hat{\theta}) \ . \tag{2.6.5}$$

Thus, mean-squared error incorporates both variance and bias and hence provides a more complete measure of the quality of $\hat{\theta}$ than either measure alone.

d) Consistency

Intuitively, a desirable property for any estimator is that the more observations used, the closer the parameter estimate $\hat{\theta}$ should be to the parameter θ. We can formalize this property as follows:

An estimation procedure producing an estimate $\hat{\theta}$ is consistent if $\hat{\theta}$ converges to θ as $M \to \infty$.

Here, by convergence of the estimate to θ we infer

$$\lim_{M \to \infty} \{\hat{\theta}\} \to \theta \ ,$$

in probability [6].[10] A sufficient condition for convergence in probability is that the mean-squared error (2.6.5) should converge to zero as $M \to \infty$

$$\lim_{M \to \infty} \{mse(\hat{\theta})\} \to 0 \ .$$

(see also Problem 2.33).

e) Sufficiency

Formally, we say that $\hat{\theta} = T(x(n))$ is a sufficient statistic[11] for θ if the distribution of the observed data $x(0), x(1), ..., x(M-1)$ conditioned on $\hat{\theta}$ is not dependent on θ. This may appear rather

---

10. $\hat{\theta}$ converges to θ in probability if for any $\epsilon > 0$ and any δ such that $0 < \delta < 1$, we can find an M_0 such that

$p(|\hat{\theta} - \theta| \geq \epsilon) < \delta$ for all $M \geq M_0$ [6].

In words, $\hat{\theta}$ converges in probability to θ if, for sufficiently large M, the probability that $\hat{\theta}$ is far from θ can be made arbitrarily small.

11. In estimation theory, any function of the observed data is referred to as a **statistic** [1].

confusing but means simply that $\hat{\theta}$ contains all of the information in the observations which is relevant to θ. We may interpret this further as follows: The observed data only give information about θ if their probability density functions depend on θ. If the estimation procedure is such that all relevant information from $x(0), x(1),...,x(M-1)$ is included in $\hat{\theta}$, then the density of the data conditioned on the estimate will not depend on θ, and $\hat{\theta}$ will be sufficient. (See also Problem 2.24).

These general properties of estimators may seem somewhat abstract at this stage. Their interpretation in the context of random signal analysis will become clearer in the following discussion on the estimation of moments, and again in Chapter 3 when we discuss optimal estimation procedures.

2.6.2 Estimation of Moments − Time Averages

As we have already suggested, the problem usually encountered in practical situations is how to estimate moments such as the mean and correlation using only a single time-series, one member of the ensemble. Consider, for example, the mean value of a random signal x at sample index n. Recall from Section 2.1 that the relative frequency interpretation gave a result for μ_n as

$$\mu_n = \lim_{N \to \infty} \frac{1}{N} \{x_1(n) + x_2(n) + ... + x_N(n)\} , \qquad (2.1.8)$$

where $x_i(n)$ is the i-th repetition (ensemble member) associated with index n. Given only a single realization, however, evaluating μ_n using (2.1.8) is clearly not an option.

If the sequence is stationary an alternative might be to replace the ensemble average by a **time average**. That is, given a sequence of M points from a stationary sequence, μ could be estimated by

$$\hat{\mu} = \frac{1}{M} \sum_{n=0}^{M-1} x(n) . \qquad (2.6.6)$$

It is apparent that such an estimate is unbiased since

$$E\{\hat{\mu}\} = \frac{1}{M} \sum_{n=0}^{M-1} E\{x(n)\} = \mu . \qquad (2.6.7)$$

It would seem reasonable that stationarity would be required before such time averaging can be used. In fact, time averaging cannot always be used even if $x(n)$ *is* stationary. That is, stationarity is a necessary but not a sufficient condition for a time average of the form (2.6.6) to converge to μ. When such a time average can be employed the sequence is said to be **ergodic**. Various forms of ergodicity can be defined including ergodicity in the mean, ergodicity in mean-square and so on. Each referring to a particular moment. For example, $x(n)$ is ergodic in the mean if the estimator (2.6.6) is consistent:

$$\lim_{M \to \infty} \left\{ \frac{1}{M} \sum_{n=0}^{M-1} x(n) \right\} \to \mu \ . \tag{2.6.8}$$

Given the unbiased nature of $\hat{\mu}$, equation (2.6.8) is satisfied provided

$$\lim_{M \to \infty} \left\{ Var\{\hat{\mu}\} \right\} \to 0 \ . \tag{2.6.9}$$

Now,

$$Var\{\hat{\mu}\} = E\{\hat{\mu}^2\} - E^2\{\hat{\mu}\} = E\left\{ \left[\frac{1}{M} \sum_{n=0}^{M-1} x(n) \right]^2 \right\} - \mu^2 \ ,$$

$$= \frac{1}{M^2} \sum_{n=0}^{M-1} \sum_{m=0}^{M-1} E\{x(n)x(m)\} - \mu^2 \ ,$$

$$Var\{\hat{\mu}\} = \frac{1}{M^2} \sum_{n=0}^{M-1} \sum_{m=0}^{M-1} C(n,m) \ , \tag{2.6.10}$$

where $C(n,m) = E\{x(n)x(m)\} - \mu_m \mu_n$ is the covariance defined by equation (2.1.17). For stationary $x(n)$ this may be written

$$Var\{\hat{\mu}\} = \frac{1}{M^2} \sum_{n=0}^{M-1} \sum_{m=0}^{M-1} C(n-m) \ . \tag{2.6.11}$$

We require that $Var\{\hat{\mu}\} \to 0$ as $M \to \infty$, and one may show [7] that a sufficient condition for this is

$$\lim_{i \to \infty} \left\{ C(i) \right\} = 0 \ , \tag{2.6.12}$$

and provided this holds $x(n)$ is ergodic in the mean. Most physically occurring signals which are stationary are also ergodic with respect to the first and second moments. In common with the usual engineering practice we shall assume that the condition holds for all stationary signals we encounter (though see Problem 2.25 for a simple counterexample).

Now let us consider the estimation of the correlation. A relative frequency interpretation of the correlation for a stationary signal is given by:

$$r(m)= \lim_{N\to\infty} \{\frac{1}{N}\sum_{i=1}^{N} x_i(n)x_i(n+m)\} , \qquad (2.6.13)$$

where $x_i(n)$ is the n-th sample of the i-th realization of $x(n)$, and where the averaging takes place across the ensemble, (see also equation (2.1.14)). Usually we are given only a finite set of samples $x(n)$ for $n=0,1,...,M-1$, say, from a single realization. As with estimation of the mean we replace the ensemble average of (2.6.13) with an estimate obtained by time averaging. One such estimate is given by:

$$\hat{r}(m)=\frac{1}{M-|m|} \sum_{n=0}^{M-|m|-1} x(n)x(n+|m|) , \qquad (2.6.14)$$

where the upper limit of the summation arises because of the restriction to $x(n)$ for $n=0,1,...,M-1$, and where the inclusion of the modulus ensures that $\hat{r}(m)$ satisfies

$$\hat{r}(m)=\hat{r}(-m) . \qquad (2.6.15)$$

Also the divisor $M-|m|$ is chosen so that

$$E\{\hat{r}(m)\}=\frac{1}{M-|m|} \sum_{n=0}^{M-|m|-1} r(m)=E\{\hat{r}(m)\}=r(m) , \qquad (2.6.16)$$

and $\hat{r}(m)$ is unbiased. From equation (2.6.14) however, as m increases each successive $\hat{r}(m)$ is based on less data and the estimate becomes less and less reliable. In the extreme case $m=M-1$ and $\hat{r}(m)$ is derived from a single data product. This effect is quantified via the variance of the estimate, which for Gaussian data and sufficiently large M, is approximately given by [15]

$$Var\{\hat{r}(m)\} \approx \frac{1}{[M-|m|]} \sum_{i=-\infty}^{\infty} [r^2(i)+r(i-m)r(i+m)] \ . \quad (2.6.17)$$

Note that as $M \to \infty$, $Var\{\hat{r}(m)\} \to 0$, so that the estimate $\hat{r}(m)$ is consistent. On the other hand, as m increases relative to M, the variance of the estimate increases.

As a general principle when computing $\hat{r}(m)$, to avoid very high variability it is important to ensure that $M \gg m$ ($M > m/4$ is an empirical rule often given). A further problem associated with $\hat{r}(m)$ is that the positive semi-definite property of the correlation matrix R, constructed using the true correlation values, is lost when the elements of R are the $\hat{r}$'s.

An alternative estimate for $r(m)$, designed to overcome the problems of high variability, is provided by

$$r'(m) = \frac{1}{M} \sum_{n=0}^{M-|m|-1} x(n)x(n+|m|) \ . \quad (2.6.18)$$

This estimate also has the advantage of retaining the positive semi-definite property for the correlation matrix. However, applying expectations to both sides of equation (2.6.18) we have

$$E\{r'(m)\} = \frac{M-|m|}{M} r(m) \ , \quad (2.6.19)$$

and $r'(m)$ is biased. Actually, from (2.6.19) we see that $r'(m)$ is asymptotically unbiased

$$\lim_{M \to \infty} \{E\{r'(m)\}\} \to r(m) \ .$$

This suggests that for $M \gg m$, bias is not a major problem. Additionally for Gaussian data and large M, the variance of $r'(m)$, is approximately given by [15]

$$Var\{r'(m)\} \approx \frac{1}{M} \sum_{i=-\infty}^{\infty} [r^2(i)+r(i-m)r(i+m)] \ . \quad (2.6.20)$$

As with $\hat{r}(m)$, we see that $Var\{r'(m)\} \to 0$ as $M \to \infty$, so that since $r'(m)$ is also asymptotically unbiased it is a consistent estimate. Comparing (2.6.17) and (2.6.20) we see that for Gaussian signals at least, $Var\{r'(m)\}$ is lower than $Var\{\hat{r}(m)\}$.

As we have indicated, when comparing estimators that are biased it is better to compare the mean-squared error rather than the variance. Priestley [6] records that, while the precise values of the mean-squared error for the two estimates will depend on the data length M and sequence $r(m)$, the *biased estimate* $r'(m)$ *generally has a lower mean-squared error and is therefore preferred in most cases.* Practical experience in digital signal processing problems such as deconvolution (discussed in the next Chapter), supports this conclusion. The biased estimate improves the **numerical condition** of the matrix R, and this leads to generally better results. Similar observations have led to the widespread, though not universal use of the biased estimator. In either case the important practical point is that the lag should be kept small relative to the data window length. For $\hat{r}(m)$ this keeps the variance small, for $r'(m)$ it keeps the bias small, and generally leads to very similar estimates for both formulas.

Appendix 2A — Calculating the Correlation via Contour Integration

We have seen that the general linear process $x(n)$ admits the spectral representation

$$R(z)=H(z)H(z^{-1})\sigma_w^2 , \tag{A1}$$

where we may take $h(n)$ as a stable, causal system and where σ_w^2 is the variance of the generating white input.[12] The inverse transform $r(m)$ may be calculated from (A1) via the contour integral:

$$r(m)=\frac{1}{2\pi j}\int_C R(z)z^{m-1}dz , \tag{A2}$$

$$=\frac{1}{2\pi j}\int_C H(z)H(z^{-1})z^{m-1}\sigma_w^2 dz , \tag{A3}$$

---

12. For background on inverse z-transformation using this approach see Oppenheim and Schafer [9].

where C represents a closed contour in the z-plane which lies entirely within the Region Of Convergence (ROC) for the transform $R(z)$.[13] The integral may be evaluated using **Cauchy's residue theorem** which reduces (A2) to a sum of **residues** as:

$$r(m) = \sum \text{Residues of } R(z)z^{m-1} \text{ at the poles within } C. \quad (A4)$$

The residue at an individual k-th order pole at $z=a$ is given by

$$\{\text{Residue at } z=a\} = \lim_{z \to a} \left\{ \frac{1}{(k-1)!} \frac{d^{k-1}}{dz^{k-1}} \left[(z-a)^k R(z) z^{m-1} \right] \right\}. \quad (A5)$$

We shall illustrate the procedure by examining first and second order AR examples:

Example: AR(1) Model

From Section 2.4, equation (2.4.12), the AR(1) sequence has generating system:

$$H(z) = \frac{1}{1 - a_1 z^{-1}} \quad ; \ |a_1| < 1 , \quad (A6)$$

with ROC $|z| > |a_1|$. Hence

$$R_{xx}(z) = \frac{1}{(1 - a_1 z^{-1})(1 - a_1 z)} \sigma_w^2 , \quad (A7)$$

with ROC $|a_1| < |z| < |1/a_1|$. We now substitute (A7) into (A2) giving

$$r(m) = \frac{1}{2\pi j} \int_C \frac{z^m \sigma_w^2}{(z - a_1)(1 - a_1 z)} dz , \quad (A8)$$

where the contour is chosen to lie entirely within the ROC of $R(z)$. For $m \geq 0$, the only pole enclosed by such a contour is at $z = a_1$, and we may write

$$r(m) = \text{Residue of } \left\{ \frac{z^m \sigma_w^2}{(z - a_1)(1 - a_1 z)} \right\} \text{ at } z = a_1 \quad ; \ m \geq 0 . \quad (A9)$$

Using the residue formula (A5), we have

$$r(m)=\{Residue\ at\ z=a_1\ \}=\frac{a_1^m \sigma_w^2}{1-a_1^2}\quad ;\ m\geq 0\ . \qquad (A10)$$

Although we can use the residue formula to obtain the correlation values for $m<0$, the task is complicated by the presence of multiple poles at the origin due to the term z^m in (A8). Fortunately this difficulty can be avoided since we need only impose symmetry to complete the result

$$r(m)=\frac{a_1^{|m|}\sigma_w^2}{1-a_1^2}\quad ;\ -\infty<m<\infty\ , \qquad (A11)$$

which agrees with the result of equation (2.4.15) obtained previously.

Example: AR(2) Model

From equation (2.4.3) the general AR(2) model has generating function

$$H(z)=\frac{1}{1-a_1 z^{-1}-a_2 z^{-2}}\ , \qquad (A12)$$

where $H(z)$ represents a stable, causal system. We may rewrite $H(z)$ as

$$H(z)=\frac{z^2}{z^2-a_1 z-a_2}=\frac{z^2}{(z-p_1)(z-p_2)}\ ,$$

where

$$p_1=\frac{a_1+\sqrt{a_1^2+4a_2}}{2}\ ,\qquad p_2=\frac{a_1-\sqrt{a_1^2+4a_2}}{2}\ , \qquad (A13)$$

and where we assume p_1, p_2 are distinct. Hence from (A1), $R(z)$ is given by

$$R(z)=\frac{z^2\sigma_w^2}{(z-p_1)(z-p_2)(1-p_1 z)(1-p_2 z)}\ . \qquad (A14)$$

Since $H(z)$ is stable and causal, p_1, p_2 lie inside $|z|=1$, and $1/p_1$, $1/p_2$, fall outside. Using (A2) we have

$$r(m)=\frac{1}{2\pi j}\int_C \frac{z^{m+1}\sigma_w^2}{(z-p_1)(z-p_2)(1-p_1 z)(1-p_2 z)}\,dz\ ,\qquad \text{(A15)}$$

where C is chosen to lie within the ROC of $R(z)$ ($|z|=1$, say). For $m\geq-1$ the poles inside the contour are at p_1 and p_2, and using equations (A4) and (A5) we may write

$$r(m)=\frac{\sigma_w^2 p_1^{m+1}}{(p_1-p_2)(1-p_1^2)(1-p_1 p_2)}+\frac{\sigma_w^2 p_2^{m+1}}{(p_2-p_1)(1-p_2^2)(1-p_1 p_2)}$$

$$;\ m\geq-1\ ,$$

or

$$r(m)=Ap_1^m+Bp_2^m\ ,\qquad\text{(A16)}$$

where

$$A=\frac{\sigma_w^2 p_1}{(p_1-p_2)(1-p_1^2)(1-p_1 p_2)}\ ,\qquad\text{(A17a)}$$

$$B=\frac{\sigma_w^2 p_2}{(p_2-p_1)(1-p_2^2)(1-p_1 p_2)}\ .\qquad\text{(A17b)}$$

Finally we may impose symmetry to obtain the complete result

$$r(m)=Ap_1^{|m|}+Bp_2^{|m|}\ ,\qquad\text{(A18)}$$

where A, B are given by (A17).

---

13. Strictly $R(z)=R_{xx}(z)$ but we ignore the subscript to simplify notation.

Problems

2.1 Find the autocorrelation sequence for each of the following processes:

i) $x(n) = a_0 w(n) + a_1 w(n-1)$,

ii) $x(n) = w(n) + 0.4w(n-1) + 0.2w(n-2)$,

iii) $x(n) = a_0 w(n) + a_1 w(n-1) + a_2 w(n-2)$,

where $w(n)$ is a zero-mean white sequence with variance σ_w^2, and where the a_i's are constants. In each case give conditions for which $x(n)$ is stationary.

2.2 Consider the signal

$$y(n) = av^2(n) ,$$

where $v(n)$ is iid with unit variance, and a is constant. Under what conditions is $y(n)$ (weakly) stationary?

2.3 a) Find the mean value, correlation and covariance for the sequence

$$x(n) = A\sin(\omega nT + \theta) ,$$

where ω is a fixed radial frequency, T is the sample interval, A is a constant, and θ is a random initial phase angle which is uniformly distributed over $[-\pi, \pi]$.

b) Repeat part a) if additionally A is a random variable (independent of θ) with zero-mean and unit variance.

2.4 A **harmonic random sequence** is defined by

$$x(n) = \sum_{i=1}^{k} A_i \sin(\omega_i n + \phi_i) ,$$

where A_i, ω_i are constants, and the ϕ_i are independent starting phases that are uniform over $[-\pi, \pi]$. For $k=2$, find the mean, correlation and covariance of $x(n)$.

2.5 a) Show that for a stationary sequence $x(n)$ with correlation $r(m)$, the spectrum

$$R(e^{j\omega}) = \sum_{i=-\infty}^{\infty} r(i)e^{-j\omega i} \ ,$$

is real and even.

b) Assuming that $x(n)$ is a linear process, infer that $R(e^{j\omega})$ is non-negative.

2.6 Given that $w(n)$ is a zero-mean, iid sequence, and $x(n) = h(n) * w(n)$, find $r_{xx}(m)$ if

i) $h(0) = 1$, $h(1) = a$.

ii) $h(n) = a^{2n}$; $n \geq 0$, $h(n) = 0$; $n < 0$.

2.7 A zero mean iid signal $w(n)$, with variance σ_w^2, is input to a system $h(n)$ with

$$h(n) = a^{2n} u(n) \ ,$$

where $u(n)$ is the unit step. Find the correlation $r(m)$ of the output from the system.

2.8 Find $R_{yy}(e^{j\omega})$ for each of the following:

i)

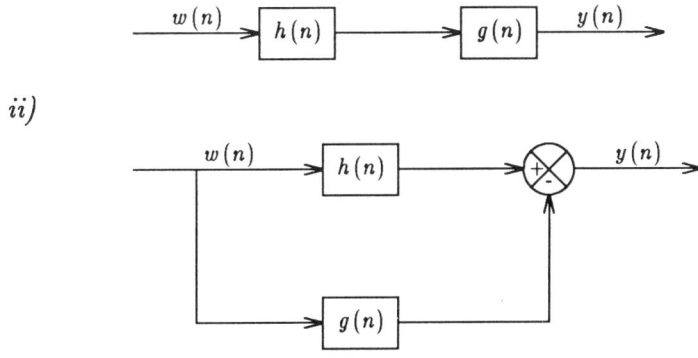

ii)

where $h(0) = 1$, $h(1) = 0.5$, and $g(0) = 1$, $g(1) = -0.5$.

2.9 For a stationary input sequence $x(n)$, and a linear shift-invariant system $h(n)$, with output $y(n)$, show that

a)
$$r_{xy}(m)=E\{x(n)y(n+m)\}= \sum_{j=-\infty}^{\infty} h(j)r_{xx}(m-j) \ .$$

b)
$$R_{xy}(e^{j\omega})=H(e^{j\omega})R_{xx}(e^{j\omega}) \ .$$

2.10 a) Suppose that a zero-mean stationary input $x(n)$ is input to a linear shift-invariant system $h(n)$ producing an output $y(n)$. Use the definition

$$r_{yx}(m)=E\{y(n)x(n+m)\} \ ,$$

to show that

$$R_{yx}(e^{j\omega})=H(e^{-j\omega})R_{xx}(e^{j\omega}) \ .$$

b) Repeat part a) using the definition of $r_{xy}(m)$ and the property

$$r_{xy}(m)=r_{yx}(-m) \ .$$

2.11 Find the power spectrum of the MA(1) process

$$x(n)=b_0 w(n)+b_1 w(n-1) \ .$$

Sketch the result if $b_0=1$, $b_1=0.5$.

2.12 Find the autocorrelation function associated with the AR(2) process

$$x(n)=x(n-1)-0.25x(n-2)+w(n) \ .$$

2.13 Given

$$y(n)=ay(n-2)+w(n) \qquad ; 0<a<1 \ ,$$

where $w(n)$ is a zero-mean white noise sequence with variance σ_w^2, find $r_{yy}(m)$ and $R_{yy}(e^{j\omega})$.

2.14 Show that the general linear process

$$x(n)=\sum_{i=0}^{\infty} h(i)w(n-i) \ ,$$

where $w(n)$ is zero-mean iid with variance σ_w^2, has correlation

$$r(m)=\sigma_w^2\sum_{i=0}^{\infty} h(i)h(i+m) \quad ; m\geq 0 \ ,$$

$$r(m)=r(-m) \ .$$

2.15 The general linear process of equation (2.4.1) is called **invertible** if it can be written:

$$x(n)=\sum_{i=1}^{\infty} b_i x(n-i)+w(n) \ ,$$

with finite coefficients b_i, where $w(n)$ is a zero-mean iid process with variance σ_w^2. For each of the following, give conditions under which the model is invertible and find the inverted representation:

i) $x(n)=w(n)+aw(n-1)$.

ii) $x(n)=w(n)+0.5w(n-1)+aw(n-2)$.

2.16 Show that the MA(M) process

$$x(n)=a_0 w(n)+a_1 w(n-1)+...+a_M w(n-M) \ ,$$

where $E\{w^2(n)\}=\sigma_w^2$, has autocorrelation $r(m)$ which for positive lags can be written

$$r(m)=(a_0 a_m+...+a_{M-m}a_M)\sigma_w^2 \quad ; m=0,1,...,M \ ,$$

$$=0 \quad ; m>M \ .$$

2.17 a) Rewrite the ARMA(1,1) process

$$x(n)=0.8x(n-1)+w(n)+0.5w(n-1) \ ,$$

as a purely MA result.

b) For the same process find a purely AR representation.

2.18 Show that the ARMA(1,1) process

$$x(n)=a_1 x(n-1)+b_0 w(n)+b_1 w(n-1) \ ,$$

has autocorrelation:

$$r(m)=\frac{(b_0^2+b_1^2+2a_1 b_1 b_0)}{(1-a_1^2)}\sigma_w^2 \quad ; \ m=0 \ ,$$

and

$$r(m)=\frac{a_1^{m-1}\sigma_w^2}{1-a_1^2}(a_1^2 b_1 b_0+a_1 b_0^2+a_1 b_1^2+b_1 b_0) \quad ; \ m\geq 1 \ ,$$

$$r(m)=r(-m) \qquad ; \ m<0 \ .$$

2.19 For the MA(1) process

$$x(n)=w(n)+b_1 w(n-1) \ ,$$

show that

$$\left|\frac{r(1)}{r(0)}\right|\leq 0.5 \ ,$$

and find the value(s) of b_1 which produce equality.

2.20 A second order AR process is generated by passing Gaussian white noise through a difference equation

$$x(n)=a_1 x(n-1)+a_2 x(n-2)+w(n) \ .$$

i) Find conditions on the coefficients a_1, a_2, to ensure stability of the signal.

ii) Show that the autocorrelation can be written in the form

$$r(i)=c_1\left[\frac{a_1+\sqrt{a_1^2+4a_2}}{2a}\right]^{|i|}+c_2\left[\frac{a_1-\sqrt{a_1^2+4a_2}}{2a}\right]^{|i|} \ ,$$

where c_1, c_2 are constants.

2.21 For each of the following, show that

$$|H(e^{j\omega})|^2 = constant .$$

i)

$$H(z) = \frac{(1 - z_1 z^{-1})}{(z^{-1} - z_1)} ,$$

where z_1 is real.

ii)

$$H(z) = \frac{(1 - z_a z^{-1})(1 - z_a^* z^{-1})}{(z^{-1} - z_a)(z^{-1} - z_a^*)} ,$$

where $*$ denotes complex conjugate.
Use these results to infer that a general $H(z)$ expressible in the form of equation (2.5.16), is all-pass.

2.22 Consider the sequence

$$x(n) = w(n) + aw(n-1),$$

where $w(n)$ is zero-mean iid with variance σ_w^2.

a) Find the correlation $r(m)$.

b) Find the power spectrum $R(e^{j\omega})$.

c) Find the bias associated with the estimate

$$r^{'}(1) = \frac{1}{N} \sum_{n=0}^{N-2} x(n)x(n+1) .$$

2.23 Using the definitions

$$B(\hat{\theta}) = E\{\hat{\theta}\} - \theta ,$$

and

$$Var(\hat{\theta}) = E\{(\hat{\theta} - \bar{\theta})^2\} ,$$

where $\bar{\theta}=E\{\hat{\theta}\}$, show that

$$mse(\hat{\theta})=E\{(\hat{\theta}-\theta)^2\}=Var(\hat{\theta})+B^2(\hat{\theta}) \ .$$

2.24 *Result:[6]* If the joint density of a sequence $x(1),x(2),...,x(M)$ and a parameter θ can be factorized into two functions

$$f\left(x(1),x(2),...,x(M)\right)=g\left(\hat{\theta},\theta\right)h\left(x(1),x(2),...,x(M)\right) \ ,$$

then the estimator $\hat{\theta}$ is a sufficient statistic for θ. Use this result to show that for Gaussian iid random variables $x(i)$; $i=1,2,...,M$, distributed as $N(\mu,\sigma)$ then

$$\bar{x}=\frac{1}{M}\sum_{i=1}^{M}x(i) \ ,$$

is a sufficient statistic for μ.

2.25 Consider the signal

$$x(n)=A\cos\theta+y(n) \ ,$$

where $y(n)$ is a zero-mean iid sequence, θ is initially distributed uniformly on $[-\pi,\pi]$, and A is a constant.

a) Compute $E\{x(n)\}$, the ensemble average.

b) By computing the time average

$$\lim_{N\to\infty}\{\frac{1}{N}\sum_{n=1}^{N}x(n)\} \ ,$$

show that $x(n)$ is *not* ergodic in the mean.

2.26 A deterministic signal is estimated by averaging M noise corrupted measurements

$$y_j(n)=x(n)+v_j(n) \qquad ; j=0,1,...,M-1 \ ,$$

where $v_j(n)$ is a zero mean iid sequence and $E\{v_j(n)v_i(n)\}=\sigma^2\delta(i-j)$. Find the variance of

$$\hat{x}(n)=\frac{1}{M}\sum_{m=1}^{M}y_m(n) \ .$$

2.27 A signal $x(n)$ is assumed to consist of independent samples drawn from a Laplacian density:

$$f\left(x(n)\right)=\frac{1}{\lambda}e^{\frac{-2|x(n)|}{\lambda}} \qquad ; \lambda>0 \ .$$

This signal is input to the second order filter:

$$H(z)=\frac{1}{(1-0.5z^{-1})(1-0.25z^{-1})} \ .$$

The output of this filter is denoted $y(n)$.

a) Find the correlation and covariance of the excitation signal $x(n)$.

b) Find $r_{yy}(m)$ and $R_{yy}(e^{j\omega})$.

Note: You may assume that

$$\int_{0}^{\infty}x^n e^{-ax}dx=\frac{n!}{a^{n+1}} \ .$$

2.28 For each of the following signals, show that the sample mean

$$\hat{\mu}=\frac{1}{M}\sum_{i=0}^{M-1}x(i) \ ,$$

is unbiased and consistent.

a) $x(n)$ iid with mean value μ, variance σ^2.

b) $x(n)=w(n)+aw(n-1)$ where $w(n)$ is zero-mean unit variance iid.

2.29 If $x(i)$; $i=0,1,...,M-1$ are iid Gaussian samples with mean μ and variance σ_x^2, evaluate the variance estimate

$$\hat{v}=\frac{1}{M}\sum_{i=0}^{M-1}(x(i)-\mu)^2 \ ,$$

for *i)* Bias and *ii)* Consistency.

Note: you may use the following result: For a zero-mean Gaussian variate x,

$$E\{x^n\}=0 \quad ; n=1,3,5,..$$

$$=1.3...(n-1)\sigma^n \quad ; n=2,4,..$$

2.30 Consider a set of observations $x(i)$; $i=0,1,...,L-1$, where the $x(i)$'s are iid Gaussian with mean μ and variance σ^2. Show that the general form for the L-dimensional joint distribution given by equation $(2.1.41)$ can be reduced to

$$f(\underline{x})=\frac{1}{(2\pi)^{L/2}\,\sigma^L}\,e^{\left[-\frac{1}{2\sigma^2}\sum_{i=0}^{L-1}(x(i)-\mu)^2\right]} \quad .$$

2.31 a) Consider two *uncorrelated* zero-mean jointly Gaussian random variables x_1, x_2 with variances σ_1^2, σ_2^2, respectively. By expanding the joint Gaussian density of equation $(2.1.41)$ show that x_1, x_2 are also *independent*.

   b) Extend the result of part a) to the case of N dimensional random vectors $\underline{x}_1$, $\underline{x}_2$ both zero-mean, jointly Gaussian with $E\{\underline{x}_1\underline{x}_2^t\}=0$, and with

$$E\{\underline{x}_i\underline{x}_i^t\}=R_i \quad ; i=1,2 \ .$$

Note: You may use the result that for a block-diagonal matrix

$$R=\begin{bmatrix} R_1 & 0 \\ 0 & R_2 \end{bmatrix} ,$$

$$Det(R)=Det(R_1)Det(R_2) \ .$$

2.32 Consider independent samples $x(i)$; $i=0,1,...,M-1$, with mean μ and variance σ_x^2. Show that

$$\sum_{i=0}^{M-1}(x(i)-\mu)^2\equiv\sum_{i=0}^{M-1}(x(i)-\overline{x})^2+M(\overline{x}-\mu)^2 \ .$$

Hence show that

$$v' = \frac{1}{M} \sum_{i=0}^{M-1} (x(i) - \bar{x})^2 ,$$

where

$$\bar{x} = \frac{1}{M} \sum_{i=0}^{M-1} (i) .$$

is a biased estimator for σ_x^2. Determine the bias as $M \to \infty$, what can you conclude about v'?

2.33 An estimator $\hat{\theta}$ is called **consistent in mean-square** if

$$\lim_{M \to \infty} \{mse(\hat{\theta})\} \to 0 .$$

Show that $\hat{\theta}$ is consistent in mean-square if and only if: *i)* $Var\{\hat{\theta}\} \to 0$ and *ii)* $\hat{\theta}$ is asymptotically unbiased.

2.34 Assume we have independent trials x_i ; $i=1,2,...,n$ drawn from the probability law:

$$prob\{x_i = 1\} = p ,$$
$$prob\{x_i = 0\} = (1-p) = q .$$

Consider

$$K = \frac{1}{N} \sum_{i=1}^{n} x_i .$$

a) Show that K is an unbiased estimator of $E\{x_i\}$ and that $(1-K)$ is an unbiased estimator of q, but that $K(1-K)$ is *not* an unbiased estimator of pq.

b) Use the result of Problem **2.33** to infer that K is a consistent estimator for p.

References

1.	A.Papoulis, *Probability, Random Variables and Stochastic Processes*, McGraw-Hill, 1984.

2.	W.B.Davenport, Jr., *Probability and Random Processes*, McGraw-Hill, 1970.

3.	H.Stark and J.W.Woods, *Probability, Random Processes and Estimation Theory for Engineers*, Prentice-Hall, 1986.

4.	G.E.P.Box and G.M.Jenkins, *Time Series Analysis, Forecasting and Control*, Holden-Day, 1976.

5.	G.M.Jenkins and D.G.Watts, *Spectral Analysis and its Applications*, Holden-Day, 1968.

6.	M.B.Priestley, *Spectral Analysis and Time Series, Vols I and II*, Academic Press, 1981.

7.	W.W.S.Wei, *Time Series Analysis: Univariate and Multivariate Methods*, Addison-Wesley, 1990.

8.	C.L.Nikias and M.R.Raghuveer, "Bispectrum estimation: A digital signal processing framework," *Proc. IEEE*, vol. 75, pp. 869-891, 1987.

9.	A.V.Oppenheim and R.W.Schafer, *Discrete-Time Signal Processing*, Prentice-Hall, 1989.

10.	S.J.Orfanidis, *Optimum Signal Processing: An Introduction*, McGraw-Hill, 1988.

11.	H.O.A.Wold, *A Study in the Analysis of Stationary Time-Series*, Almquist and Wiksell, 1938.

12.	R.E.Mickens, *Difference Equations – Theory and Applications*, Sec. Ed., Van Nostrand Reinhold, 1990.

13.	E.A.Robinson, *Statistical Communication and Detection with Special Reference to Radar and Seismic Signals*, Griffin, 1967.

14.	H.L.Van Trees, *Detection, Estimation and Modulation Theory*, Wiley, 1968.

15.	M.S.Bartlett, "On the theoretical specification of sampling properties of autocorrelated time series," *J. Roy. Statist. Soc. Suppl.*, vol. 8, pp. 27-41, 1946.

chapter three

Optimal Signal Processing

3.1 Optimal Estimation Procedures

3.1.1 Classical and Bayesian Estimation

Estimation theory plays a major role in statistical signal processing [1]-[3]. Apart from the estimation of moments which was described in Section 2.6, and of spectra (which we defer until Chapter 7), we often require estimation procedures for signals and for a wide range of signal model parameters. These problems include smoothing, filtering and prediction of noisy observations of signals, as well as the estimation of the parameters of signal generating models such as the AR, MA and ARMA models described in Section 2.4. We shall discuss these estimation problems in fairly general terms before giving detailed consideration to particular procedures which will subsequently be used extensively.

We consider the general problem of estimating one or more unknown parameters from a series of data measurements (observations) where the parameters are functionally related to the observed data in some manner. Estimation problems are often delineated into **deterministic** (or **classical**) parameter estimation, and **random** (or **Bayesian**) parameter estimation. In classical estimation, the observations are random but the parameters are thought of as unknown constant (deterministic) values. In Bayesian parameter estimation the parameters are also viewed as random, and our prior knowledge of their behavior is expressed through an **a priori density**. In terms of the 'relative frequency' interpretation of probability there is certainly a fundamental difference between the two approaches: In classical estimation, if we repeat the experiment, then although the observed data values change, the parameter is deterministic and therefore remains unchanged. By contrast, in Bayesian estimation if the experiment were repeated the parameter value would vary according to the probability law expressed by the *a priori* density. However, there is another view of the division between classical and Bayesian estimation. In particular, we may think of the parameters as random only prior to any measurements or observations. In this view the *a priori* density is simply a reflection of our prior knowledge of (or belief about) the likely behavior of the parameters. In either view the essential distinction is one of *prior information* − classical estimation procedures operate without *a priori* knowledge of the parameters whereas Bayesian estimation methods exploit *a priori* knowledge of the parameter behavior to produce an estimate based on both prior and posterior (after observation) information.

Both classical and Bayesian estimation have found application in signal processing. In particular, three estimation procedures have been used widely:

1. Maximum A Posteriori (MAP) Estimation
2. Maximum Likelihood (ML) Estimation
3. Minimum Mean-Squared Error (MMSE) Estimation

In the sequel we shall consider each of these separately. In each case we shall classify the form of the estimator (classical or Bayesian), define the properties, and finally use examples to illustrate the relation between estimators. We begin by examining MAP and ML estimation in the context of a particularly simple problem; that of estimating a single parameter given one or more measurements.

Maximum A Posteriori (MAP) and Maximum Likelihood (ML) Estimation of a Single Parameter

1. MAP Estimation of a Single Random Parameter

Suppose we are given a set of M samples from a discrete random sequence $x(n)$ for $n=0,1,...,M-1$. Suppose further that from this data we need to estimate the value of a random parameter θ, which is dependent on $x(n)$ in some way. We assume that the parameter θ is characterized by a probability density function $f(\theta)$ called the *a priori* density. $f(\theta)$ corresponds to the assumed probability distribution for the parameter in the absence of other information, that is prior to any measurements. At this stage we will not be too specific about the parameter θ but simply note that in one example it might represent a constant signal contained within noisy measurements $x(n)$, or in another case θ may be a parameter of a signal generating model and so on. As a notational convenience we represent the observed data in vector form $\underline{x} = [x(0), x(1), ..., x(M-1)]^t$.

The probability density function $f(\theta/\underline{x})$ is the probability density of θ conditioned on the observed values of $x(n)$, and is referred to as the **a posteriori** density for θ. The Maximum A Posteriori (MAP) estimate of θ is the peak value of this density. This maximum corresponds to the *most probable* value of θ given the observed data. The maximum can generally be found by taking the partial derivative of $f(\theta/\underline{x})$ with respect to θ and equating to zero[1]

$$\frac{\partial}{\partial \theta} \{ f(\theta/\underline{x}) \} = 0 . \tag{3.1.1}$$

This represents a turning point in the posterior density, and provided the density is unimodal[2] then the parameter obtained from the solution of this equation is the MAP estimate, θ_{MAP}. Given the random model for θ we can classify the procedure as Bayesian.

It is often more convenient to work in terms of the logarithm of the posterior density $\ln[f(\theta/\underline{x})]$. The same value of θ maximizes both $f(\theta/\underline{x})$ and $\ln[f(\theta/\underline{x})]$ because of the monotonicity of the logarithm. An equivalent procedure for the estimation is therefore

$$\frac{\partial}{\partial \theta} \{ \ln[f(\theta/\underline{x})] \} = 0 . \tag{3.1.2}$$

Once again the solution yields θ_{MAP}.

Determination of $f(\theta/\underline{x})$, or $\ln[f(\theta/\underline{x})]$ is often problematic. Applying Bayes rule [4], however, we may rewrite the conditional density as

$$f(\theta/\underline{x}) = \frac{f(\underline{x}/\theta) f(\theta)}{f(\underline{x})} , \tag{3.1.3}$$

---

1. Note that $f(\theta/\underline{x})$ is a one-dimensional function of the scalar parameter θ. The vector $\underline{x}$ simply fixes the location of the density.

2. Strictly (3.1.1) gives only a necessary condition for the maximum [1] and we should check to ensure that θ_{MAP} corresponds to the global maximum. We

where, as above, $f(\theta)$ is the *a priori* density of θ. Applying the logarithm to both sides of equation (3.1.3) we have

$$\ln[f(\theta/\underline{x})]=\ln[f(\underline{x}/\theta)]+\ln[f(\theta)]-\ln[f(\underline{x})] \ .$$

The MAP procedure of equation (3.1.2) may thus be written

$$\frac{\partial}{\partial\theta}\{\ln[f(\underline{x}/\theta)]+\ln[f(\theta)]-\ln[f(\underline{x})]\}=0 \ ,$$

but, since $f(\underline{x})$ is not dependent on θ we may write

$$\frac{\partial}{\partial\theta}\{\ln[f(\underline{x}/\theta)]+\ln[f(\theta)]\}=0 \ . \qquad (3.1.4)$$

Hence θ_{MAP} may equivalently be obtained from the solution of (3.1.1), (3.1.2) or (3.1.4) as convenience dictates.

2. ML Estimation of a Single Unknown Parameter

Let us now consider a slightly different estimation problem. As before assume that we are given a sequence of measurements $x(n)$ for $n=0,1,...,M-1$ and that we wish to estimate a parameter θ. Now however, we assume that θ is an unknown (deterministic) constant. The **Maximum Likelihood (ML)** estimate of θ is obtained by selecting the value θ_{ML} which maximizes the density of the observed data $\underline{x}$ with respect to the parameter θ. That is, θ_{ML} is obtained by maximizing the function L_θ where

$$L_\theta=f(\underline{x};\theta) \ . \qquad (3.1.5)$$

L_θ is the density of the observations $\underline{x}$, and the notation $f(\underline{x};\theta)$ reflects the dependence of the density on θ. In ML estimation we view $f(\underline{x};\theta)$ as a function of θ, that is with $\underline{x}$ as a fixed set of observations. The function $L_\theta=f(\underline{x};\theta)$ is called the **likelihood function** of θ. Note that if

$$f(\underline{x};\theta_1)>f(\underline{x};\theta_2) \ , \qquad (3.1.6)$$

---

simplify things here by assuming that the posterior density *is* unimodal.

then θ_1 is a 'more plausible' value of θ than θ_2. The method of Maximum Likelihood is based on the principle that θ be estimated by its most plausible value given the observations $\underline{x}$. That is, θ_{ML} is chosen to maximize L_θ. Given the model of θ as an unknown deterministic constant we may classify ML as a classical estimation method.[3]

As with MAP estimation it is often more convenient to work with the logarithm, and the **log-likelihood function** is defined by

$$\ln L_\theta = \ln[f(\underline{x};\theta)] \ .$$

The ML estimate can generally be obtained by differentiating and equating to zero:

$$\frac{\partial}{\partial \theta}\{\ln L_\theta\}=0 \ . \tag{3.1.7}$$

Assuming $f(\underline{x};\theta)$ is unimodal, the solution of (3.1.7) yields the ML estimate θ_{ML}. Note that L_θ and $\ln L_\theta$ are functionally dependent on the random data, and are therefore also random variables.

The power of the ML estimation process is reflected in a number of important properties:

a) Sufficiency – If a sufficient statistic for θ exists, then the ML estimate will be a sufficient statistic.

b) Efficiency – As we shall see shortly, there exists a lower bound on the variance attainable by an unbiased estimator for an unknown θ. An estimator whose variance attains this lower bound, and is sufficient is called **fully efficient**. If a fully efficient statistic exists then the ML estimate will be that statistic. Even if a fully efficient estimate does not exist for finite M, in almost all practical cases ML estimates are asymptotically fully efficient [5].

c) Gaussianity – ML estimates are asymptotically Gaussian [1].

Contrasting ML and MAP estimation, we see that the MAP estimate is derived using a combination of prior and posterior knowledge. In MAP estimation we assume something is known

---

3. There is nothing absolute about this classification, however, it is possible to derive the ML estimator via the MAP estimator by considering θ to be random *a priori* and uniformly distributed over an infinite range [2].

about θ and we formulate this knowledge into the prior density $f(\theta)$. ML estimation is potentially more useful than MAP estimation in practical problems because no prior knowledge of the parameter is required. It must be remembered, however, that both procedures require knowledge of the joint posterior density of the observations.

Example — Noisy Measurement (1 Parameter, 1 Observation)

A simple example will help to illustrate the MAP and ML estimation procedures: Consider a single measurement x consisting of the sum of a parameter θ and a random variable w which is zero-mean Gaussian with variance σ_w^2:

$$x = \theta + w \ . \tag{3.1.8}$$

(We may think of x as a single sample, noise corrupted measurement consisting of signal θ and additive noise w.)
Find:

   i) The ML estimate if θ is an unknown constant.

   ii) The MAP estimate if θ is random with *a priori* density $N(\bar{\theta}, \sigma_\theta)$.

i) ML Estimate — The likelihood function is

$$L_\theta = f(x;\theta) \ , \tag{3.1.5}$$

where, as before the notation reflects the functional dependence on θ but where in this single parameter, single observation case, $f(x;\theta)$ is a scalar function of a single unknown variable. From equation (3.1.8), x is a Gaussian signal with mean θ and variance σ_w^2 and the likelihood function L_θ reflects that dependence

$$L_\theta = f(x;\theta) = \frac{1}{\sqrt{2\pi}\,\sigma_w} e^{-\frac{1}{2\sigma_w^2}(x-\theta)^2} \ . \tag{3.1.9}$$

The log-likelihood function is then

$$\ln L_\theta = \ln[f(x;\theta)] = -\frac{1}{2}\ln(2\pi\sigma_w^2) - \frac{1}{2\sigma_w^2}(x-\theta)^2 \ . \tag{3.1.10}$$

Differentiating with respect to θ and equating to zero we have

$$\frac{1}{\sigma_w^2}(x - \theta_{ML}) = 0 ,$$

which yields

$$\theta_{ML} = x . \tag{3.1.11}$$

Thus, the best estimate in the ML sense is the raw data. This is an intuitively satisfying result − without any information about θ we cannot refine the measurement in any way.

The variance associated with this estimate will be of interest in the sequel. It is evaluated as

$$Var\{\theta_{ML}\} = E\{\theta_{ML}^2\} - E\{\theta_{ML}\}^2$$
$$= E\{x^2\} - E^2\{x\} ,$$

which for $x = \theta + w$ gives

$$Var\{\theta_{ML}\} = \theta^2 + \sigma_w^2 - \theta^2 = \sigma_w^2 . \tag{3.1.12}$$

ii) MAP Estimate − We now have $x = \theta + w$ with w zero-mean Gaussian with variance σ_w^2, and with the prior density $f(\theta)$ distributed as $N(\bar{\theta}, \sigma_\theta)$. The MAP estimate is obtained from equation (3.1.4)

$$\frac{\partial}{\partial \theta}\{\ln[f(x/\theta)] + \ln[f(\theta)]\} = 0 . \tag{3.1.4}$$

The density of x given the value θ is Gaussian with mean θ and variance σ_w^2. Hence, the logarithm of the density is

$$\ln[f(x/\theta)] = -\frac{1}{2}\ln(2\pi\sigma_w^2) - \frac{1}{2\sigma_w^2}(x - \theta)^2 . \tag{3.1.13}$$

The *a priori* density $f(\theta)$ is given by

$$f(\theta) = \frac{1}{\sqrt{2\pi}\,\sigma_\theta} e^{-\frac{1}{2\sigma_\theta^2}(\theta - \bar{\theta})^2} ,$$

and hence

$$\ln[f(\theta)] = -\frac{1}{2}\ln(2\pi\sigma_\theta^2) - \frac{1}{2\sigma_\theta^2}(\theta-\overline{\theta})^2 \ . \tag{3.1.14}$$

Substituting (3.1.13) and (3.1.14) into (3.1.4) gives

$$\frac{\partial}{\partial\theta}\left[-\frac{1}{2}\ln(2\pi\sigma_w^2) - \frac{1}{2\sigma_w^2}(x-\theta)^2 - \frac{1}{2}\ln(2\pi\sigma_\theta^2) - \frac{1}{2\sigma_\theta^2}(\theta-\overline{\theta})^2\right] = 0 \ .$$

Differentiating we have

$$\frac{(x-\theta_{MAP})}{\sigma_w^2} - \frac{(\theta_{MAP}-\overline{\theta})}{\sigma_\theta^2} = 0 \ .$$

Rearranging yields the MAP estimate

$$\theta_{MAP} = \frac{x\sigma_\theta^2 + \overline{\theta}\sigma_w^2}{\sigma_w^2 + \sigma_\theta^2} \ ,$$

or

$$\theta_{MAP} = \left[\frac{x + \overline{\theta}(\sigma_w^2/\sigma_\theta^2)}{1 + (\sigma_w^2/\sigma_\theta^2)}\right] \ . \tag{3.1.15}$$

Comparing this to the ML estimate $\theta_{ML}=x$ of equation (3.1.11) we see that θ_{MAP} can be viewed as a weighted sum of the ML estimate x and the prior mean $\overline{\theta}$. In (3.1.15), the ratio $(\sigma_w^2/\sigma_\theta^2)$ can be viewed as a measure of confidence in the *a priori* model of θ. The smaller the value of σ_θ^2 the larger the ratio of σ_w^2 to σ_θ^2, the greater our confidence in $\overline{\theta}$ and the less weight we give to the observation x. In the limit as $(\sigma_w^2/\sigma_\theta^2)\rightarrow\infty$ the MAP estimate is simply given by $\overline{\theta}$, the *a priori* mean. At the other extreme as σ_θ^2 becomes large, then $\theta_{MAP}\rightarrow x$ which is the ML estimate.

Example – Noisy Measurements (1 Parameter, M Observations)

We may easily extend the previous example by considering M observations $x(0), x(1), ..., x(M-1)$ with

$$x(n) = \theta + w(n) \quad ; \ n = 0, 1, ..., M-1 \ , \tag{3.1.16}$$

where the samples $w(n)$ are iid Gaussian random variables distributed as $N(0,\sigma_w)$. Here we may think of $x(n)$ as noisy measurements of a signal which is constant over the observations. As in the previous example we consider two cases:

  i) The ML estimate if θ is an unknown constant.

  ii) The MAP estimate if θ is random with *a priori* density $N(\bar{\theta},\sigma_\theta)$.

i) The ML estimate — The observed variables form a set of independent Gaussian random variables with mean θ. The likelihood function is now equal to the M-dimensional density of the observations, which includes the functional dependence on the mean value θ:

$$L_\theta = f\left(\underline{x};\theta\right) = \frac{1}{(2\pi\sigma_w^2)^{M/2}} e^{\frac{-1}{2\sigma_w^2}\sum_{i=0}^{M-1}(x(i)-\theta)^2} , \qquad (3.1.17)$$

where the form of the multidimensional Gaussian density follows from equation (2.1.41) of Section 2.1. The log-likelihood function now becomes

$$\ln L_\theta = \ln[f\left(\underline{x};\theta\right)] = -\frac{M}{2}\ln(2\pi\sigma_w^2) - \frac{1}{2\sigma_w^2}\sum_{i=0}^{M-1}\left(x(i)-\theta\right)^2 . \qquad (3.1.18)$$

Hence, performing the maximization we have

$$\frac{\partial \ln L_\theta}{\partial \theta} = \sum_{i=0}^{M-1}\left(x(i)-\theta_{ML}\right) = 0 ,$$

or

$$\theta_{ML} = \frac{1}{M}\sum_{i=0}^{M-1}x(i) . \qquad (3.1.19)$$

So that the best estimate in the maximum likelihood sense is the sample mean of the observations. Again, this seems entirely consistent with our intuition. Note, however, that this solution holds for Gaussian observation noise but the simple average is *not* the maximum likelihood estimate if the noise density is non-Gaussian (see Problem 3.3).

ii) The MAP estimate – The MAP estimate is obtained from equation (3.1.4) with the posterior density given by the multidimensional Gaussian function of equation (3.1.17). Substituting and performing the differentiation yields

$$-\sigma_\theta^2 \left[\sum_{i=0}^{M-1} \left(x(i) - \theta_{MAP} \right) \right] + \sigma_w^2 (\theta_{MAP} - \bar{\theta}) = 0 \ ,$$

which, upon rearrangement gives:

$$\theta_{MAP} = \frac{\dfrac{1}{M} \sum_{i=0}^{M-1} x(i) + \bar{\theta} \left(\dfrac{\sigma_w^2}{M\sigma_\theta^2} \right)}{1 + \left(\dfrac{\sigma_w^2}{M\sigma_\theta^2} \right)} \ . \tag{3.1.20}$$

The MAP estimate can be viewed as equal to a weighted sum of the ML estimate (driven by the observations) and the prior estimate $\bar{\theta}$. As in the single observation case, the relative weight given to prior and posterior components depends on the ratio of σ_w^2 to σ_θ^2. Comparing with the single observation estimate (equation (3.1.15)), we see that the effect of using M independent observations is to reduce the dependence on the prior density by a factor of M. This is entirely reasonable – each observation reduces the overall variance of the observations and thus reduces our dependence on the prior model.

MAP and ML Estimation of Multiple Parameters

The ML and MAP estimation methods generalize to the case of multiple parameters in a straightforward fashion. If we have observations $x(n)$ for $n = 0, 1, ..., M-1$, and we seek estimates of parameters $\underline{\theta} = [\theta(0), \theta(1), ..., \theta(L-1)]^t$, then:

1. MAP Estimation of L Random Parameters

By analogy with the single parameter problem, the MAP estimator is obtained by maximizing the posterior density $f(\underline{\theta}/\underline{x})$ or equivalently $\ln[f(\underline{\theta}/\underline{x})]$, with respect to θ. This is achieved by differentiating with respect to each component of θ and equating the result to zero. That is

$$\frac{\partial}{\partial \theta(i)}\{\ln[f(\underline{\theta}/\underline{x})]\}=0 \quad ; i=0,1,...,L-1 \; . \tag{3.1.21}$$

Hence we have a separate partial derivative, and thus a separate equation, for each parameter $\theta(i)$. Thus, for parameters $\theta(0),\theta(1),...,\theta(L-1)$ we have a set of L equations in L unknowns, the solution of which yields the MAP estimates. An alternative notation for (3.1.21) is

$$\nabla_{\underline{\theta}}\{\ln[f(\underline{\theta}/\underline{x})]\}=\underline{0} \; ,$$

where $\nabla = [\dfrac{\partial}{\partial \theta(0)}, \dfrac{\partial}{\partial \theta(1)},...,\dfrac{\partial}{\partial \theta(L-1)}]^t$.

As in the single parameter case, it is usually helpful to rewrite the density $f(\underline{\theta}/\underline{x})$ using Bayes rule as

$$f(\underline{\theta}/\underline{x})=\frac{f(\underline{x}/\underline{\theta})f(\underline{\theta})}{f(\underline{x})} \; , \tag{3.1.22}$$

or, in terms of the logarithm

$$\ln[f\,(\underline{\theta}/\underline{x})]=\ln[f\,(\underline{x}/\underline{\theta})]+\ln[f\,(\underline{\theta})]-\ln[f\,(\underline{x})] \; .$$

As in the single parameter case, $\ln[f(\underline{x})]$ has no dependence on $\underline{\theta}$ so that substituting into (3.1.21) we may write

$$\frac{\partial}{\partial \theta(i)}\{\ln[f(\underline{x}/\underline{\theta})]+\ln[f\,(\underline{\theta})]\}=0 \quad ; i=0,1,...,L-1 \; , \tag{3.1.23}$$

and the solution of this set of L simultaneous equations yields $\underline{\theta}_{MAP}$.

2. ML Estimation of L Unknown Parameters

The ML estimator for the unknown parameter vector $\underline{\theta}$ is obtained by maximizing the multidimensional likelihood function

$$L_{\theta}=f(\underline{x};\underline{\theta}) \; , \tag{3.1.24}$$

or equivalently the log-likelihood

$$\ln L_{\theta}=\ln[f\,(\underline{x};\underline{\theta})] \; . \tag{3.1.25}$$

The maximum of this function is obtained from the solution of

$$\frac{\partial[\ln L_\theta]}{\partial\theta(i)}=0 \quad ; \ i=0,1,...,L-1 \ . \tag{3.1.26}$$

Once again we have a set of L equations in L unknowns and the solution yields $\underline{\theta}_{ML}$.

Having specified the general form for the ML estimator we now pose the question: How good is any particular estimator? The following result gives a valuable clue and also demonstrates the power of the ML estimator:

Result – Cramer-Rao Lower Bound (CRLB)

If $\underline{\theta}$ is an $(L\times1)$ unknown (deterministic) constant parameter vector, and if $\hat{\underline{\theta}}$ is any unbiased estimator of $\underline{\theta}$ then[4]

$$Cov\{\hat{\underline{\theta}}\}\geq J^{-1} \ , \tag{3.1.27}$$

where $Cov\{\hat{\underline{\theta}}\}$ is the $(L\times L)$ covariance matrix for the parameter vector with

$$C=Cov\{\hat{\underline{\theta}}\}=E\{(\underline{\theta}-\hat{\underline{\theta}})(\underline{\theta}-\hat{\underline{\theta}})^t\} \ , \tag{3.1.28}$$

(see Section 2.1), and where J is the $(L\times L)$ matrix with elements

$$J(i,j)=-E\{\frac{\partial^2\ln[f(\underline{x};\theta)]}{\partial\theta(i)\partial\theta(j)}\} \quad ; \ i,j=0,2,...,L-1 \ . \tag{3.1.29}$$

J is known as the **Fisher information matrix.**

Notes

i) The inequality (3.1.27) is known as the **Cramer-Rao Lower Bound (CRLB).** It provides a fundamental lower limit on the variance attainable with an unbiased estimator for an unknown parameter vector.[5] An estimator which has these properties is referred to as the Minimum Variance Unbiased Estimator

---

4. See, for example, Van Trees [1] for a proof of this result.
5. This result is subject to certain 'regularity' conditions on the distribution [6].

(MVUE) for the problem. Note that the matrix inequality $Cov\{\hat{\theta}\} \geq J^{-1}$ is interpreted as meaning that the matrix $[C - J^{-1}] \geq 0$ is positive semi-definite. The best any unbiased estimator can do is attain equality in (3.1.27) for the variance. Such an estimator is then called fully efficient. As we have already observed, it is possible that such an estimator does not exist. However, if a fully efficient statistic exists, then the ML estimator is that statistic.

ii) Equation (3.1.27) gives a general relation for the bound on the covariances of the parameters. It is often more helpful to be able to bound the variances of the individual parameter estimates. These correspond to the diagonal elements of $C = Cov\{\hat{\theta}\}$. Now, the matrix $[C - J^{-1}]$ is positive semidefinite. A property of such matrices is that the diagonal elements are all non-negative [7]. Hence

$$Var\{\theta(i)\} \geq J^{-1}(i,i) \quad ; \quad i = 0, 1, ..., L-1 \ ,$$

where $J^{-1}(i,i)$ is the *i*th diagonal element of the *inverse* matrix J^{-1}.

iii) For the special case of scalar parameter estimation, the CRLB of (3.1.27) takes on a simpler form: For an unbiased estimate of an unknown deterministic parameter θ we have

$$Var\{\theta\} \geq \frac{1}{J} \ , \tag{3.1.30}$$

or

$$Var\{\hat{\theta}\} \geq \frac{1}{-E\{\frac{\partial^2}{\partial \theta^2}\{\ln[f(\underline{x};\theta)]\}\}} \ . \tag{3.1.31}$$

We will illustrate the CRLB by returning to the simple single parameter, single observation example described earlier. We observe x with

$$x = \theta + w \ ,$$

with θ unknown and w distributed as $N(0, \sigma_w)$. As before we have $f(x;\theta)$ Gaussian with mean θ, and variance σ_w^2 and the log-likelihood function is therefore

$$L_\theta = \ln[f(x;\theta)] = -\frac{1}{2}\ln(2\pi\sigma_w^2) - \frac{1}{2\sigma_w^2}(x-\theta)^2 \ . \qquad (3.1.10)$$

The CRLB is thus given by equation (3.1.31) with this likelihood function. That is

$$Var\{\hat{\theta}\} \geq \frac{1}{-E\{\frac{\partial^2}{\partial\theta^2}\ln[f(x;\theta)]\}} = \frac{1}{-E\{\frac{\partial}{\partial\theta}\left[\frac{1}{\sigma_w^2}(x-\theta)\right]\}} \ . \qquad (3.1.32)$$

Simplifying, we see that the Cramer-Rao Lower Bound is given by

$$Var\{\hat{\theta}\} \geq \sigma_w^2 \ . \qquad (3.1.33)$$

The lower limit coincides with the variance for the ML estimator as given by equation (3.1.12). Hence we can conclude that, in this example, the ML estimate achieves the CRLB even for finite M.

3.1.2 Minimum Mean-Squared Error (MMSE) Estimation

Suppose that we wish to estimate a random parameter θ as some function of an observed random variable x. We shall construct an estimate $\theta_{MS} = h(x)$ which minimizes the mean-squared error between the estimate and the true or 'desired' parameter θ. This is achieved by choosing θ_{MS} to minimize the **quadratic cost functional**, J, with

$$J = E\{[\theta - h(x)]^2\} \ . \qquad (3.1.34)$$

The cost function J can be evaluated via the expected value:

$$J = \int\limits_{-\infty}^{\infty} \int\limits_{-\infty}^{\infty} [\theta - h(x)]^2 f(x,\theta)\,d\theta\,dx \ .$$

The joint density of x and θ may be further expanded as

$$f(x,\theta) = f(\theta/x)f(x) \ ,$$

so that we have

$$J = \int\limits_{-\infty}^{\infty} f(x)\left[\int\limits_{-\infty}^{\infty} [\theta - h(x)]^2 f(\theta/x)\,d\theta\right]dx \ . \qquad (3.1.35)$$

Both the inner and outer integrals are positive everywhere (because all the quantities within the integrals are non-negative everywhere). Moreover, the outer integral is unaffected by the choice of $h(x)$. Accordingly, we may equivalently minimize just the inner integral

$$\int\limits_{-\infty}^{\infty} [\theta - h(x)]^2 f(\theta/x)\, d\theta \ .$$

The minimization is achieved by differentiating with respect to $h(x)$ and equating to zero giving

$$-2 \int\limits_{-\infty}^{\infty} [\theta - h(x)] f(\theta/x)\, d\theta = 0 \ ,$$

or

$$h(x) \int\limits_{-\infty}^{\infty} f(\theta/x)\, d\theta = \int\limits_{-\infty}^{\infty} \theta f(\theta/x)\, d\theta \ .$$

But

$$\int\limits_{-\infty}^{\infty} f(\theta/x)\, d\theta = 1 \ ,$$

hence

$$\theta_{MS} = h(x) = E\{\theta/x\} \ . \tag{3.1.36}$$

Thus, the Minimum Mean-Squared Error (MMSE) estimator is obtained when $h(x)$ is chosen as the expectation of θ conditioned on the observed data. Note that unlike the MAP and ML methods, this estimator requires the conditional mean value of the posterior density but does *not* rely on explicit knowledge of the density itself. Observe also that $\theta_{MS} = E\{\theta/x\}$ is in general a nonlinear function of the data. An important exception occurs when the posterior density is Gaussian, in this case θ_{MS} is a *linear* function of x (see Problem 3.25).

It is interesting to compare this MMSE estimator with the MAP estimator described above. Both deal with the estimation of a random parameter θ, thus both may be classified as Bayesian. Both estimators produce estimates that are a function of the posterior density for θ. The distinction between the estimators lies in the nature of that function. For MAP it is the peak, while for MMSE it is the mean value. It is interesting to note that for symmetric densities the peak and the mean, and therefore the MAP and MMSE estimates, coincide. This class includes the important special case of Gaussian posterior densities.

Contrasting classical and Bayesian estimation; in classical estimation the quality of an estimator is assessed relative to the properties of bias, consistency, efficiency etc. In Bayesian estimation, the random nature of θ makes such performance indicators inappropriate. Instead we evaluate the performance of the estimator via a cost function J. Here, we focus on the MMSE cost function, though there is nothing unique about this choice. In principle, we could choose any number of functionals. For example, one possibility is the Minimum Absolute Error (MAE) criterion:

$$J = E\{|\theta - h(x)|\} .\tag{3.1.37}$$

In fact, even the MAP estimator can be derived from consideration of a form of cost function (for details see, for example, [2]).

As we illustrate in Problem 3.11, the optimal estimate obtained from the MAE criterion is the **median** of the posterior density. For symmetric densities this too coincides with the MAP and MMSE estimates. In fact, for symmetric unimodal densities the optimum obtained from any of a large class of functionals including all symmetric convex functionals will coincide with θ_{MS}.[6]

As with MAP and ML estimates we may easily extend the MMSE estimation procedure to the general problem of estimating L parameters $\theta = [\theta(0), \theta(1), ..., \theta(L-1)]^t$ given M observations $\underline{x} = [x(0), x(1), ..., \overline{x}(M-1)]^t$. By direct extension of the above anlaysis we find that

$$\underline{\theta}_{MS} = E\{\underline{\theta}/\underline{x}\} .\tag{3.1.36}$$

---

6. See Van Trees [1] for a general discussion and for the required conditions on J.

Linear MMSE Estimation

We have observed that the optimal minimum mean-squared error estimator of a parameter θ is equal to the conditional mean value

$$\theta_{MS} = E\{\theta/\underline{x}\} , \tag{3.1.36}$$

and that this is generally a nonlinear function of the observed data. Suppose now, however, that the estimator is constrained to be a linear function θ^*, say, of these measured data. For a single parameter, M observation problem we have

$$\theta^* = \sum_{i=0}^{M-1} h_i x(i) , \tag{3.1.38}$$

where the h_i's are weights chosen to minimize the mean-squared error

$$J = E\{e^2\} = E\{[\theta - \sum_{i=0}^{M-1} h_i x(i)]^2\} . \tag{3.1.39}$$

The minimization is performed by separately differentiating (3.1.39) with respect to each coefficient h_j and equating to zero. This yields

$$\frac{\partial J}{\partial h_j} = E\{[\theta - \sum_{i=0}^{M-1} h_i x(i)] x(j)\} = 0 \; ; \; j=0,1,...,M-1 ,$$

or

$$E\{e x(j)\} = 0 , \tag{3.1.40}$$

where e is the estimation error:

$$e = \theta - \sum_{i=0}^{M-1} h_i x(i) .$$

Recalling the definition of orthogonality given in Section 2.1 equation (2.1.28), equation (3.1.40) says that *the error is orthogonal to the data (measurements).* This **orthogonality principle** is a fundamental property of linear minimum mean-squared error estimation (see also Section 3.2).

3.1.3 Estimation of Signals

The discussion so far has concerned the estimation of one or more parameters from a set of observations. We now seek to place estimation in a signal processing framework. Generally, given some signal $x(n)$ for $n=0,1,...,N-1$, we are concerned to estimate another signal $d(n)$ for $n=0,1,...,M-1$, say, which we may think of as a **desired** or **ideal** signal. For example, $x(n)$ might be a noise corrupted measurement and $d(n)$ might be the noise-free signal (the enhancement problem). Or $d(n)$ might be a shifted version of $x(n)$ itself (the prediction problem). The nature of these, and other similar problems is a focus of later sections of this chapter (indeed of most of this text). For now, we will simply consider $d(n)$ as a desired signal — without specifying the application or environment. The estimate of the desired signal, denoted $\hat{d}(n)$, say, will be a function of the signal $x(n)$

$$\hat{d}(n)=T\{x(n)\} \, , \tag{3.1.41}$$

where $T\{\ \}$ represents a functional mapping. The MAP, ML and MMSE estimates for this problem are obtained from:

1. MAP Estimator

$$maximize \ \ f\left(d(n)/x(n)\right) \, . \tag{3.1.42}$$

2. ML Estimator

$$maximize \ \ f\left(x(n);d(n)\right) \, . \tag{3.1.43}$$

3. MMSE Estimator

$$\hat{d}(n)=E\{d(n)/x(n)\} \, . \tag{3.1.44}$$

4. Linear MMSE Estimator

$$\hat{d}(n)=\sum_{i=0}^{N-1} h_i x(i) \, . \tag{3.1.45}$$

Comparing these four procedures we remark that the linear MMSE estimate *4.* is the least general, but is the most easily applicable. This is because *1.*, *2.* and *3.* all require either explicit knowledge of signal/parameter densities, or at least of conditional expectations. By contrast, the linear MMSE estimator *4.* may be obtained purely from knowledge of the second moments of the data and parameters. Moreover, even when such second moments are unavailable they can often be estimated directly from the data. As a further point, the linear MMSE estimation structure can be easily modified to have the form of a linear FIR filtering procedure, with obvious computational and conceptual advantages. In particular, the linear least-squares estimate of $d(n)$ at time n using only the previous L values can be written as

$$\hat{d}(n) = \sum_{i=0}^{L-1} f(i) x(n-i) \ , \tag{3.1.46}$$

where now the parameters h_i of (3.1.45) have been replaced by coefficients of a linear FIR filter $f(n)$. It is this solution which is of most practical value in optimal and adaptive processing, and it is this structure which we shall pursue in the next section.

3.2 Optimal Least-Squares Filter Design

3.2.1 Formulation of the Normal Equations

Having established the general principle of minimum mean-squared error estimation we now turn our attention to the linear problem and, in particular, to the problem of linear filter design. The filter design schemes first encountered in digital signal processing usually have as an objective some ideal frequency characteristic; lowpass, highpass etc. These design methods are typically based on well established procedures such as transformations from analog designs for IIR filters, or windowing for FIR filters [8]. An alternative approach would be to design the filter to be *optimally close to the ideal result* in some sense. We could, of course, define 'optimality' in any number of ways. There are, however, powerful reasons for the MMSE or least-squares optimization which we shall outline in this section,[7] and a

---

7. We use the terms minimum mean-squared error and the looser 'least-squares' interchangeably to refer to the filters obtained by minimizing (3.2.1a) or (3.2.1b).

considerable body of literature has been established on this important subject. Much of this has been directed at specific applications such as speech processing [9],[10], and reflection seismology [11]. More general background can be found in the tutorial papers by Makhoul [12], and by Wood and Treitel [13].

Consider the following problem: An input signal $x(n)$ is to be filtered using a linear filter with impulse response $f(n)$. We seek to design this filter in such a way that the output is as close as possible to some ideal or desired signal $d(n)$. The performance of the filter is observed through the error $e(n)$ between the desired signal and the filter output $y(n)$. This filtering process is depicted in Figure 3.2.1.

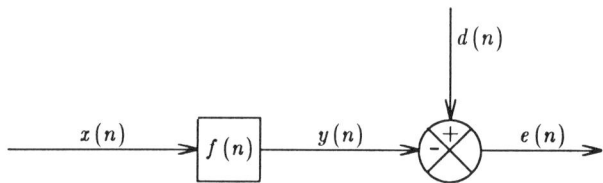

Figure 3.2.1 Least-squares optimal filtering. Filter $f(n)$ is designed to minimize the mean-squared error between desired signal $d(n)$, and output $y(n)$.

In the previous section, we developed the idea of minimum mean-squared error estimation of parameters. Here, we are concerned to use the same principle in a digital filtering framework. In this context 'mean-squared error' must be interpreted differently according to the nature of the input signals. For random input sequences, we can define an objective functional

$$J = E\{e^2(n)\} \ , \tag{3.2.1a}$$

where E is the expectation operator taken over the ensemble. For deterministic sequences J is defined by

$$J = \sum_n e^2(n) \ , \tag{3.2.1b}$$

where the range of summation for n is generally determined by the length of the input data sequence, and at this point remains to be specified. In both cases, our objective is to design $f(n)$ so as to minimize J. Now, referring to Figure 3.2.1

$$e(n) = d(n) - f(n) * x(n) , \qquad (3.2.2a)$$

or, assuming a transversal implementation

$$e(n) = d(n) - \sum_i f(i) x(n-i) . \qquad (3.2.2b)$$

Here, the range of summation over i depends on the length of the filter and for the moment we leave this also unspecified. The minimization is performed by differentiating J with respect to each coefficient $f(j)$ and equating the results to zero. Taking the random case, for example, the differentiation yields

$$\frac{\partial J}{\partial f(j)} = \frac{\partial E\{e^2(n)\}}{\partial f(j)} = 2E\{e(n) \frac{\partial e(n)}{\partial f(j)}\} , \qquad (3.2.3)$$

but using (3.2.2b)

$$\frac{\partial e(n)}{\partial f(j)} = \frac{\partial}{\partial f(j)} (d(n) - \sum_i f(i) x(n-i)) = -x(n-j) .$$

Hence combining this result with (3.2.3) we have

$$\frac{\partial J}{\partial f(j)} = 2 \left[\sum_i R(n-i, n-j) f(i) - g(n-j, n) \right] , \qquad (3.2.4)$$

where $R(n-i, n-j)$ is the autocorrelation of the input signal $x(n)$, defined by

$$R(n-i, n-j) = E\{x(n-i) x(n-j)\} , \qquad (3.2.5)$$

and $g(n-j, n)$ is the cross-correlation between the input $x(n)$ and the desired output $d(n)$, defined by

$$g(n-j, n) = E\{x(n-j) d(n)\} . \qquad (3.2.6)$$

Equating (3.2.4) to zero gives

$$\sum_i R(n-i, n-j) f(i) = g(n-j, n) . \qquad (3.2.7)$$

(3.2.7) forms a set of *linear* algebraic equations known as the **normal equations**. The solution of the normal equations yields the optimal filter. The result is known as the minimum mean-squared error solution or simply as the least-squares filter.

Note that when the inputs are deterministic, an equivalent procedure applied to (3.2.1b) yields an identical set of normal equations. For the deterministic case, however, the correlation coefficients are obtained by replacing the expectations in (3.2.5) and (3.2.6) by summation over the data.

The Orthogonality Principle

The term normal equations arises from the **orthogonality** of the input and the error, which results from equating the derivative of J to zero. From (3.2.3) when the derivative of the error is equated to zero we see that

$$-2E\{e(n)x(n-j)\}=0 ,$$

or

$$E\{e(n)x(n-j)\}=0 . \tag{3.2.8}$$

Thus, the orthogonality of input and error is confirmed and the result is seen to hold over the range of j as specified by the length of the filter. A corresponding result holds for the deterministic case but with the expectations replaced by summation over index n.

From Section 2.1.3 we note that since $E\{e(n)\}=E\{x(n)\}=0$, then the condition (3.2.8) also implies that $e(n)$ and $x(n)$ are uncorrelated. Interestingly, the result works in either direction. That is, the optimization gives rise to the orthogonality condition, but the orthogonality condition also defines the least-squares filter. This follows as a consequence of the **uniqueness** of the least-squares solution.

The least-squares solution also guarantees orthogonality of the error and output at time n because

$$E\{e(n)y(n)\}=E\{e(n)\sum_i f(i)x(n-i)\}$$

$$=\sum_i f(i)E\{e(n)x(n-i)\} .$$

Hence from (3.2.8), for the optimal filter

$$E\{e(n)y(n)\}=0 \ . \tag{3.2.9}$$

Viewed as a function of the filter coefficients, J in (3.2.1) is a quadratic functional and thus has a single turning point. We can be confident that the solution obtained at this turning point is not a maximum since, for any $f(n)$ producing a particular J, another $f(n)$ can easily be found which will increase J. No maximum for J can be found. (We shall formally demonstrate that the result is a minimum in Section 3.2.4. Such a result can also be derived from consideration of the second derivatives of J see, for example, [14].) The unimodality of the functional is one of the two major attractions of this form of optimization. The other advantage, again arising from the quadratic nature of J, is that with a transversal implementation, such a criterion gives rise to a set of *linear* equations for the optimal filter.

Autocorrelation and Autocovariance Forms

There are two distinct forms for the normal equations (3.2.7). These are known as the **autocorrelation** and the **autocovariance** formulations [12]:[8]

a) Autocorrelation Formulation

The autocorrelation formulation occurs when the input signals are stationary random processes, or when they are deterministic with the minimization in (3.2.1b) taken over all n. In this case for stationary inputs, as we have seen in Chapter 2

$$R(n-i,n-j)=R(j-i)=r(j-i) \quad ; \ g(n-j,n)=g(j) \ ,$$

and equations (3.2.7) may be written

$$\sum_i r(j-i)f(i)=g(j) \ . \tag{3.2.7a}$$

A similar form arises in the deterministic case (see Problem 3.12).

---

8. The titles Autocorrelation and Autocovariance forms are very confusing because they are not directly related to the definitions of correlation and covariance given in Section 2.1. As Parsons [15] points out, perhaps **stationary** and **nonstationary** formulations would be more appropriate.

b) Autocovariance Formulation

The autocovariance formulation occurs when the signals are non-stationary random, or when they are deterministic with the minimization taken over a finite interval for n. In this case, the equations are as specified by (3.2.7)

$$\sum_i R(n-i,n-j)f(i)=g(n-j,n) \ . \tag{3.2.7}$$

Note, however, that the autocorrelation form may be obtained even when the minimization is taken over a finite interval by windowing the input signal appropriately In adaptive filtering we are usually more concerned with the autocorrelation form for the least-squares filter and we shall largely restrict attention to this formulation.

3.2.2 Solution of the Normal Equations

a) Infinite Two-Sided Filter

Consider firstly an infinite length two-sided (acausal) filter $f(n)$. Equations (3.2.7a) become

$$\sum_{i=-\infty}^{\infty} r(j-i)f(i)=g(j) \qquad ; \ -\infty<j<\infty \ .$$

Hence, for inputs which give rise to the autocorrelation formulation, recognizing that the left hand side of these equations is the convolution of the filter with the input autocorrelation, and z transforming both sides gives

$$R(z)F(z)=G(z) \ , \tag{3.2.7b}$$

where $R(z)$, $F(z)$, and $G(z)$ are the two-sided z transforms of $r(n)$, $f(n)$ and $g(n)$, respectively. Rearranging we have

$$F(z)\rightarrow F^{*}(z)=\frac{G(z)}{R(z)} \ , \tag{3.2.10}$$

where we have used the superscript $*$ to denote the optimal solution.

Note that if the filter is restricted to causality, the solution cannot be obtained by simple spectral division as in (3.2.10), but may still be found by employing a spectral factorization technique [16].

b) Finite Causal Filter

If, in addition to causality, the filter is also constrained to be finite (of length L, say) then the normal equations (3.2.7a) become

$$\sum_{i=0}^{L-1} r(j-i)f(i)=g(j) \quad ; 0\leq j\leq L-1 \;.$$

This represents a set of L linear equations for the optimal L point impulse response function $f^*(i)$. It is appropriate at this point to introduce the vector notation for the optimal filter which is used almost universally in least-squares and adaptive filter design. For an L point FIR filter we define a vector of filter impulse response coefficients $\underline{f}$ as:

$$\underline{f}=[f(0),f(1),...,f(L-1)]^t \;,$$

and a data vector $\underline{x}_n$ as:

$$\underline{x}_n=[x(n),x(n-1),...,x(n-L+1)]^t.$$

$\underline{x}_n$ is a vector whose elements are the L most recent samples of the input signal. Using this notation the convolution

$$y(n)=\sum_{i=0}^{L-1} f(i)x(n-i) \;, \qquad (3.2.11a)$$

can be written as

$$y(n)=\underline{f}^t\underline{x}_n=\underline{x}_n^t\underline{f} \;. \qquad (3.2.11b)$$

Also the gradient of J as expressed by equations (3.2.4) (with $R(n-i,n-j)=r(j-i)$) can be written

$$\nabla J=2(R\underline{f}-\underline{g}) \;, \qquad (3.2.4a)$$

where R is the matrix of autocorrelation coefficients which for the autocorrelation formulation is given by equation (2.1.35) of Section 2.1:

$$R = \begin{bmatrix} r(0) & r(1) & r(2) & . & r(L-1) \\ r(1) & r(0) & r(1) & . & . \\ . & r(1) & . & . & . \\ . & . & . & . & . \\ r(L-1) & . & . & . & r(0) \end{bmatrix}, \qquad (2.1.35)$$

where $\underline{g}$ is given by

$$\underline{g} = [g(0), g(1), ..., g(L-1)]^t ,$$

and where

$$\nabla J = [\frac{\partial J}{\partial f(0)}, \frac{\partial J}{\partial f(1)}, ..., \frac{\partial J}{\partial f(L-1)}]^t ,$$

is the gradient of J with respect to $\underline{f}$. With this notation the normal equations (3.2.7a) become

$$R\underline{f} = \underline{g} , \qquad (3.2.7c)$$

and the solution is given by

$$\underline{f}^* = R^{-1}\underline{g} . \qquad (3.2.12)$$

Here again the superscript * is used to denote the fact that we have obtained the optimal solution.

As we indicated in Section 2.1, R is a symmetric Toeplitz matrix and has a total of only L distinct elements. As we shall see, this has a significant impact on the computation of $\underline{f}^*$. Also, if R is non-singular then the inverse R^{-1} exists and is unique. This confirms that $\underline{f}^*$ is the unique solution of this minimization problem. R is certainly non-singular if it is positive definite. As we show in Appendix 3A, however, R is only positive semi-definite. This leaves open the possibility of a singular correlation matrix R. In practice, for most random signals R is indeed non-singular. For deterministic inputs this does not always hold. One simple example occurs when

$$x(n) = A\cos(\omega_0 nT + \phi) .$$

That is, when the input is a sinusoid. With such an input for any $L > 2$, R is singular and the solution (3.2.12) is not meaningful (see also Problem 3.14).

3.2.3 Computation of the Least-Squares Solution

There are two stages in the computation of the least-squares filter f^*: *i)* Computation of auto and cross-correlation coefficients and *ii)* Solution of the normal equations. For analytic examples, the correlation coefficients are available *a priori* or may be precisely calculated using analytic models of the signals. For practical computational problems we are usually provided with only a data sequence $x(n), d(n)$ for $n = 0, 1, ..., M-1$. In this case, the auto and cross-correlation coefficients must be estimated using time averages. We have already discussed the general procedures for this time averaging in Section 2.6. Taking the biased estimator for example, we may interpret equation (2.6.18) to give

$$r^{'}(i) = \frac{1}{M} \sum_{n=0}^{M-i-1} x(n)x(n+i) \quad ; i = 0, 1, ..., L-1 , \qquad (3.2.13)$$

and

$$g^{'}(i) = \frac{1}{M} \sum_{n=0}^{M-i-1} x(n)d(n+i) \quad ; i = 0, 1, ..., L-1 , \qquad (3.2.14)$$

where L is the filter length.

As we have already indicated, this same autocorrelation formulation can be used for deterministic signals $x(n)$ and $d(n)$.[9] In fact, a further advantage of the biased estimator is that these expressions are used to form the correlations in both cases. (The division by M does not represent a significant difference since it occurs on both sides of (3.2.7).) The important point is that the use of these estimators is appropriate for either deterministic or random data. Consequently it is not necessary to make any decision regarding the form of the data — the computation is identical in both cases. In fact, the only difference in the results

---

9. To obtain an autocorrelation form with a finite M point data segment with index from 0 to $M-1$ say, it is necessary to apply a window of length $M+L$. With this condition equation (3.2.5) becomes

$$R(i,j) = \sum_{n=0}^{M+L-1} x(n-i)x(n-j) .$$

It is easy to show that this reduces to a function of the difference $i-j$, and is equivalent to (3.2.13) except for the scale factor.

will be the repeatability of the optimal solution.

Once the correlation coefficients have been estimated, the normal equations may be solved. The numerical solution of a general set of linear equations of order L requires on the order of $L^3/2$ operations using standard techniques. For symmetric coefficient matrices this can be reduced by approximately a factor of three using the Cholesky factorization [9]. However, due to the highly structured nature of the autocorrelation matrix, the solution of (3.2.12) for the optimal filter $\underline{f}^*$ can be obtained in approximately $2L^2$ operations using a technique known as the **Levinson recursion** [17]. This is a very significant reduction in the computational load, and is another important attraction of least-squares optimization.[10] The Levinson recursion is one of the most important algorithms in all of digital signal processing, and is outlined in Appendix 3B.

3.2.4 Error Energy in Finite Length Least-Squares Filtering

The Minimum Mean-Squared Error Energy

We note firstly that any filter vector $\underline{f}$ will result in some value for error energy J as given by (3.2.1a) or (3.2.1b), and this error energy is a measure of the performance of the filter $\underline{f}$. We may obtain an expression for this error energy by expanding (3.2.1a) or (3.2.1b) as appropriate. Consider $x(n)$, $d(n)$ jointly stationary and zero-mean (a similar analysis may be conducted for deterministic inputs). We have,

$$J = E\{e^2(n)\} \ . \tag{3.2.1a}$$

Using vector notation the error may be written

$$e(n) = d(n) - \underline{f}^t \underline{x}_n \ , \tag{3.2.2c}$$

so that

---

10. In spite of the computational gain derived from the Levinson recursion, we should keep in mind that the major portion of the computation usually lies in estimating the auto and cross-correlation coefficients to set up the equations, rather than in computing the solution.

$$J = E\{d^2(n)\} + \underline{f}^t E\{\underline{x}_n \underline{x}_n^t\} \underline{f} - 2\underline{f}^t E\{\underline{x}_n d(n)\} \ .$$

Hence we have

$$J = \sigma_d^2 + \underline{f}^t R \underline{f} - 2\underline{f}^t \underline{g} \ , \tag{3.2.15}$$

where σ_d^2 is the mean-squared value of $d(n)$. The error energy J is thus a second order function of the filter coefficient vector. That is, J is a 'bowl-shaped' function of the filter coefficients. J is known as the **error performance surface** or simply as the **performance surface** for the filter $\underline{f}$. For an L point filter the bowl is actually L-dimensional – a hyperparaboloid. The minimum value for the mean-squared error, $J_{\min}$ say, occurs at the bottom of this bowl when $\underline{f} \rightarrow \underline{f}^*$ as given by (3.2.12). Substituting this solution into equation (3.2.15) and simplifying gives

$$J_{\min} = \sigma_d^2 - \underline{f}^{*t} \underline{g} \ , \tag{3.2.16a}$$

or

$$J_{\min} = \sigma_d^2 - \underline{f}^{*t} R \underline{f}^* \ , \tag{3.2.16b}$$

where $\underline{f}^*$ is given by equation (3.2.12).

The mean-squared error J can also be expressed in terms of the difference between the filter $\underline{f}$ and the optimal result $\underline{f}^*$. In particular, defining an **error vector** $\underline{v}$ as:

$$\underline{v} = \underline{f} - \underline{f}^* \ , \tag{3.2.17}$$

then the expression $\underline{v}^t R \underline{v}$ may be expanded as

$$\underline{v}^t R \underline{v} = (\underline{f} - \underline{f}^*)^t R (\underline{f} - \underline{f}^*)$$

$$= \underline{f}^t R \underline{f} + \underline{f}^{*t} R \underline{f}^* - 2\underline{f}^{*t} R \underline{f} \ ,$$

but, using (3.2.12) we have

$$\underline{v}^t R \underline{v} = \underline{f}^t R \underline{f} + \underline{f}^{*t} \underline{g} - 2\underline{f}^t \underline{g} \ . \tag{3.2.18}$$

Now, using (3.2.15) and (3.2.16) we have

$$J - J_{\min} = \sigma_d^2 + \underline{f}^t R \underline{f} - 2\underline{f}^t \underline{g} - \sigma_d^2 + \underline{f}^{*t} \underline{g} \ .$$

Hence

$$J = J_{\min} + \underline{f}^t R \underline{f} + \underline{f}^{*t} g - 2\underline{f}^t g ,$$

or, from (3.2.18)

$$J = J_{\min} + \underline{v}^t R \underline{v} . \tag{3.2.19}$$

This equation is especially useful as it expresses the excess above the minimum mean-squared error in terms of the error between any filter $\underline{f}$ and the optimal filter $\underline{f}^*$. Note also that since R is positive semi-definite (see Appendix 3A) then

$$\underline{v}^t R \underline{v} \geq 0 . \tag{3.2.20}$$

This demonstrates that $\underline{f} = \underline{f}^*$ is a minimum of J since, from (3.2.19) any change in $\underline{f}$ away from $\underline{f}^*$ can only increase J. The equality in (3.2.20) which is associated with a singular R allows the possibility of a non-unique minimum. Conversely, if R were strictly positive definite then

$$\underline{v}^t R \underline{v} > 0 ,$$

and the minimum is unique.

Monotonicity of $J_{\min}$ with L

Let us denote the error energy associated with the L point filter by $J_{\min}^{(L)}$. We have the following result:

   The error energy $J_{\min}^{(L)}$ is a monotonically non-increasing function of L.

This says simply that the performance of the least-squares filter cannot get worse as the filter length is increased. This result is intuitively rather obvious because any change in the coefficients from $\underline{f}^{*(L)}$ to $\underline{f}^{*(L+1)}$ which produced $J_{\min}^{(L+1)} > J_{\min}^{(L)}$ could not be minimum. This is because the simple extension $\underline{f}^{(L+1)} = [f^{*(L)}(0), ..., f^{*(L)}(L-1), 0]^t$ produces $J^{(L+1)} = J_{\min}^{(L)}$. Hence the optimum solution $\underline{f}^{*(L+1)}$ must do at least this well if it is to correspond to the minimum error energy. Hence, $J_{\min}^{(L+1)} \leq J_{\min}^{(L)}$ and the result follows.

Example: Identification of an Unknown System

Consider a zero-mean iid sequence with variance σ_w^2, input to a simple system with three point impulse response $h(0)=1$, $h(1)=0.5$, $h(2)=0.7$. This response produces a third order MA process as indicated by equation (2.4.10) of Section 2.4. Suppose we are given only the input $x(n)$ and the output $y(n)$, that is we assume that the system $h(n)$ is unknown. We shall apply the least-squares optimization to identify $h(n)$. This is achieved by using $y(n)$ as the desired output for the filter and $x(n)$ as input. The setup is depicted in Figure 3.2.2.

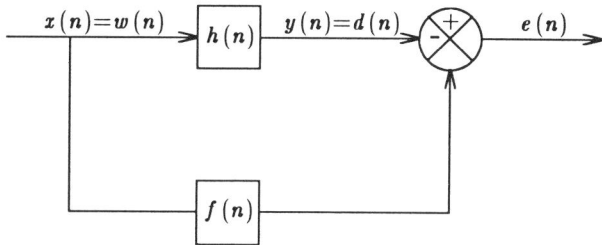

Figure 3.2.2 Block diagram for system identification example. A filter $f(n)$ models an unknown system $h(n)$ using input $x(n)$ and using the system output $y(n)$ as desired signal.

We seek to construct the least-squares solution $\underline{f}^*$ given by

$$\underline{f}^* = R^{-1}\underline{g} \ . \qquad (3.2.12)$$

Since $x(n)$ is zero-mean iid with variance σ_w^2, then R is diagonal with elements σ_w^2. Also

$$g(i) = E\{x(n-i)d(n)\} \ ,$$

but

$$d(n) = y(n) = x(n) + 0.5x(n-1) + 0.7x(n-2) \ ,$$

so that

$$\underline{g} = [1, 0.5, 0.7, 0, 0, ..., 0]^t \sigma_w^2 \ .$$

Note that we have calculated the exact correlations. We have been able to do this because of our prior knowledge of $h(n)$. In practice, given only signals $x(n)$, $y(n)$ the correlations would have to be estimated via the usual time averaging formulae. The result would only approximately equal that given for R and g. However, neglecting this effect, the result is easily obtained from (3.2.12) as

$$\underline{f}^* = \frac{1}{\sigma_w^2} \begin{bmatrix} 1 & 0 & 0 & . & . & 0 \\ 0 & 1 & 0 & . & . & 0 \\ 0 & 0 & 1 & . & . & \\ . & . & . & . & . & . \\ . & . & . & . & 1 & 0 \\ 0 & 0 & 0 & 0 & . & 1 \end{bmatrix} \begin{bmatrix} 1 \\ 0.5 \\ 0.7 \\ 0 \\ . \\ 0 \end{bmatrix} \sigma_w^2 \ ,$$

or

$$\underline{f}^* = [1, 0.5, 0.7, 0, 0, ..., 0]^t \ . \tag{3.2.21}$$

Note that for filter lengths $L \geq 3$, the result is an exact identification of $h(n)$. For $L=1$ or $L=2$ the result is a truncated version of the unknown system.

The mean-squared error associated with this solution can be calculated from

$$J_{\min} = \sigma_d^2 - \underline{f}^{*t} \underline{g} \ , \tag{3.2.16a}$$

but

$$\sigma_d^2 = E\{d^2(n)\} = [1 + (0.5)^2 + (0.7)^2]\sigma_w^2 \ .$$

As before, if we denote the minimum error energy $J_{\min}$ corresponding to filter length L by $J_{\min}^{(L)}$, we have

$$J_{\min}^{(1)} = 1.74\sigma_w^2 - 1\sigma_w^2 = 0.74\sigma_w^2 \ ,$$

$$J_{\min}^{(2)} = 1.74\sigma_w^2 - 1.25\sigma_w^2 = 0.49\sigma_w^2 \ ,$$

$$J_{\min}^{(n)} = 0 \quad ; n = 3, 4, ... $$

This demonstrates the expected result that filter lengths of three and above give perfect identification of the unknown system. We also see the expected monotonically non-increasing behavior of $J_{\min}^{(L)}$ with increasing L.

This example can be generalized as follows: Let $h(n)$ be a causal shift-invariant system. Assume measurements $x(n)$ and $y(n)$ are obtained, where $x(n)$ is a zero-mean iid sequence and

$$y(n) = \sum_{i=0}^{n} h(i)x(n-i) \ ,$$

then the least-squares filter obtained from (3.2.12) with $x(n)$ as the input and $y(n)$ as the desired signal is a truncated version of $h(n)$

$$\underline{f}^* = [h(0), h(1), ..., h(L-1)]^t \ , \tag{3.2.22}$$

and from (3.2.16a)

$$J_{\min}^{(L)} = \sum_{i=L}^{\infty} h^2(i)\sigma_w^2 \ . \tag{3.2.23}$$

This result follows as a simple extension of the above three point example and is considered in Problem 3.16.

The interpretation of the least-squares solution (3.2.12) is not usually so easy. The case examined above is the simplest possibility having a diagonal correlation matrix so that the result is shaped purely by the cross-correlation between $x(n)$ and $d(n)$. We cannot in general expect things to be this simple. In all cases, however, $\underline{f}^*$ is crucially influenced by the cross-correlation. If this quantity is zero then $\underline{f}^*$ will be zero. More generally, $\underline{f}^*$ can loosely be thought of as being shaped by that part of $x(n)$ which is correlated with $d(n)$.

Normalized Mean-Squared Error and Performance

A normalized version of the minimum mean-squared error, dubbed $J'_{\min}$ may be defined by dividing equation (3.2.16) by σ_d^2:

$$J'_{\min} = 1 - \frac{1}{\sigma_d^2} \underline{f}^{*t} \underline{g} \ , \tag{3.2.24a}$$

$$= 1 - \frac{1}{\sigma_d^2} \underline{f}^{*t} R \underline{f}^* \ . \tag{3.2.24b}$$

The normalized mean-squared error then satisfies

$$0 \leq J'_{\min} \leq 1 \ . \tag{3.2.25}$$

The lower limit of (3.2.25) follows simply because

$$J'_{min} = \frac{E\{e^2_{min}(n)\}}{\sigma^2_d} ,$$

must be a positive quantity. The upper limit follows from (3.2.24a) because the positive semi-definite property for R requires that $\underline{f}^t R \underline{f} \geq 0$ for any $\underline{f}$. Also, we may define the **performance of the filter** P, say, in terms of this normalized error energy simply as:

$$P = 1 - J'_{min} . \qquad (3.2.26)$$

Hence from (3.2.24)

$$P = \frac{1}{\sigma^2_d} \underline{f}^{*t} \underline{g} , \qquad (3.2.27a)$$

or

$$P = \frac{1}{\sigma^2_d} \underline{f}^{*t} R \underline{f}^* . \qquad (3.2.27b)$$

It follows from the bound on J'_{min} that

$$0 \leq P \leq 1 . \qquad (3.2.28)$$

So that $P=1$ implies that the least-squares filter perfectly maps $x(n)$ onto $d(n)$. At the other extreme $P=0$ reflects total failure of the least-squares minimization to model $d(n)$.

It should be noted that J_{min} and P are bounded as indicated by (3.2.25) and (3.2.28) *only* when the equations are formed using the theoretical values for R as in the analytic examples in this Chapter. In practice, when these correlation elements are replaced by estimates derived from time-averages, the bounds may fail. In particular it is not uncommon, when computing the solution, for the performance indicator P to become negative. This is generally an indication that the matrix R is poorly conditioned. This is particularly likely to occur when using the unbiased estimator for the correlation which, as observed in Chapter 2 is not necessarily positive semi-definite. This particular phenomenon can be overcome by the device of applying an exponential weighting that decays with index to the data sequence prior to calculating the correlation coefficients, or by the method of 'prewhitening' discussed in Section 3.3.2 below.

3.3 Applications of Least-Squares Filters

3.3.1 Linear Prediction

One of the most useful applications of the least-squares filtering method is to linear prediction. Indeed, predicting a signal using a linear combination of its past samples is a classic problem in time-series analysis. In signal processing, the procedure has been employed in a number of areas but most notably in geophysical exploration and in digital speech processing. Linear prediction is interesting both because of these practical applications, and because the least-squares predictor has properties which give fundamental insights into signal modeling and analysis. The reader should be aware that there are several texts devoted exclusively to this subject (see, for example, [10],[18]). The topic is also covered in some depth in most speech processing texts [9],[15] and in some books on adaptive filtering [19],[20]. An excellent review paper is the 1975 article by Makhoul [12].

Consider a zero-mean, stationary signal $x(n)$. A **linear prediction** of this signal takes the form

$$\hat{x}(n) = \sum_{i=1}^{p} c_i x(n - n_0 - i + 1) \ . \tag{3.3.1}$$

This is referred to as the **p-th order predictor for prediction distance n_0** and the coefficients c_i are called **predictor coefficients**. The term linear prediction arises from the fact that the estimate or prediction $\hat{x}(n)$ is formed as a linear combination of previous values of $x(n)$. A special case is the single-step ($n_0=1$) predictor:

$$\hat{x}(n) = \sum_{i=1}^{p} c_i x(n - i) \ . \tag{3.3.2}$$

Associated with any prediction we can define a **prediction error** as

$$e(n) = x(n) - \hat{x}(n) = x(n) - \sum_{i=1}^{p} c_i x(n - n_0 - i + 1) \ . \tag{3.3.3}$$

The predictor coefficients are selected using the least-squares minimization technique described in the previous section. Thus, for the zero-mean stationary data assumed we minimize

$$J=E\{e^2(n)\}=E\{(x(n)-\hat{x}(n))^2\}\ ,$$

or

$$J=E\{(x(n)-\sum_{i=1}^{p}c_ix(n-n_0-i+1))^2\}\ . \qquad (3.3.4)$$

This procedure is essentially the same as that of Section 3.2 for the finite causal least-squares filter. In this case, however, the input is $x(n-n_0)$, the desired signal is $x(n)$, and the filter coefficients $f(0), f(1),...,f(L-1)$ are replaced by predictor coefficients $c_1, c_2,...,c_L$ (compare (3.3.4) to (3.2.1a) and (3.2.2a), and see Figure 3.3.1).

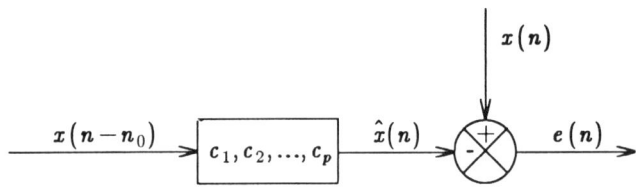

Figure 3.3.1 Least-squares optimal prediction for prediction distance n_0.

To minimize J, we differentiate and equate to zero. The procedure is identical to that employed in Section 3.2 and the details are left as an exercise. As in Section 3.2, we obtain a set of normal equations:

$$R\underline{c}=\underline{g}\ , \qquad (3.3.5)$$

where R is the p-th crder Toeplitz autocorrelation matrix (see Section 2.1, equation (2.1.35)) with elements

$$r(i-j)=E\{x(n-n_0-i+1)x(n-n_0-j+1)\}\quad ;\ i,j=1,2,...,p\ ,$$

$$=E\{x(n-i)x(n-j)\}\ . \qquad (3.3.6)$$

Similarly $\underline{g}$ is the cross-correlation between input and desired signals

$$\underline{g}=[E\{x(n)x(n-n_0)\},,...,E\{x(n)x(n-n_0-p+1)\}]^t\ ,$$

or

$$\underline{g}=[r(n_0),r(n_0+1),...,r(n_0+p-1)]^t\ ,\qquad(3.3.7)$$

and where the coefficient vector $\underline{c}$ consists of the predictor coefficients

$$\underline{c}=[c_1,c_2,...,c_p]^t\ .$$

Apart from replacing the filter coefficients $f(i)$ with predictor coefficients c_j, the distinction between (3.3.5) and the general normal equations (3.2.7c) lies in the form of g. Here, the cross-correlation vector consists purely of autocorrelation coefficients. For the special case of the single-step predictor $(n_0=1)$ the solution is obtained from (3.3.5) with the elements of R obtained from (3.3.6) but where g becomes

$$\underline{g}=[r(1),r(2),...,r(p)]^t\ .\qquad(3.3.8)$$

The Prediction Error Filter

As we have seen, associated with any prediction is a prediction error

$$e(n)=x(n)-\sum_{i=1}^{p}c_i x(n-n_0-i+1)\ .\qquad(3.3.3)$$

We can define a new operator, the **Prediction Error Filter** (PEF) as

$$\underline{f}=[1,0,...,0,-c_1,-c_2,...,-c_p]^t\ ,\qquad(3.3.9)$$

where the operator $\underline{f}$ contains a total of (n_0-1) zeros. The application of this filter to the data $x(n)$ produces the prediction error directly, that is

$$e(n)=\underline{f}^t\underline{x}_n\ .\qquad(3.3.10)$$

In particular, for the single-step predictor

$$\underline{f}=[1,-c_1,-c_2,...,-c_p]^t\ .$$

To compute this operator we may set up the normal equations as in (3.3.5), solve for the predictor coefficients, and then form $\underline{f}$ using (3.3.9). Alternatively, we can set up an augmented set of normal equations whose solution will yield the PEF directly. This alternative solution is of importance for both theoretical and practical reasons. It can be obtained by firstly writing out equations (3.3.5) for the single-step case as:

$$r(0) \ c_1 + r(1) \ c_2 + \cdots + r(p-1) \ c_p = r(1)$$
$$r(1) \ c_1 + r(0) \ c_2 + \cdots + r(p-2) \ c_p = r(2)$$

$$\begin{matrix} \cdot & & \cdot & & \cdot \\ \cdot & & \cdot & & \cdot \end{matrix}$$

$$r(p-1) \ c_1 + r(p-2) c_2 + \cdots + r(0) c_p = r(p) \ . \qquad (3.3.11)$$

This set can be rearranged as

$$r(1) \ - \ r(0) \ c_1 - r(1) \ c_2 \ - \cdots - \ r(p-1) \ c_p = 0$$
$$r(2) \ - \ r(1) \ c_1 - r(0) \ c_2 - \cdots - \ r(p-2) \ c_p = 0$$

$$\begin{matrix} \cdot & & \cdot & & \cdot \\ \cdot & & \cdot & & \cdot \end{matrix}$$

$$r(p) - r(p-1) c_1 - r(p-2) c_2 - \cdots - r(0) c_p = 0 \ . \qquad (3.3.12)$$

To these we add a further equation $-$ the expression for the minimum mean-squared error. For the finite L point causal least-squares filter $\underline{f}$ this was obtained in Section 3.2, equation (3.2.16a) as

$$J_{\min} = \sigma_d^2 - \underline{f}^{*t} \underline{g} \ , \qquad (3.2.16a)$$

or

$$J_{\min} = E\{d^2(n)\} - \sum_{i=0}^{L-1} f^*(i) g(i) \ .$$

For the single-step predictor, expanding the modified functional (3.3.4) we obtain (see Problem 3.17)

$$J_{\min} = r(0) - \sum_{i=1}^{p} c_i r(i) \ . \qquad (3.3.13)$$

This equation is added to the set (3.3.12) and in matrix form the augmented equations (3.3.12) and (3.3.13) can be written

$$
\begin{bmatrix}
r(0) & r(1) & . & . & r(p) \\
r(1) & r(0) & . & . & r(p-1) \\
. & . & . & . & . \\
. & . & . & . & . \\
. & . & . & . & . \\
r(p) & r(p-1) & . & . & r(0)
\end{bmatrix}
\begin{bmatrix}
1 \\
-c_1 \\
-c_2 \\
. \\
-c_p
\end{bmatrix}
=
\begin{bmatrix}
J_{\min} \\
0 \\
0 \\
. \\
0
\end{bmatrix} . \qquad (3.3.14)
$$

This augmented (order $p+1$) set of normal equations can be solved directly using the Levinson recursion, to give the prediction error filter (see Appendix 3B). However, we see that the right-hand side vector g has a greatly simplified form having just a single non-zero element. This can be exploited using a modification of the Levinson recursion. This modified algorithm is known as Durbin's recursion [21] and can be used to solve the set even more efficiently. Also, we see that in the augmented equations, $J_{\min}$ is obtained directly from the solution of the normal equations. An exposition of Durbin's algorithm, which is used for the solution of the prediction error filter, is given in Appendix 3C.

Linear Prediction and the AR Model

Recall the AR(M) model defined in Section 2.4, with generating system

$$
H(z) = \frac{1}{1 - a_1 z^{-1} - a_2 z^{-2} - \dots - a_M z^{-M}} , \qquad (2.4.3)
$$

producing a stationary zero-mean signal $x(n)$ as:

$$
x(n) = a_1 x(n-1) + \dots + a_M x(n-M) + w(n) , \qquad (2.4.9)
$$

where $w(n)$ is a zero-mean white noise with variance σ_w^2. Now, let us apply linear prediction to a signal $x(n)$ generated via this AR model. A single-step linear prediction of order p takes the form

$$
\hat{x}(n) = c_1 x(n-1) + c_2 x(n-2) + \dots + c_p x(n-p) \qquad (3.3.1)
$$

with error

$$
e(n) = x(n) - \hat{x}(n) . \qquad (3.3.3)
$$

If the order of the predictor equals that of the model $(M=p)$ then we may combine (2.4.9) and (3.3.1) to obtain

$$e(n)=(a_1-c_1)x(n-1)+(a_2-c_2)x(n-2)+...$$
$$+(a_p-c_p)x(n-p)+w(n) \ . \qquad (3.3.15)$$

It seems apparent from (3.3.15) that the solution which minimizes

$$J=E\{e^2(n)\} \ ,$$

is given by

$$a_i=c_i \quad ; \ i=1,2,...,p \ . \qquad (3.3.16)$$

This can be confirmed if we recognize that with this solution

$$J_{\min}=E\{e^2(n)\}=E\{w^2(n)\}=\sigma_w^2 \ . \qquad (3.3.17)$$

Clearly, no predictor we could construct could produce a lower value for J, since $w(n)$ is by definition unpredictable. Hence, $\underline{a}\equiv\underline{c}$ must be the solution and we may conclude that the least-squares linear predictor exactly obtains the parameters $\underline{a}$ of the AR(M) model. Furthermore, if $p>M$ it is easy to demonstrate that the predictor is given by

$$\underline{c}=[a_1,...,a_p,0,0,...,0]^t \ . \qquad (3.3.18)$$

Hence we conclude that by choosing a sufficiently large order p we can always obtain the exact parameter values when $x(n)$ is an AR process. Of course, in practice the result is obtained from the solution of the normal equations (3.3.5) using **estimates** of the correlation coefficients. Consequently, the solution will not be identical to (3.3.18)

The close relationship between the AR model and the linear predictor extends to the prediction error filter. Recall from Section 2.4 the synthesis procedure for generating an AR process $x(n)$ (depicted in Figure 2.4.1a)). The generating model $H(z)$ for such a process is given by

$$H(z)=\frac{1}{A(z)}=\frac{1}{1-a_1z^{-1}-a_2z^{-2}-...-a_Mz^{-M}}=\frac{1}{A(z)} \ , \qquad (2.4.3)$$

where $A(z)=1-a_1 z^{-1}-a_2 z^{-1}-...-a_M z^{-M}$. Consequently the corresponding analysis filter is (see Figure 2.4.1b))

$$G(z)=\frac{1}{H(z)}=A(z) \ . \tag{2.4.11}$$

Note that for the all-pole AR sequence, the analysis filter is all-zero and is therefore stable for any finite coefficients a_i. Now if, as described above, the linear prediction process produces an **exact** set of coefficients $\underline{c}=\underline{a}$, then the prediction error filter

$$\underline{f}=[1,-c_1,-c_2,...,-c_p]^t \ ,$$

$$=[1,-a_1,-a_2,...,-a_p]^t \ ,$$

or

$$F(z)=A(z) \ . \tag{3.3.19}$$

where $A(z)$ is the analysis filter for the AR model. Thus, for finite AR processes, not only does the least-squares linear predictor produce an exact match to the predictor coefficients, but the prediction error filter is identical to the analysis filter for the process. Consequently the output of the least-squares prediction error filter applied to the AR process is a white noise sequence.

We note that because of the FIR nature of the prediction error filter $(1,-c_1,-c_2,...,-c_p)$, the filter response is characterized by zeros only. Consequently, the stability of the operator is guaranteed for any set of finite coefficients c_i. For the inverse filter $1/F(z)$, the requirement is that the zeros of $F(z)$ should lie within $|z|=1$. It is a remarkable property of the least-squares PEF that the operator $F(z)$ *always* satisfies this condition whatever the input. It will be recalled from Section 2.5 that this condition on a finite length sequence (that its zeros all lie inside $|z|=1$) is called minimum-phase. Hence, *the prediction error filter $F(z)$ is always minimum-phase irrespective of the phase structure of $x(n)$* and in consequence the least-squares inverse filter is always stable. A proof of this important property is given in Appendix 3D.

Case Study: Application of Linear Prediction to
Geophysical Exploration

One area where linear prediction has been widely employed is in **geophysical prospecting** [13]. In particular, prediction error filters have been used for the purpose of removing, or at least attenuating, predictable components from seismic signals. This predictable energy represents reverberant, or multiple energy, which is superimposed on the desired information. The problem is particularly prevalent in marine seismic data. The principle is explained with reference to Figure 3.3.2.

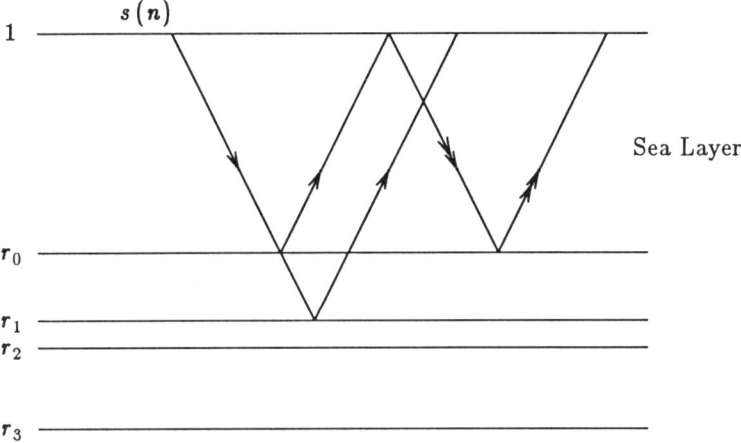

Figure 3.3.2 Reflections in an idealized marine seismic system. Single arrows show primary reflections. Double arrows show multiple reflections. $s(n)$ is a (downward propagating) acoustic energy source. In practice, the arrows would be vertical, giving coincident source and receiver. Here, they are drawn with offsets for clarity. Also for clarity, only a single multiple and two primary arrivals are shown.

An air gun or similar source transmits a pulse (or other waveform) of acoustic energy downwards through the sea from a surface vessel. On encountering the sea-bed the acoustic energy is partially transmitted through to the sub-layers of mud and rock, and partially reflected. The reflected energy returns to the surface where it is received by a hydrophone.[11] The transmitted energy

---

11. Note that in practice, an array of hydrophones towed behind the vessel is

propagates down through the first sub sea-bed layer until it encounters a second interface, marking a change between two rock types and characterized by a change of acoustic impedance. On encountering this interface, energy is again partly reflected back upwards and is partly transmitted into the next layer, and so on. Each rock type has a characteristic wave velocity and impedance [22]. Consequently, the response measured at the surface comprises a series of reflections whose strengths and arrival times assist in determining the rock-types and layer thicknesses of the sub-surface structure.

A number of factors conspire to complicate this picture; one of the main problems arises from the fact that energy returning to the surface is itself reflected back towards the sea-bed. This arises because the air-sea interface is an almost perfect reflector (see Figure 3.3.2). Taken together with the fact that the sea-bed may have a very high reflection coefficient, this leads to a considerable portion of the transmitted energy being trapped within the sea layer. This trapped energy reverberates between surface and sea-bed and is recorded repeatedly at the hydrophones together with the desired ('primary') reflections which it overlays. The presence of these 'multiple' reflections in the recorded seismic signal makes correct interpretation of the desired primary information difficult and their removal, or at least attenuation, is an important aim of processing.

One technique which can be employed to attenuate the multiples is prediction error filtering. In most applications of linear prediction the prediction distance or 'gap' is unity. Here, the filter is given a predictive gap which is approximately equal to the two-way travel time of energy in the sea-layer k_0, samples, say. This is fairly easily determined since the sea depth is usually at least approximately known, as is the velocity of sound in water.

We attempt to predict the multiples using a least-squares prediction filter with prediction distance n_0 as:

$$\underline{f} = [1, 0, ..., 0, -c_1, -c_2, ..., -c_p]^t , \qquad (3.3.20)$$

and this operator is applied to the data. One may argue that the

---

generally employed. A single hydrophone is considered here in the interests of simplifying the discussion.

location of the sub sea-bed layers is essentially random and hence unpredictable, whereas the multiples repeat at (more or less) fixed intervals and are thus largely predictable [13]. As an illustration, Figure 3.3.3a) shows a synthetic response (trace).

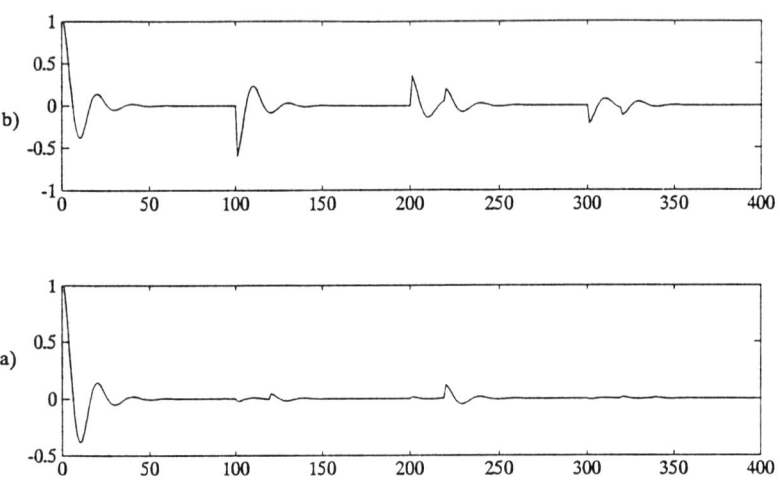

Figure 3.3.3 Synthetic seismic response (trace) and predictive deconvolution. a) Synthetic Response: Consists of primary arrivals at 0 and 220, and multiples at 100,200, 300 and 320 and 320. b) Prediction Error Filtered response showing attenuation of multiple components.

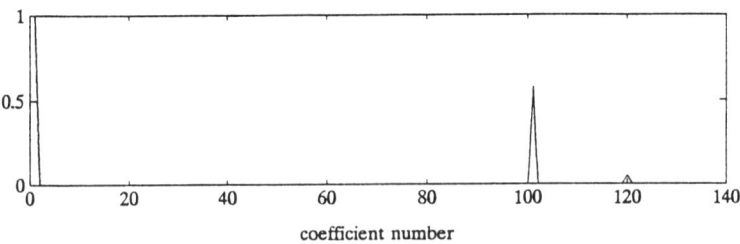

Figure 3.3.4 Prediction error filter applied to data in Figure 3.3.3.

The synthetic response illustrated was constructed as follows: The primary arrivals $s_p(n)$ are modeled as the convolution of the source waveform $s(n)$ with the impulse response of the earth $r(n)$, say

$$s_p(n) = s(n) {}^* r(n) \ . \tag{3.3.21}$$

The source is modeled as a simple damped cosine

$$s(n)=e^{-\lambda n}\cos\omega_0 n \ , \tag{3.3.22}$$

and the earth by an impulse train

$$r(n)=1+r_0\delta(n-k_0)+r_1\delta(n-k_1)+... \tag{3.3.23}$$

where the r_i's represent reflection strengths and the k_i's the arrival times corresponding to the i-th layer. The unit response at time $n=0$ represents the direct arrival (no reflection).

The measured response $s_r(n)$ consists of the primary response $s_p(n)$ convolved with a multiple response $m(n)$

$$s_r(n)=s(n)\,{}^*r(n)\,{}^*m(n) \ . \tag{3.3.24}$$

This multiple response is due to the fact that all primary arrivals at the surface are reflected back towards the subsurface. We assume that the reverberant energy in the sea-layer is dominant. This is not an unreasonable assumption most of the time. Then each impulse which returns to the surface is subsequently reflected towards the sea-bed (with reflection coefficient -1) and back from the sea-bed (with strength r_0) and returns to the surface k_0 later, only to be reflected again. The response shown in Figure 3.3.3a) includes the sea-bed primary signal (recorded at time index $n=0$) and a single primary at $n=220$, with reflection coefficient 0.2. Superimposed are multiple arrivals at intervals $k_0=100$ samples. The reflection coefficient for the multiples was $r_0=0.6$. The source waveform in this case was a simple damped cosine. The nominal sample rate was $f_s=100$ Hz so that $k_0=100$ corresponds to a 1 second travel time. The speed of sound in water is approximately 1500 m/s and thus in this example, the water depth is approximately 750m.

For this response we may compute prediction operators using the general structure (3.3.5), where the correlation coefficients are estimated using the sample correlation.[12] As an example, Figure 3.3.4 shows an $L=40$ point prediction error filter example for a

---

12. Here we used the raw correlation of (2.6.18), that is without any 'bias correction'. Note that for the data used in this example the impact of the correction is small

prediction distance of $n_0=92$. The result of applying this PEF to the data of Figure 3.3.3a) is shown in Figure 3.3.3b). We see that the predictor is almost completely successful in removing the multiple arrivals at $n=100$, 200, 300 and 320 from the response while the primaries at $n=0$ and $n=220$ have not been attenuated. Overall, the PEF may be considered successful and this in spite of the fact that n_0 was set at 92, which is not equal to the true interval between multiples of 100. In fact, a little thought shows that for such a simple example the predictor will be effective for any choice of prediction distance in the range

$$k_0-L+1\leq n_0\leq k_0 . \qquad (3.3.25)$$

This **predictive deconvolution** technique was originally proposed by Robinson [23]. In spite of the obvious flaws in the model described, the method generally works well in practice [13] and is widely used in seismic exploration processing.

3.3.2 Deconvolution

Deconvolution, or inverse filtering, is the name given to the process of unravelling the effects of the convolution of two signals (or of a system and a signal) so as to restore one of the signals. For example, if a signal $x(n)$ is input to a linear shift-invariant system $h(n)$ producing an output $y(n)$, then we seek to design and apply some operator which will restore $x(n)$ (see Figure 3.3.5).

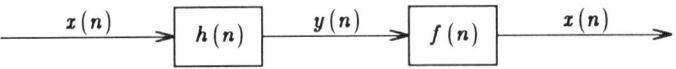

Figure 3.3.5 Deconvolution using an inverse filter $f(n)$.

It is assumed here that the signal $y(n)$ and the system $h(n)$ are known (measured) quantities. Moreover, we restrict attention to deconvolution using a linear time-invariant filter $f(n)$. $f(n)$ is the 'inverse filter' for $h(n)$ and may be denoted $h^{-1}(n)$. As we shall see, however, it is usually not possible to realize this inverse exactly, in which case $f(n)$ will be an approximation to $h^{-1}(n)$.

The most straightforward approach to deconvolution is simply to calculate $F(z)$ by inverting the system transform $H(z)$. For the sake of simplicity, let us assume that the system $h(n)$ is stable, causal, and has a z transform which may be represented in the rational form:

$$H(z) = \frac{\displaystyle\prod_{i=1}^{N}(1-a_i z^{-1})}{\displaystyle\prod_{j=1}^{M}(1-b_j z^{-1})} , \qquad (3.3.27)$$

where, as a consequence of stability and causality, the poles b_j satisfy

$$|b_j| < 1 \quad ; j=1,2,...,M .$$

However, no such restriction applies to the numerator, and in general we may order and separate the zeros such that

$$|a_i| < 1 \quad ; i=1,2,...,k$$
$$|a_i| > 1 \quad ; i=k+1,k+2,...,N .$$

Thus we may write $H(z)$ as

$$H(z) = \frac{\displaystyle\prod_{i=1}^{k}(1-a_i z^{-1}) \prod_{i=k+1}^{N}(1-a_i z^{-1})}{\displaystyle\prod_{j=1}^{M}(1-b_j z^{-1})} . \qquad (3.3.28)$$

Referring to Figure 3.3.6, the deconvolution problem can equivalently be phrased as the determination of $f(n)$ such that

$$h(n) * f(n) = \delta(n) , \qquad (3.3.29)$$

or equivalently

$$F(z) = \frac{1}{H(z)} . \qquad (3.3.30)$$

In principle, the inverse can be found simply by taking the reciprocal of the transform of the system as indicated by equation (3.3.30). From (3.3.30), however, the zeros of (3.3.28) become the poles of $F(z)$ and vice-versa. Hence, $F(z)$ has zeros at b_j for

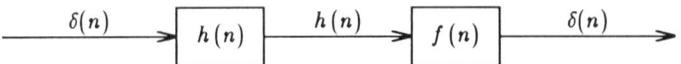

Figure 3.3.6 Deconvolution as inverse filtering of a system $h(n)$.

$j=1,2,...,M$ and poles at a_i for $i=1,2,...,N$. Unfortunately, from the definition, some of the poles of $F(z)$ lie outside the unit circle (specifically, those at $z=a_i$ for $i=k+1,k+2,...,N$). Clearly, $F(z)$ can only be implemented as a *stable* causal system if all the zeros of $H(z)$ lie within $|z|=1$. We recall from Section 2.5 that this condition corresponds to $h(n)$ being **minimum phase**. We can overcome this restriction if the possibility of an acausal or anticipatory solution is admitted. To illustrate, consider a simple two-point system example [24]

$$h(0)=1 , \quad h(1)=a , \qquad\qquad (3.3.31a)$$

so that

$$H(z)=1+az^{-1} , \qquad\qquad (3.3.31b)$$

which has a single zero at $z=-a$. Now, inverting $H(z)$ as in (3.3.30) and expanding using long-division gives

$$F(z)=\frac{1}{H(z)}=\frac{1}{1+az^{-1}}=1-az^{-1}+a^2z^{-2}-... \qquad (3.3.32)$$

providing $|a|<1$ this sequence converges and thus represents a stable transform (see Figure 3.3.7a)). If $|a|>1$ on the other hand, the magnitude of the terms of the sequence increase without bound and $F(z)$ is unstable (see Figure 3.3.7b)). This condition on $|a|$ is, of course, nothing more than the familiar constraint on the poles of a causal transfer function. However, we could have expanded $1/H(z)$ in powers of z rather than z^{-1}, that is, as an acausal sequence:

$$F(z)=\frac{1}{H(z)}=\frac{1}{a}z-\frac{1}{a^2}z^2+\frac{1}{a^3}z^3-... \qquad (3.3.33)$$

By contrast with (3.3.32), for this expansion the sequence converges provided $|a|>1$. The resulting filter is, however, purely anticipatory (see Figure 3.3.7c)). Hence, we could produce a stable solution if $|a|>1$ but only by using an acausal filter. More

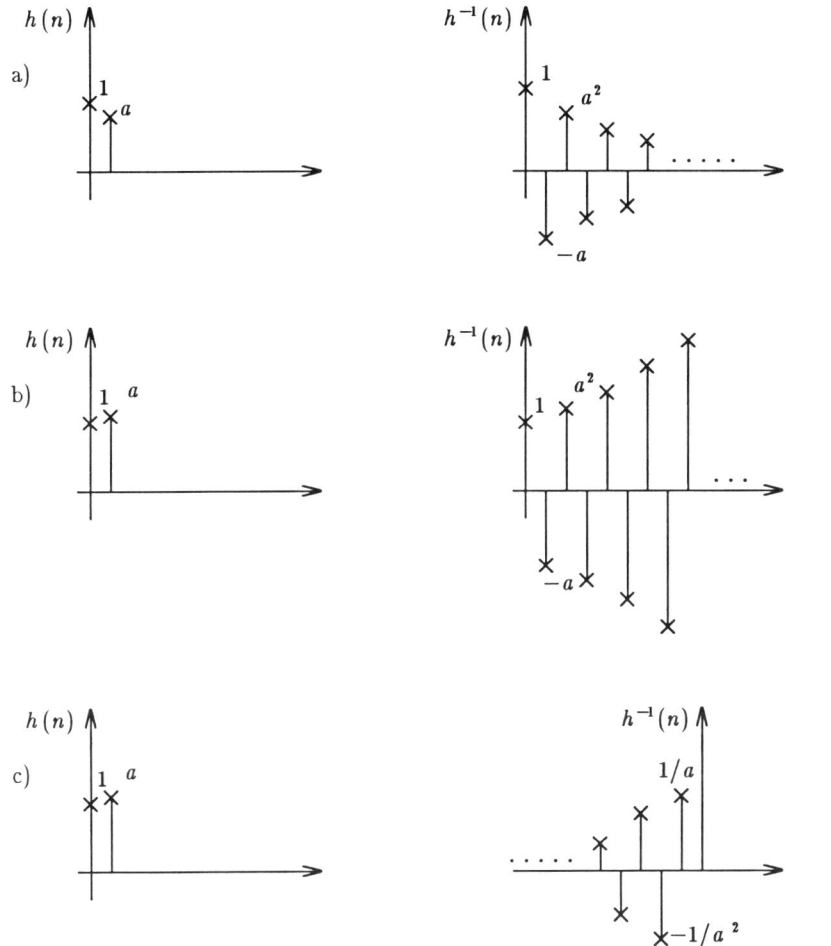

Figure 3.3.7 Inversion of the two-point sequence $h(0)=1$, $h(1)=a$. a) $|a|<1$, $h^{-1}(n)$ is stable causal. b) $|a|>1$, $h^{-1}(n)$ is unstable causal. c) $|a|>1$, $h^{-1}(n)$ is stable acausal.

generally, $H(z)$ contains zeros both inside and outside $|z|=1$. Such a mixed-phase system only has a stable inverse as a two-sided filter.

A further problem with direct inversion arises from the fact that the inverse of $h(n)$ may be infinite in length, even if $h(n)$ itself is finite (in fact this occurs in the two-point example described above). Consequently, construction of the inverse using a Discrete Fourier Transform (DFT) causes problems because an infinite sequence is represented by a finite length transform.

Optimal Least Squares Inverse Filter Design

In view of the above discussion, if a stable, finite, and causal inverse is required, it will generally only be possible to find an approximate solution. Adopting the minimum mean-squared error philosophy for filter design, the approximation process in the inversion can be viewed as depicted in Figure 3.3.8.

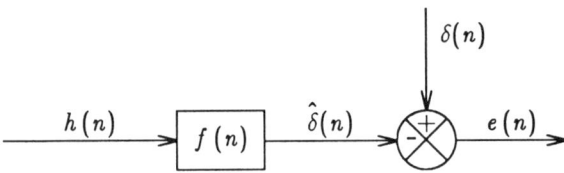

Figure 3.3.8 Inverse filter design by least-squares approximation.

As usual, the problem is to minimize the difference between the filter output and the desired signal. In this case the desired signal is simply an impulse and we have

$$J= \sum_{n=-\infty}^{\infty} e^2(n)= \sum_{n=-\infty}^{\infty} (\delta(n)-\hat{\delta}(n))^2 ,$$

$$= \sum_{n=-\infty}^{\infty} (\delta(n)-f(n)*h(n))^2 , \qquad (3.3.34)$$

where $\hat{\delta}(n)$ is the output from the inverse filter. By differentiating with respect to $f(j)$ we have

$$\sum_i r(i-j)f(i)=g(j) , \qquad (3.3.35)$$

where

$$r(i-j)= \sum_{n=-\infty}^{\infty} h(n-i)h(n-j) , \qquad (3.3.36)$$

and

$$g(j)= \sum_{n=-\infty}^{\infty} \delta(n)h(n-j) =h(-j) . \qquad (3.3.37)$$

In view of the causality of $h(n)$, $g(j)$ has only a single non-zero value. Note that (3.3.35) corresponds to the normal equations

(3.2.7) of Section 3.2. Also, applying a change of variables to (3.3.36) we have

$$r(i-j)= \sum_{n=-\infty}^{\infty} h(n)h(n+i-j) \, ,$$

or

$$r(m)= \sum_{n=-\infty}^{\infty} h(n)h(n+m) \, . \qquad (3.3.36a)$$

This, of course, corresponds to the sample correlation first encountered in Section 2.3 (see equation (2.3.5)). As in Section 3.2, various solutions to (3.3.35) are obtained depending on any length and causality constraints placed on the filter. We consider three cases:

a) Infinite Unconstrained Filter — If the inverse filter is not constrained to be causal, then the optimal $F(z)$ can be obtained simply by z transforming both sides of the normal equations (3.3.35):

$$\sum_{i=-\infty}^{\infty} r(j-i)f(i)=g(j) \, . \qquad (3.2.7)$$

Recognizing that the left-hand side of this equation is simply the convolution of the filter with the input autocorrelation, we have

$$R(z)F(z)=H(z^{-1}) \, , \qquad (3.3.38)$$

where the right-hand side is obtained by transforming the cross-correlation of (3.3.37). From (3.3.36a) the transform $R(z)$ is

$$R(z)=H(z)H(z^{-1}) \, ,$$

so that combining with (3.3.38), we obtain

$$F(z)=\frac{1}{H(z)} \, . \qquad (3.3.39)$$

As expected, the solution is exact because the filter has no causality or finite length restrictions.

b) Infinite Causal Filter — If the inverse filter is constrained to be causal, the normal equations (3.3.35), with $g(j)$ given by (3.3.37), become

$$\sum_{i=0}^{\infty} r(j-i)f(i) = h(0) \qquad ; j=0 ,$$

$$= 0 \qquad ; j \neq 0 .$$

The solution to this set can be shown to have the form [24]

$$F(z) = \frac{C}{H_{eqmin}(z)} , \qquad (3.3.40)$$

where $H_{eqmin}(z)$ is the equivalent minimum phase transfer function corresponding to $H(z)$ (see Section 2.5) and C is a constant. For $H(z)$ if the form (3.3.28), $H_{eqmin}(z)$ is

$$H_{eqmin}(z) = \frac{\prod_{i=1}^{k} (1 - a_i z^{-1}) \prod_{i=k+1}^{N} (z^{-1} - a_i)}{\prod_{j=1}^{M} (1 - b_j z^{-1})} , \qquad (3.3.41)$$

so that $H_{eqmin}(z)$ has all its poles and zeros within the unit circle, as expected. Hence

$$F(z) = \frac{C \prod_{j=1}^{M} (1 - b_j z^{-1})}{\prod_{i=1}^{k} (1 - a_i z^{-1}) \prod_{i=k+1}^{N} (z^{-1} - a_i)} . \qquad (3.3.42)$$

It follows that *the causal, infinite least-squares filter is minimum phase irrespective of the system* $H(z)$. This is an extremely important property. Also, the filter output $Y(z)$ is given by

$$Y(z) = F(z)H(z)$$

or, using (3.3.28)

$$Y(z) = \frac{C \prod\limits_{j=1}^{M} (1-b_j z^{-1}) \prod\limits_{i=1}^{k} (1-a_i z^{-1}) \prod\limits_{i=k+1}^{N} (1-a_i z^{-1})}{\prod\limits_{i=1}^{k} (1-a_i z^{-1}) \prod\limits_{i=k+1}^{N} (z^{-1}-a_i) \prod\limits_{j=1}^{M} (1-b_j z^{-1})} ,$$

or

$$Y(z) = \frac{C \prod\limits_{i=k+1}^{N} (1-a_i z^{-1})}{\prod\limits_{i=k+1}^{N} (z^{-1}-a_i)} , \qquad (3.3.43)$$

which is all-pass (see Section 2.5). We see that the infinite causal least-squares inverse filter only produces the ideal result if $k=N$, that is, if $h(n)$ is minimum phase. In other cases the resulting output will be spectrally equivalent, but the phase will not be restored.

To illustrate this phenomenon let us return to the two-point example given above

$$h(0)=1 , \quad h(1)=a , \qquad (3.3.31a)$$

where we assume $h(n)$ is non-minimum phase, that is $|a|>1$. The infinite causal least-squares inverse filter $F(z)$ is given by equation (3.3.40). In this example, the equivalent minimum phase version of $h(n)$ is given by

$$h_{eqmin}(0)=a , \quad h_{eqmin}(1)=1 . \qquad (3.3.44)$$

This gives

$$H_{eqmin}(z)=a+z^{-1} , \qquad (3.3.45)$$

with zero at $z=-1/a$ as expected. From (3.3.40), the inverse filter has the form

$$F(z) = \frac{C}{a+z^{-1}} , \qquad (3.3.46)$$

where C is a constant and thus influences only the scale of the result. The output from the filter is then given by

$$Y(z) = C \left[\frac{1 + az^{-1}}{a + z^{-1}} \right] , \tag{3.3.47}$$

which is all-pass as expected. The pole-zero plots for $H(z)$, $F(z)$ and $Y(z)$ are shown in Figure 3.3.9.

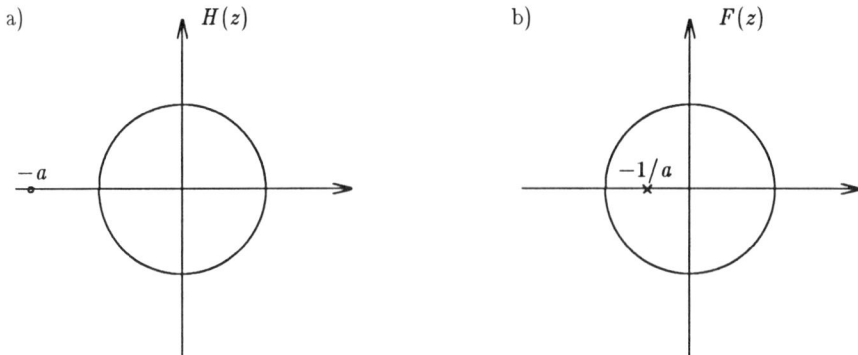

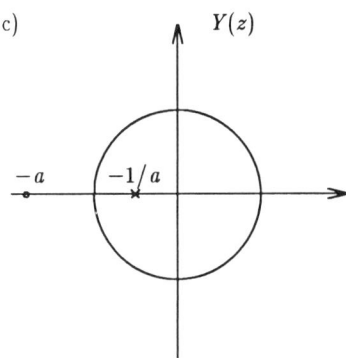

Figure 3.3.9 Pole-zero plots for causal least-squares inverse filtering of the two point sequence $h(0) = 1$, $h(1) = a$. a) System $H(z)$. b) Inverse $F(z)$. c) Filter output $Y(z)$.

c) Finite Causal Filter — If, in addition to causality, the inverse filter is also constrained to have finite length, then the normal equations take the form of (3.2.7c)

$$R\underline{f} = \underline{g} , \tag{3.2.7c}$$

where R is a symmetric Toeplitz matrix with elements consisting of the correlation coefficients of $h(n)$ obtained from (3.3.36a), where

$$\underline{g} = [h(0),0,0,...,0]^t ,$$

and where

$$\underline{f} = [f(0),f(1),...,f(L-1)]^t .$$

The equations (3.2.7c) may be solved using the Levinson recursion as described earlier in Section 3.2. However, the simplified form of the cross-correlation vector g allows further simplification using Durbin's recursion (see Appendix 3C). In fact, comparing (3.2.7c) (given the particular form for g), with (3.3.14) for the Prediction Error Filter (PEF), we see that the two have an identical structure. Hence, as with the PEF, it can be shown that the finite least-squares inverse filter f will be minimum phase, *independent of the phase structure of the input* (see Appendix 3D). Consequently, as for the infinite filter, the inversion is most effective when the input sequence *is* minimum phase.

The Use of Delay in Inverse Filtering

As we have seen, a non-minimum phase system cannot be completely deconvolved using a stable, causal inverse filter. Such a filter can effectively flatten the spectrum, but will leave a residual phase distortion, the all-pass component. There are some applications where such spectral equalization is sufficient, but there are others where the phase structure is also important.

Consider again the example $h(0)=1$, $h(1)=a$. The expansion of the inverse of the z transform as given by (3.3.32) is unstable if $|a|>1$. As we noted earlier, the long-division which gave rise to (3.3.32) could have been expanded in powers of z rather than z^{-1} giving

$$F(z) = \frac{1}{a}z - \frac{1}{a^2}z^2 + \frac{1}{a^3}z^3 - ... \tag{3.3.33}$$

This is a stable (convergent) sequence for $|a|>1$. However, as z^{-1} represents a delay, z represents an advance and the filter $F(z)$ is an anticipatory or acausal response. However, such a filter can be implemented using delay. For, if the sequence of equation

(3.3.33) is now delayed by l points to form a new filter $f_1(n)$, say, then the transform becomes

$$F_1(z) = z^{-l} F(z)$$
$$= \frac{1}{a} z^{-l+1} - \frac{1}{a^2} z^{-l+2} + \dots + (-1)^{l-1} \frac{1}{a^l} + (-1)^l \frac{1}{a^{l+1}} z + \dots$$

This response is now two-sided, having l causal components. Now, if the acausal component of $F_1(z)$ is truncated to give $\hat{F}(z)$, say, where

$$\hat{F}(z) = \frac{1}{a} z^{-l+1} - \frac{1}{a^2} z^{-l+2} + \dots + (-1)^l \frac{1}{a^l} , \qquad (3.3.48)$$

then $\hat{F}(z)$ is a causal approximation to the inverse $F(z)$. Consequently, by incorporating delay into the inversion we can improve the modelling of the inverse system. The result of using delay becomes clear if we consider the effect of convolving the inverse with the original impulse response. Ignoring the effect of the approximation caused by truncating the response, the output should be an impulse. A moments thought confirms that the position of the impulse is at sample index l. Thus, the use of delay may improve the inversion but at the cost of delaying the overall system output. Looked at another way, a perfect inversion produces an impulse at time index zero. This represents a flat spectrum and no phase distortion. A non-minimum phase system cannot be successfully inverted by a causal filter — the spectrum can be equalized but an all-pass component (with a nonlinear phase distortion) remains. Even with delay we cannot produce a perfect result, but we can produce a flat spectrum and a *linear* phase curve.

Least-Squares Inverse Filtering with Delay

As we have seen, using a finite delay can improve the inversion of a non-minimum phase sequence. On the other hand, for any finite delay the result will remain approximate. Furthermore, the design approach adopted in the example considered above, that is simply truncating the ideal inverse operator, is somewhat arbitrary. However, it is possible to design an inverse operator which incorporates a delay using the least-squares method. This modified least-squares design problem corresponds to replacing the desired signal $\delta(n)$ by a delayed version $\delta(n-l)$, see Figure 3.3.10. Mathematically, we replace

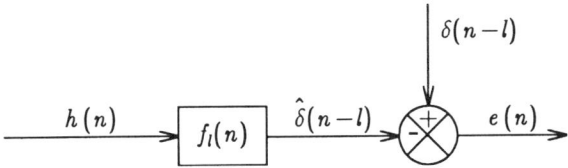

Figure 3.3.10 Least-squares inverse filtering with delay l.

the zero-delay minimization of (3.3.34) by

$$J_l = \sum_{n=-\infty}^{\infty} (\delta(n-l) - f_l(n) * h(n))^2 , \qquad (3.3.49)$$

where $f_l(n)$ refers to the filter associated with delay l. Differentiating J_l and equating to zero gives a set of normal equations of the form (3.3.35) as usual. The correlation elements $r(i)$ are unchanged by the incorporation of delay, but the cross-correlation $g(j)$ is now given by

$$g(j) = \sum_{n=-\infty}^{\infty} \delta(n-l) h(n-j) = h(l-j) . \qquad (3.3.50)$$

a) Infinite Causal Filter — For the infinite causal filter it can be shown [24] that

$$F_l(z) = \frac{h_{ap}(0) z^{-l} + h_{ap}(1) z^{-l+1} + \ldots + h_{ap}(l)}{H_{eqmin}(z)} , \qquad (3.3.51)$$

where the $h_{ap}(i)$ are time domain coefficients obtained from the inverse transform of

$$H_{ap}(z) = C \frac{\displaystyle\prod_{i=k+1}^{N} (1 - a_i z^{-1})}{\displaystyle\prod_{i=k+1}^{N} (z^{-1} - a_i)} ,$$

which is all-pass. From this one may obtain two results [24]:

i) The infinite causal least-squares filter perfectly inverts $H(z)$ if the output is infinitely delayed.

ii) The error energy J_l is monotonically non-increasing with l.

b) Finite Causal Filter — If the causal least-squares filter with delay l is constrained to be finite, then the normal equations (3.2.7c) become

$$Rf_l = [h(l), h(l-1), ..., h(0), 0, 0, ..., 0]^t . \qquad (3.3.52)$$

The results stated above for the causal infinite filter are not generally applicable for the causal finite least-squares filter. In particular, the filter performance is *not* in general a monotonic function of delay l. It follows, therefore, that without some *a priori* knowledge of the delay required to produce a desired accuracy, the above formulation is not of much value. Fortunately, this is not a serious practical problem since R is unchanged as the delay varies. Hence, an algorithm which evaluates f_l as l changes need only invert R once. Simpson [25] produced a highly efficient algorithm known as the 'Simpson Sidewards Recursion'. Starting from the zero-delay solution obtained from the Levinson recursion, the algorithm sequentially computes the filters for delay $l=1,2,...$ up to any desired delay. Furthermore, it is possible to monitor the performance at each step in the recursion and halt the process according to some pre-determined threshold.

Deconvolution of Noisy Measured Signals

In practice, signals are typically corrupted by measurement noise. For example, consider the system shown in Figure 3.3.11, where $h(n)$ is a known system, subjected to a known input $x(n)$, producing an output which is corrupted by measurement noise $v(n)$. In this problem, even if $f(n)$ is ideal in the sense that $f(n)=h^{-1}(n)$, the input is not recovered exactly due to the measurement noise. Moreover, in areas of the spectrum where the power of the system response is low relative to the noise power, the effect of an ideal inverse will be to significantly amplify already high levels of measurement noise. We mention two approaches to

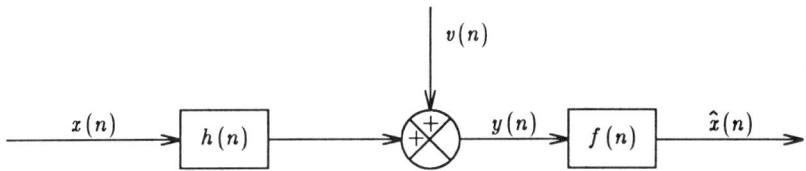

Figure 3.3.11 Deconvolution of a noisy measured signal.

mitigate this problem:

a) Prewhitening [26] — Is the term used to describe the practice of adding a small positive quantity to each element of the leading diagonal of the system autocorrelation matrix (that is, the zeroth lag of the autocorrelation) in the normal equations. That is equations (3.2.7c)

$$R\underline{f}=\underline{g} \;, \qquad\qquad (3.2.7c)$$

are replaced by

$$(R+\sigma^2 I)\underline{f}=\underline{g} \;. \qquad\qquad (3.3.53)$$

where σ^2 is a small positive constant. The term prewhitening arises from the fact that adding a number to the diagonal term of the correlation matrix (that is to $r(0)$) is equivalent to adding white noise to the input. However, this whitening of the spectrum is purely hypothetical as far as the input time-series is concerned, since σ^2 is added only to the correlation, and then only for the purposes of computing the inverse operator. In the inversion, the raising of the input spectral level causes a corresponding decrease in the level of amplification in the inverse operator.

From a numerical point of view, prewhitening can be seen as a means of improving the stability of the solution. The determination of the filter becomes problematic when the system of normal equations becomes ill-conditioned (when the autocorrelation matrix is almost singular). A number of measures of the condition of a system have been defined. Among these is the idea of **condition number, C**

$$C=\frac{|\lambda_{\max}|}{|\lambda_{\min}|} \;, \qquad\qquad (3.3.54)$$

where $\lambda_{\max}$ and $\lambda_{\min}$ denote the maximum and minimum eigenvalues respectively of the autocorrelation matrix. The larger the value of C, the worse the condition of the system, and in the limit as $C \to \infty$, then $\lambda_{\min} \to 0$ and the matrix is singular. Since R is symmetric positive-definite then the eigenvalues of R are real and non-negative (see also Section 4.2). Hence

$$C = \frac{\lambda_{\max}}{\lambda_{\min}} . \qquad (3.3.55)$$

It is easy to demonstrate that if R has eigenvalues $\lambda_1, \lambda_2, ..., \lambda_L$, then $R + \sigma^2 I$ has eigenvalues $\lambda_1 + \sigma^2$, $\lambda_2 + \sigma^2, ..., \lambda_L + \sigma^2$. Hence, the condition number for $R + \sigma^2 I$ is

$$C' = \frac{\lambda_{\max} + \sigma^2}{\lambda_{\min} + \sigma^2} , \qquad (3.3.56)$$

and $C' \leq C$ for all σ^2, $\lambda_{\min}$, $\lambda_{\max}$, with equality holding if and only if $\lambda_{\max} = \lambda_{\min}$, that is if the input is white.

The level of prewhitening applied to a system, that is the value of σ^2, is usually given as a percentage of the zero lag autocorrelation coefficient. If σ^2 is the prewhitening applied to the system, then the **Percentage Prewhitening (PPW)** is

$$PPW = \frac{100\sigma^2}{r(0)} . \qquad (3.3.57)$$

The selection of a suitable level for PPW is ad-hoc, however, and values anywhere between 0.5% and 5% of $r(0)$ have been suggested. The main advantage of prewhitening is its simplicity, and the fact that well-conditioned systems are insensitive to its application. Hence, a small level of prewhitening can be applied in all deconvolution problems without a large performance penalty.

b) Shaping – Another approach to the reduction of noise in inverse filtering is to relax the output impulse requirement in favor of a broader pulse as desired function. This approach has been proposed by Senmoto and Childers [27], who employed a 'generalized inverse' of the form

$$F(e^{j\omega}) = \left[\frac{D(e^{j\omega})}{H(e^{j\omega})} \right] , \qquad (3.3.58)$$

where $D(e^{j\omega})$ is the desired output pulse. The method provides a way to control which areas of the spectrum are completely equalized and which are not. This can be used as a mechanism to avoid high noise amplification in areas of low signal content. This assumes, of course, that such signal-to-noise information is available *a priori*.

Appendix 3A − Positive Semi-Definite Form of the Autocorrelation Matrix

For a stationary signal $x(n)$, the $(L \times L)$ correlation matrix R may be written

$$R = E\{\underline{x}_n \underline{x}_n^t\} \ , \tag{A1}$$

where $\underline{x}_n = [x(n), x(n-1), ..., x(n-L+1)]^t$. For any $(L \times 1)$ vector $\underline{a}$, say, we may define

$$y = \underline{a}^t \underline{x}_n \ , \tag{A2}$$

so that

$$y^2 = \underline{a}^t \underline{x}_n \underline{x}_n^t \underline{a} \ , \tag{A3}$$

and

$$E\{y^2\} = \underline{a}^t R \underline{a} \geq 0 \ . \tag{A4}$$

We have thus demonstrated that R is positive semi-definite as was asserted in Section 2.1.

Appendix 3B − The Levinson Recursion

The determination of the L point least-squares optimal filter is obtained from the solution of the linear algebraic equations (3.2.7). Classical techniques for the solution of such a set require $O(L^3/2)$ operations and $O(L^2)$ for storage. However, for this problem it is possible to take advantage of the symmetry of the coefficient matrix to reduce the computational burden. Firstly, in the case of the covariance formulation we can use the Cholesky factorization which requires $O(L^3/6)$ operations and $O(L^2)$ storage. This represents about a third of the requirement for standard techniques. However, our interest lies primarily with the autocorrelation formulation of equations (3.2.7c)

$$\sum_{i=0}^{L-1} r(i-j)f(i)=g(j) \quad ; j=0,1,...,L-1 \ ,$$

or

$$R\underline{f}=\underline{g} \ , \tag{3.2.7c}$$

where $\underline{f}=[f(0),f(1),...,f(L-1)]^t$, $\underline{g}=[g(0),g(1),...,g(L-1)]^t$, and where R has elements $R(i,j)=r(i-\overline{j})$. We may obtain significant reductions in the computational requirement for the solution of this set by exploiting the Toeplitz structure of R. Levinson [17] has produced an algorithm which solves equations (3.2.7c) in approximately $2L^2$ operations and with a reduced storage requirement as compared to that of classical algorithms.

The algorithm begins with the (trivial) solution to the scalar (1×1) set:

$$r(0)f_0^{(0)}=g(0) \ ,$$

and iteratively increases the order of the equations up to the full $(L\times L)$ system. We use the notation $f_j^{(i)}$ to denote the j-th component of the filter vector corresponding to the intermediate system of order $(i+1)$.[13] Rather than giving a formal statement of the iteration equations, we will illustrate the algorithm by examining the iteration at a single step. Consider the iteration from $i=2$ to 3. That is, assume that we have computed from the previous step the filter coefficients $f_0^{(2)}$, $f_1^{(2)}$, $f_2^{(2)}$, where

$$\begin{bmatrix} r(0) & r(1) & r(2) \\ r(1) & r(0) & r(1) \\ r(2) & r(1) & r(0) \end{bmatrix} \begin{bmatrix} f_0^{(2)} \\ f_1^{(2)} \\ f_2^{(2)} \end{bmatrix} = \begin{bmatrix} g(0) \\ g(1) \\ g(2) \end{bmatrix} \ . \tag{B1}$$

---

13. In this Appendix, to help simplify notation, components of the filter coefficient and auxiliary coefficient vector are denoted via subscripts.

Also, assume that as a by-product of the computation we have calculated **auxiliary coefficients** $a_0^{(2)}$, $a_1^{(2)}$, $a_2^{(2)}$ and α_3, where

$$
\begin{bmatrix} r(0) & r(1) & r(2) \\ r(1) & r(0) & r(1) \\ r(2) & r(1) & r(0) \end{bmatrix} \begin{bmatrix} a_0^{(2)} \\ a_1^{(2)} \\ a_2^{(2)} \end{bmatrix} = \begin{bmatrix} \alpha_3 \\ 0 \\ 0 \end{bmatrix} . \tag{B2}
$$

As a first step we extend the system (B1) by augmenting the order of the correlation matrix and by adding a zero to the filter vector:

$$
\begin{bmatrix} r(0) & r(1) & r(2) & r(3) \\ r(1) & r(0) & r(1) & r(2) \\ r(2) & r(1) & r(0) & r(1) \\ r(3) & r(2) & r(1) & r(0) \end{bmatrix} \begin{bmatrix} f_0^{(2)} \\ f_1^{(2)} \\ f_2^{(2)} \\ 0 \end{bmatrix} = \begin{bmatrix} g(0) \\ g(1) \\ g(2) \\ \gamma_3 \end{bmatrix} , \tag{B3}
$$

where the term γ_3 is computed directly from (B3) as

$$
\gamma_3 = r(3) f_0^{(2)} + r(2) f_1^{(2)} + r(1) f_2^{(2)} . \tag{B4}
$$

We next extend the auxiliary system (B2) as

$$
\begin{bmatrix} r(0) & r(1) & r(2) & r(3) \\ r(1) & r(0) & r(1) & r(2) \\ r(2) & r(1) & r(0) & r(1) \\ r(3) & r(2) & r(1) & r(0) \end{bmatrix} \begin{bmatrix} a_0^{(2)} \\ a_1^{(2)} \\ a_2^{(2)} \\ 0 \end{bmatrix} = \begin{bmatrix} \alpha_3 \\ 0 \\ 0 \\ \beta_3 \end{bmatrix} , \tag{B5}
$$

where β_3 may be computed directly from (B5) as

$$
\beta_3 = r(3) a_0^{(2)} + r(2) a_1^{(2)} + r(1) a_2^{(2)} . \tag{B6}
$$

Now, the Toeplitz structure of the coefficient matrix allows us to *reverse* the equations (B5) without changing the coefficient matrix R. That is

$$
\begin{bmatrix} r(0) & r(1) & r(2) & r(3) \\ r(1) & r(0) & r(1) & r(2) \\ r(2) & r(1) & r(0) & r(1) \\ r(3) & r(2) & r(1) & r(0) \end{bmatrix} \begin{bmatrix} 0 \\ a_2^{(2)} \\ a_1^{(2)} \\ a_0^{(2)} \end{bmatrix} = \begin{bmatrix} \beta_3 \\ 0 \\ 0 \\ \alpha_3 \end{bmatrix} . \tag{B7}
$$

We next subtract some constant multiple k_3, say, of equations (B7) from (B5) as

$$
\begin{bmatrix} r(0) & r(1) & r(2) & r(3) \\ r(1) & r(0) & r(1) & r(2) \\ r(2) & r(1) & r(0) & r(1) \\ r(3) & r(2) & r(1) & r(0) \end{bmatrix} \left\{ \begin{bmatrix} a_0^{(2)} \\ a_1^{(2)} \\ a_2^{(2)} \\ 0 \end{bmatrix} - k_3 \begin{bmatrix} 0 \\ a_2^{(2)} \\ a_1^{(2)} \\ a_0^{(2)} \end{bmatrix} \right\} =
$$

$$
= \left\{ \begin{bmatrix} \alpha_3 \\ 0 \\ 0 \\ \beta_3 \end{bmatrix} - k_3 \begin{bmatrix} \beta_3 \\ 0 \\ 0 \\ \alpha_3 \end{bmatrix} \right\} . \tag{B8}
$$

We choose k_3 so that the resulting system will have only one non-zero element on the right-hand side. This may be achieved by selecting

$$
k_3 = \frac{\beta_3}{\alpha_3} , \tag{B9}
$$

so that from (B8) we have

$$
\begin{bmatrix} r(0) & r(1) & r(2) & r(3) \\ r(1) & r(0) & r(1) & r(2) \\ r(2) & r(1) & r(0) & r(1) \\ r(3) & r(2) & r(1) & r(0) \end{bmatrix} \begin{bmatrix} a_0^{(3)} \\ a_1^{(3)} \\ a_2^{(3)} \\ a_3^{(3)} \end{bmatrix} = \begin{bmatrix} \alpha_4 \\ 0 \\ 0 \\ 0 \end{bmatrix} , \tag{B10}
$$

where $\alpha_4 = \alpha_3 - k_3 \beta_3$, and

$$
a_0^{(3)} = a_0^{(2)} ,
$$
$$
a_1^{(3)} = a_1^{(2)} - k_3 a_2^{(2)} ,
$$
$$
a_2^{(3)} = a_2^{(2)} - k_3 a_1^{(2)} ,
$$
$$
a_3^{(3)} = -k_3 a_0^{(2)} . \tag{B11}
$$

We can also reverse the system (B10) because of the autocorrelation matrix structure. This yields

$$\begin{bmatrix} r(0) & r(1) & r(2) & r(3) \\ r(1) & r(0) & r(1) & r(2) \\ r(2) & r(1) & r(0) & r(1) \\ r(3) & r(2) & r(1) & r(0) \end{bmatrix} \begin{bmatrix} a_3^{(3)} \\ a_2^{(3)} \\ a_1^{(3)} \\ a_0^{(3)} \end{bmatrix} = \begin{bmatrix} 0 \\ 0 \\ 0 \\ \alpha_4 \end{bmatrix} . \tag{B12}$$

We now subtract some multiple, c_3, say of (B12) from (B3)

$$\begin{bmatrix} r(0) & r(1) & r(2) & r(3) \\ r(1) & r(0) & r(1) & r(2) \\ r(2) & r(1) & r(0) & r(1) \\ r(3) & r(2) & r(1) & r(0) \end{bmatrix} \left\{ \begin{bmatrix} f_0^{(2)} \\ f_1^{(2)} \\ f_2^{(2)} \\ 0 \end{bmatrix} - c_3 \begin{bmatrix} a_3^{(3)} \\ a_2^{(3)} \\ a_1^{(3)} \\ a_0^{(3)} \end{bmatrix} \right\} = \left\{ \begin{bmatrix} g(0) \\ g(1) \\ g(2) \\ \gamma_3 \end{bmatrix} - c_3 \begin{bmatrix} 0 \\ 0 \\ 0 \\ \alpha_4 \end{bmatrix} \right\} .$$

Finally, we select c_3 so that

$$\gamma_3 - c_3 \alpha_4 = g(3) ,$$

or

$$c_3 = \frac{\gamma_3 - g(3)}{\alpha_4} , \tag{B13}$$

and we obtain

$$\begin{bmatrix} r(0) & r(1) & r(2) & r(3) \\ r(1) & r(0) & r(1) & r(2) \\ r(2) & r(1) & r(0) & r(1) \\ r(3) & r(2) & r(1) & r(0) \end{bmatrix} \begin{bmatrix} f_0^{(3)} \\ f_1^{(3)} \\ f_2^{(3)} \\ f_3^{(3)} \end{bmatrix} = \begin{bmatrix} g(0) \\ g(1) \\ g(2) \\ g(3) \end{bmatrix} , \tag{B14}$$

where

$$\begin{aligned} f_0^{(3)} &= f_0^{(2)} - c_3 a_3^{(3)} , \\ f_1^{(3)} &= f_1^{(2)} - c_3 a_2^{(3)} , \\ f_2^{(3)} &= f_2^{(2)} - c_3 a_1^{(3)} , \\ f_3^{(3)} &= - c_3 a_0^{(3)} . \end{aligned} \tag{B15}$$

Hence, we have computed the filter coefficients, auxiliary coefficients, and constant α_4 required for the next step of the iteration. At this point the process is repeated, starting with the augmentation as in (B3) to produce the coefficients for the next iteration and so on.

Since we have completely specified the set up, it only remains to specify the initial conditions for the algorithm. We have

$$r(0)f_0^{(0)}=g(0) \; ,$$

from which

$$f_0^{(0)}=g(0)/r(0) \; . \tag{B16}$$

For the auxiliary vector we have

$$r(0)a_0^{(0)}=\alpha_1 \; .$$

We may select $a_0^{(0)}$ as an arbitrary non-zero value. $a_0^{(0)}=1$ is a common choice, and this gives

$$\alpha_1=r(0) \; . \tag{B17}$$

Appendix 3C − Durbin's Algorithm

Consider the simplified normal equations arising in prediction error filtering:

$$
\begin{bmatrix}
r(0) & r(1) & r(2) & . & r(L-1) \\
r(1) & r(0) & . & . & . \\
. & . & . & . & . \\
. & . & . & . & . \\
r(L-1) & r(L-2) & . & . & r(0)
\end{bmatrix}
\begin{bmatrix}
f(0) \\
f(1) \\
. \\
. \\
f(L-1)
\end{bmatrix}
=
\begin{bmatrix}
J_{min} \\
0 \\
. \\
. \\
0
\end{bmatrix}
, \tag{C1}
$$

where

$$\underline{f}=[f(0),f(1),...,f(L-1)]^t=[1,-c_1,-c_2,...,-c_p]^t \; ,$$

where c_i is the i-th least-squares predictor. We shall describe a modification of the Levinson algorithm of Appendix 3B for the solution of these equations. This modified method is known as

Durbin's Algorithm [21]. The algorithm is similar to the Levinson recursion but exploits the simplified form of the right-hand side in (C1) and, for systems of the same order, provides a further computational reduction compared to Levinson, by about a factor of two.

As with the Levinson recursion, Durbin's algorithm begins with the (trivial) solution to the scalar (1×1) set:

$$r(0)f_0^{(0)} = J_{\min}^{(0)} \ ,$$

and iteratively obtains the solution for orders 2,3,... up to the full $(L \times L)$ system. As in Appendix 3B, we use the notation $f_j^{(i)}$ to denote the j-th component of the filter vector corresponding to the intermediate system of order $(i+1)$. We begin by considering a single step in the solution from $i=2$ to 3. That is, we assume that we have computed from the previous iteration the filter coefficients $f_0^{(2)}, f_1^{(2)}, f_2^{(2)}$, where

$$\begin{bmatrix} r(0) & r(1) & r(2) \\ r(1) & r(0) & r(1) \\ r(2) & r(1) & r(0) \end{bmatrix} \begin{bmatrix} f_0^{(2)} \\ f_1^{(2)} \\ f_2^{(2)} \end{bmatrix} = \begin{bmatrix} J_{\min}^{(2)} \\ 0 \\ 0 \end{bmatrix} . \qquad (C2)$$

The approach to the solution is similar to that for the Levinson algorithm, but the result is even simpler in this case because no auxiliary vector is required. We begin by augmenting equations (C2) as

$$\begin{bmatrix} r(0) & r(1) & r(2) & r(3) \\ r(1) & r(0) & r(1) & r(2) \\ r(2) & r(1) & r(0) & r(1) \\ r(3) & r(2) & r(1) & r(0) \end{bmatrix} \begin{bmatrix} f_0^{(2)} \\ f_1^{(2)} \\ f_2^{(2)} \\ 0 \end{bmatrix} = \begin{bmatrix} J_{\min}^{(2)} \\ 0 \\ 0 \\ \alpha_3 \end{bmatrix} , \qquad (C3)$$

where the term α_3 is computed directly from (C3) as

$$\alpha_3 = r(3)f_0^{(2)} + r(2)f_1^{(2)} + r(1)f_2^{(2)} \ . \qquad (C4)$$

Now, the Toeplitz structure of the coefficient matrix allows us to *reverse* the equations (C3) without changing the coefficient matrix R. That is

$$
\begin{bmatrix} r(0) & r(1) & r(2) & r(3) \\ r(1) & r(0) & r(1) & r(2) \\ r(2) & r(1) & r(0) & r(1) \\ r(3) & r(2) & r(1) & r(0) \end{bmatrix} \begin{bmatrix} 0 \\ f_2^{(2)} \\ f_1^{(2)} \\ f_0^{(2)} \end{bmatrix} = \begin{bmatrix} \alpha_3 \\ 0 \\ 0 \\ J_{\min}^{(2)} \end{bmatrix} . \tag{C5}
$$

We may subtract some multiple k_3 of (C5) from (C3) as

$$
\begin{bmatrix} r(0) & r(1) & r(2) & r(3) \\ r(1) & r(0) & r(1) & r(2) \\ r(2) & r(1) & r(0) & r(1) \\ r(3) & r(2) & r(1) & r(0) \end{bmatrix} \left\{ \begin{bmatrix} f_0^{(2)} \\ f_1^{(2)} \\ f_2^{(2)} \\ 0 \end{bmatrix} - k_3 \begin{bmatrix} 0 \\ f_2^{(2)} \\ f_1^{(2)} \\ f_0^{(2)} \end{bmatrix} \right\} =
$$

$$
= \left\{ \begin{bmatrix} J_{\min}^{(2)} \\ 0 \\ 0 \\ \alpha_3 \end{bmatrix} - k_3 \begin{bmatrix} \alpha_3 \\ 0 \\ 0 \\ J_{\min}^{(2)} \end{bmatrix} \right\} , \tag{C6}
$$

where we select k_3 so that

$$
\alpha_3 - k_3 J_{\min}^{(2)} = 0 ,
$$

or

$$
k_3 = \frac{\alpha_3}{J_{\min}^{(2)}} . \tag{C7}
$$

Hence we obtain

$$
\begin{bmatrix} r(0) & r(1) & r(2) & r(3) \\ r(1) & r(0) & r(1) & r(2) \\ r(2) & r(1) & r(0) & r(1) \\ r(3) & r(2) & r(1) & r(0) \end{bmatrix} \begin{bmatrix} f_0^{(3)} \\ f_1^{(3)} \\ f_2^{(3)} \\ f_3^{(3)} \end{bmatrix} = \begin{bmatrix} J_{\min}^{(3)} \\ 0 \\ 0 \\ 0 \end{bmatrix} , \tag{C8}
$$

where

$$
f_0^{(3)} = f_0^{(2)} ,
$$
$$
f_1^{(3)} = f_1^{(2)} - k_3 f_2^{(2)} ,
$$

$$f_2^{(3)} = f_2^{(2)} - k_3 f_1^{(2)} \; , \tag{C9}$$
$$f_3^{(3)} = -k_3 f_0^{(2)} \; ,$$

and where

$$J_{\min}^{(3)} = J_{\min}^{(2)} - k_3 \alpha_3 \; . \tag{C10}$$

Now from (C7)

$$\alpha_3 = k_3 J_{\min}^{(2)} \; ,$$

so that

$$J_{\min}^{(3)} = (1 - k_3^2) J_{\min}^{(2)} \; . \tag{C11}$$

Hence, we have computed the filter coefficients, and minimum mean-squared error required for the next step.

Since we have completely specified the iterative update, it only remains to specify the initial conditions for the algorithm. We have

$$r(0) f_0^{(0)} = J_{\min}^{(0)} \; ,$$

from which

$$f_0^{(0)} = J_{\min}^{(0)} / r(0) \; . \tag{C12}$$

We may now give the equations for the complete algorithm:

Initialization:

$$r(0) f_0^{(0)} = J_{\min}^{(0)} = E\{x^2(n)\} \; . \tag{C13}$$

For $j = 1, 2, ..., L$:

$$k_j = \frac{\alpha_j}{J_{\min}^{(j-1)}} = \frac{r(j) f_0^{(j-1)} + r(j-1) f_1^{(j-1)} + ... + r(1) f_{j-1}^{(j-1)}}{J_{\min}^{(j-1)}} \; , \tag{C14}$$

$$J_{\min}^{(j)} = (1 - k_j^2) J_{\min}^{(j-1)} \; , \tag{C15}$$

$$f_i^{(j)} = f_i^{(j-1)} - k_j f_{j-i}^{(j-1)} \quad ; \; i = 1, 2, ..., j \tag{C16}$$

where

$$f_j^{(j-1)} \equiv 0 \ , \quad f_0^{(j)} \equiv 1 \ . \tag{C17}$$

This completes the algorithm. Equation (C15) is particularly interesting because it shows how $J_{\min}$ is directly computed at each step of the iteration. Also, we note that as a consequence of the monotonic property for the error energy with increasing filter length, from (C15)

$$0 \leq k_j \leq 1 \ .$$

In practice $k_j < 1$, since the prediction will not be perfect. The values k_i are called **reflection coefficients** and they will reappear in Chapter 6 when we discuss **lattice filters**. Inspection of the reflection coefficients during computation of the least-squares prediction error filter provides a ready check on stability during the computation, since values of k_j outside the range of the inequality can only result from numerical problems in the iteration.

Appendix 3D — Minimum Phase Property for the PEF

One of the most remarkable properties of the least-squares prediction error filter (and hence of the the inverse filter) is that for any order p, the single-step PEF is *always* minimum phase. There are several ways this result may be demonstrated (see, for example, [16],[20]). One of the simplest is to apply **Rouche's Theorem** from complex variable theory to the Prediction Error Filter (PEF) solution developed in Appendix 3C. We begin by restating Rouche's Theorem [28]:

If functions $G(z)$, $H(z)$ defined in the complex plane are analytic inside and on a closed contour C, and if

$$|G(z)| > |H(z)| \ , \tag{D1}$$

then $G(z)$ and $G(z)+H(z)$ have the same number of zeros within C.

To apply this theorem, we recall the defining equations for the PEF as given by Durbins's algorithm outlined in the previous Appendix:

For $j=1,2,...,L$:

$$J_{\min}^{(j)}=(1-k_j^2)J_{\min}^{(j-1)} , \tag{C15}$$

$$f_i^{(j)}=f_i^{(j-1)}-k_j f_{j-i}^{(j-1)} \quad ; \quad i=1,2,...,j , \tag{C16}$$

where

$$f_j^{(j-1)}\equiv 0 , \qquad f_0^{(j)}\equiv 1 . \tag{C17}$$

For the $(p+1)$-th order PEF to be minimum phase then

$$F^{(p)}(z)=\sum_{i=0}^{L-1} f_i^{(p)} z^{-i} , \tag{D2}$$

must have no zeros *outside* $|z|=1$. For our purposes, it is more convenient to use the equivalent result that $F(z^{-1})$ must have no zeros *inside* $|z|=1$. We shall use the Durbin recursion, and Rouche's theorem to show that such a result holds for all intermediate solutions $f_i^{(j)}$ as expressed through their transforms $F^{(j)}(z)$.

We begin with $f_0^{(0)}=1$, and consider $j=1$. We have

$$f_0^{(1)}=1 , \quad f_1^{(1)}=-k_1 ,$$

so that

$$F^{(1)}(z)=1-k_1 z^{-1} , \tag{D3}$$

and

$$F^{(1)}(z^{-1})=1-k_1 z .$$

Since $0\leq k_1<1$, the zero of $F^{(1)}(z^{-1})$ lies outside $|z|=1$ and the result holds for $j=1$.

Now, we assume the result holds for the filter with coefficients $f_i^{(j-1)}$, and consider $f_i^{(j)}$. We have

$$F^{(j)}(z)=\sum_{i=0}^{j} f_i^{(j)} z^{-i} . \tag{D4}$$

Now, using (C16)

$$F^{(j)}(z)=F^{(j-1)}(z)-k_j\sum_{i=0}^{j}f_{j-i}^{(j-1)}z^{-i}\ ,$$

or

$$F^{(j)}(z)=F^{(j-1)}(z)-k_jz^{-j}F^{(j-1)}(z^{-1})\ .$$

Hence

$$F^{(j)}(z^{-1})=F^{(j-1)}(z^{-1})-k_jz^jF^{(j-1)}(z)\ . \qquad \text{(D5)}$$

Now, with reference to Rouche's theorem we choose

$$G(z)=F^{(j-1)}(z^{-1})\ , \qquad \text{(D6)}$$

and

$$H(z)=-k_jz^jF^{(j-1)}(z)\ . \qquad \text{(D7)}$$

We see that both $G(z)$ and $H(z)$ are analytic within $|z|=1$. Moreover, on $|z|=1$

$$|F^{(j-1)}(z)|=|F^{(j-1)}(z^{-1})|\ ,$$

and since $0\leq k_j<1$, then

$$|G(z)|>|H(z)|\ .$$

Hence, we conclude that the conditions of the theorem are satisfied, and that $G(z)+H(z)=F^{(j)}(z^{-1})$ has the same number of zeros inside $|z|=1$ as does $G(z)=F^{(j-1)}(z^{-1})$, which by assumption has none. Hence the result is proved.

Problems

3.1 a) Given N iid Gaussian observations $x(1),x(2),...,x(N)$ with unknown mean μ and variance σ^2, find the ML estimates $\hat{\mu}$ and $\hat{\sigma}^2$ of these quantities.

   b) Find the Fisher information matrix associated with the likelihood function in a), and hence show that $\hat{\mu}$ meets the

Cramer-Rao Lower Bound (CRLB) and that $\hat{\sigma}^2$ asymptotically meets the CRLB.

c) If $\mu=0$, show that the ML estimate for the variance is given by

$$\hat{\sigma}^2 = \frac{1}{N} \sum_{n=1}^{N} x^2(n) \, ,$$

and that this estimate is fully efficient.

3.2 Suppose we have N independent observations $x_1, x_2, ..., x_N$ drawn from a distribution x with

$$p\,(x{=}1){=}a \, ,$$
$$p\,(x{=}0){=}(1{-}a){=}b \, .$$

Find the maximum likelihood estimate of a.

3.3 Consider N observations x_i of the form

$$x_i{=}s{+}v_i \, ,$$

where s is constant and where v_i are iid samples drawn from a Laplacian (double-exponential) distribution:

$$f(v_i){=}\frac{\alpha}{2}e^{-\alpha|v_i|} \, ,$$

where α is a parameter of the density. Show that the ML estimator of s is the sample median of x_i ; $i{=}1,2,...,N$. (For simplicity take N odd).

3.4 Consider iid observations x_i ; $i{=}1,2,...,N$ each drawn from an exponential distribution with:

$$f\,(x_i){=}\frac{1}{\theta_1}e^{-(x_i-\theta_2)/\theta_1} \quad ; \; x{\geq}\theta_2$$

$$=0 \quad ; \; x_i{<}\theta_2 \, ,$$

$-\infty{<}\theta_2{<}\infty$, $\theta_1{>}0$. Find the maximum likelihood estimates of θ_1, θ_2.

3.5 Consider a single noisy observation x of a random parameter θ as

$$x=\theta+w \ ,$$

where w is zero-mean Gaussian, and where the parameter θ has prior density $N(\bar{\theta},\sigma_\theta)$. The MAP estimate for θ is given by

$$\theta_{MAP}=\frac{x\sigma_\theta^2+\bar{\theta}\sigma_w^2}{\sigma_w^2+\sigma_\theta^2} \ . \qquad (3.1.15)$$

Suppose we view the Bayesian estimation process in terms of the prior density $f(\theta)$ as reflecting our 'prior belief' about a deterministic θ. Show that

$$Var\{\theta_{MAP}\}=\frac{\sigma_w^2}{1+(\sigma_w^2/\sigma_\theta^2)} \ .$$

3.6 Consider the single noisy measurement

$$x=\theta+w \ ,$$

with w Gaussian with zero-mean and variance σ_w.

a) Find the ML estimate if θ is unknown.

b) Find the MAP estimate if θ is uniformly distributed over $[-a,a]$.

3.7 Consider M measurements of a known signal $s(n)$ embedded in zero-mean iid Gaussian noise samples $w(n)$ as

$$x(n)=As(n)+w(n) \quad ; \ n=0,1,...,M-1 \ ,$$

where A is an amplitude parameter to be estimated and where $w(n)$ has variance σ_w^2.

a) Find the ML amplitude estimate if A is unknown.

b) Find the MAP estimate if A is $N(\bar{A},\sigma_A)$.

c) Find the MAP estimate if the prior density of A is

Rayleigh:

$$f(A) = \frac{A}{\sigma_A^2} e^{-\frac{A^2}{2\sigma_A^2}} \quad ; A \geq 0 ,$$

$$= 0 \quad ; A < 0 .$$

3.8 Given a measurement

$$x = s + w ,$$

where w is zero-mean Gaussian with variance σ_w^2, and where s has Rayleigh distribution

$$f(s) = \left(\frac{s}{\sigma_s^2} \right) e^{\frac{-s^2}{2\sigma_s^2}} \quad ; s \geq 0 ,$$

$$= 0 \quad ; s < 0 .$$

Find the MAP estimate of s.

3.9 Consider the posterior density

$$f(x/\theta) = \theta e^{-\theta x} \quad ; \theta \geq 0, \ x \geq 0 ,$$

and prior density

$$f(\theta) = \alpha e^{-\alpha \theta} \quad ; \theta > 0, \ \alpha \geq 0 .$$

a) Find the MAP estimate θ_{MAP}.

b) By explicitly identifying the density $f(\theta/x)$, show that the mean-square estimate θ_{MS} is

$$\theta_{MS} = \frac{2}{(x+\alpha)} .$$

3.10 Consider an observation x, whose density conditioned on a random parameter θ is exponential with

$$f(x/\theta) = \theta e^{-\theta x} \quad ; x \geq 0 ,$$

$$= 0 \quad ; x < 0 ,$$

where θ has prior density

$$f(\theta)=\alpha e^{-\alpha\theta} \quad ; \theta \geq 0 ,$$
$$=0 \quad ; \theta<0 ,$$

where $\alpha>0$ is a known constant. Find the MMSE estimate of θ.

Note: You may assume that for a constant a,

$$\int_0^\infty y^n e^{-ay} dx=\frac{n!}{a^{n+1}} .$$

3.11 Consider the estimation of a random parameter θ as a function of an observed random variable x. Show that the estimate $\hat{\theta}=h(x)$ which minimizes the mean absolute error (MAE):

$$J=E\{|\theta-h(x)|\} ,$$

is given by the **median** of the posterior density $f(\theta/x)$.

Note: You may assume that for a continuous random variable u, the median u_{med} satisfies $F(u_{med})=0.5$.

3.12 Given a sequence of data $x(n)$ with non-zero values restricted to some finite range $n=0,1,...,N-1$, show that minimization of

$$J= \sum_{n=-\infty}^{\infty} e^2(n) ,$$

with respect to an FIR filter with coefficients $\underline{f}=[f(0),f(1),...,f(L-1)]^t$, produces a set of normal equations

$$R\underline{f}=\underline{g} ,$$

where R is a symmetric Toeplitz matrix with elements

$$r(i)= \sum_{n=-\infty}^{\infty} x(n)x(n-i) ,$$

and $\underline{g}$ is the cross-correlation vector with elements

$$g(i)= \sum_{n=-\infty}^{\infty} d(n)x(n-i) \ .$$

3.13 Given

$$x(n)=X\delta(n-n_0) \ ,$$

$$d(n)=D\delta(n-n_0-n_1) \ ,$$

find the MMSE filter as given in Problem **3.12** and the corresponding minimum mean-squared error for the conditions:

i) $n_1<0$.

ii) $0\leq n_1 \leq L-1$.

iii) $n_1>L$.

3.14 If $x(n)=\cos(\omega_0 n)$, use

$$R(i,j)= \lim_{N\to\infty} \{\frac{1}{N} \sum_{k=-N/2}^{N/2} \cos(\omega_0(i+k))\cos(\omega_0(j+k))\} \ ,$$

to show that

$$R(i,j)\to r(i-j)=\cos(\omega_0(i-j)) \ .$$

Find the inverse of R for $L=2$. Show that for $L=3$, R is singular.

3.15 Show that for stationary zero-mean inputs $x(n)$, $d(n)$, the mean-squared error associated with the FIR MMSE filter $\underline{f}^{*}=R^{-1}\underline{g}$ is

$$J_{\min}=\sigma_d^2-\underline{f}^{*t}\underline{g}=\sigma_d^2-\underline{f}^{*t}R\underline{f}^{*} \ ,$$

where $\sigma_d^2=E\{d^2(n)\}$, $R=E\{\underline{x}_n \underline{x}_n^t\}$ and $\underline{g}=E\{\underline{x}_n d(n)\}$.

3.16 Consider inputs $x(n)=w(n)$ a zero-mean iid sequence, and $d(n)=h(n)*w(n)$.

a) Show that if $h(n)$ is causal and shift-invariant then the L

point MMSE filter $\underline{f}^{*}=R^{-1}\underline{g}$ is

$$\underline{f}^{*}=[h(0),h(1),...,h(L-1)]^{t} ,$$

with error energy

$$J_{\min}=\sum_{i=L}^{\infty} h^{2}(i)\sigma_{w}^{2} .$$

b) If $h(n)$ is two-sided with

$$\underline{h}=[h(-N+1),...,h(0),h(1),...]^{t} , ,$$

find corresponding expressions for $\underline{f}^{*}$ and $J_{\min}$.

3.17 For the single-step predictor of equation (3.3.2), expand the functional

$$J=E\{e^{2}(n)\} ,$$

and use the optimal solution obtained from equation (3.3.5) to show that

$$J_{\min}=r(0)-\sum_{i=1}^{p} c_{i}r(i) .$$

3.18 Given the AR(1) signal

$$x(n)=a_{1}x(n-1)+w(n) ,$$

where $w(n)$ is a zero-mean white signal with variance $\sigma_{w}^{2}=1$. Set up the normal equations (3.3.5) for the second-order predictor $[c_{1},c_{2}]$ and solve to show that $c_{1}=a_{1}$, $c_{2}=0$.

3.19 Assuming the solution for the second order predictor as obtained in question **3.18** above:

a) Find the corresponding minimum error energy $J_{\min}$.

b) Use the Durbin recursion to find the third-order predictor. (Use the prediction error filter $[1,-c_{1},-c_{2},-c_{3}]$). Calculate the corresponding $J_{\min}$.

3.20 For the first order MA process

$$x(n)=w(n)+aw(n-1) ,$$

where $w(n)$ is a zero-mean iid sequence with unit variance, find the first and second order predictors c_1, $[c_1, c_2]$. Find the corresponding $J_{\min}$ values.

3.21 Consider a zero-mean stationary signal $x(n)$ input to a two-point sparse filter $f(0), f(1)$ such that

$$y(n)=f(0)x(n)+f(1)x(n-i)=\underline{f}^t\underline{x} ,$$

where $\underline{f}=[f(0), f(1)]^t$, and $\underline{x}=[x(n), x(n-i)]^t$. Show that the correlation matrix $R=E\{\underline{x}\underline{x}^t\}$ is positive semi-definite. Use this result to prove the general correlation sequence property:

$$r(0)\geq |r(i)| .$$

3.22 In each of the following cases state whether R is an allowable correlation matrix corresponding to a zero-mean stationary signal. Give reasons for your answers.

i)

$$\begin{bmatrix} 2 & 1 & 0 \\ 1 & 1 & 1 \\ 0 & 1 & 2 \end{bmatrix} .$$

ii)

$$\begin{bmatrix} 2 & 1 & 0 \\ -1 & 2 & -1 \\ 0 & 1 & 2 \end{bmatrix} .$$

iii)

$$\begin{bmatrix} 2 & 1 & -3 \\ 1 & 2 & 1 \\ -3 & 1 & 2 \end{bmatrix} .$$

3.23 For each of the following cases state whether or not the eigenvalues may be derived from a valid autocorrelation matrix (assuming a zero-mean stationary signal).

i) $\lambda_1=1,\ \lambda_2=2,\ \lambda_3=\lambda_4=0.$

ii) $\lambda_1=-1,\ \lambda_2=\lambda_3=2.$

iii) $\lambda_1=1+j2,\ \lambda_2=1+j3.$

iv) $\lambda_1=1+j2,\ \lambda_2=1-j2.$

3.24 Consider the noisy system identification problem shown in the diagram:

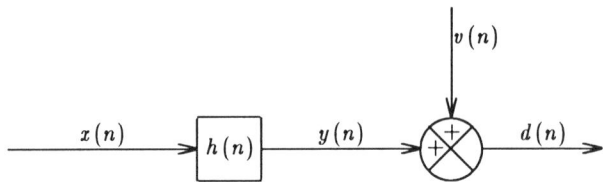

where $h(n)$ is an unknown linear shift-invariant system. Assuming that $v(n)$ is iid zero-mean Gaussian with variance σ_v^2, and given observations $d(n)$ for $n=0,1,...,M-1$, and $x(n)$ for $n=-L+1,-L+2,...,M-1$, show that the Maximum Likelihood estimate of the parameter vector $\underline{h}=[h(0),h(1),...,h(L-1)]^t$ is obtained from the solution of the linear equations

$$R\underline{h}=\underline{g}\ ,$$

and find explicit forms for R and $\underline{g}$.

3.25 Show that the minimization of the functional

$$J=E\{(\theta-\sum_{i=0}^{M-1}h_i x(i))^2\}=E\{e^2\}\ ,$$

with respect to h_i satisfies

$$E\{ex(i)\}=0\ \ ;\ i=0,1,...,M-1\ .$$

Use this result to show that if θ and the $x(i)'s$ are zero-mean, jointly Gaussian, then the *linear* MMSE is the *globally* optimal MMSE estimate.

Note: You may use the result that a linear combination of jointly Gaussian variates is Gaussian.

3.26 Consider the seismic water layer multiple train

$$m(n)=1-r_0\delta(n-n_0)+r_0^2\delta(n-2n_0)-\cdots$$

Find an inverse filter which deconvolves $m(n)$. Is the resulting operator minimum phase?

3.27 Consider a zero-mean unit variance white noise signal $w(n)$ corrupted by a single additive 'echo' with amplitude a arriving after a delay of 100 samples:

$$x(n)=w(n)+aw(n-100) .$$

It is desired to attenuate the echo using a prediction error filter. Find the $L=20$ point prediction filters corresponding to

i) prediction distance, $\hat{n}=85$,

ii) prediction distance, $\hat{n}=100$,

iii) prediction distance, $\hat{n}=105$,

and find the corresponding prediction error in each case. What can you conclude about the appropriate range of $\hat{n}$ for a 'successful prediction'?

3.28 Consider a two point impulse response $h(0)=1$, $h(1)=0.5$.

a) Calculate the $L=2$ least-squares inverse filter for this system.

b) Find the associated error energy J_{min} .

c) Show that J_{min} is lower than that for the filter obtained by simply truncating the long-division expansion of $1/H(z)$.

3.29 Repeat **3.28** parts a) and b) if the system is replaced by $h(0)=1$, $h(1)=2$.

3.30 Consider the problem of finding an L point least-squares inverse filter for a stable causal system $h(n)$. Assume that the least-squares operator is obtained by minimizing:

$$J= \sum_{n=-\infty}^{\infty} e^2(n)= \sum_{n=-\infty}^{\infty} (\delta(n)-f(n)*h(n))^2 \ .$$

Find an expression for the minimum mean-squared error $J_{\min}$ and show that this quantity is monotonically non-increasing with filter length.

3.31 Given a two-point system $h(0)=1$, $h(1)=0.5$:

  a) Calculate the $L=2$ least-squares inverse filter and corresponding error energy $J_{\min}$.

  b) Recalculate the filter and error energy if 1% prewhitening is used.

3.32 Given a three-point system $h(0)=1$, $h(1)=-1$, $h(2)=0.25$:

  a) Find the direct form inverse truncated to 4 terms.

  b) Set up and solve the $L=2$ zero-delay causal least-squares inverse filter for the same problem.

  c) Find the error energy associated with the filter in b) above.

References

---

1. H.L.Van Trees, *Detection, Estimation and Modulation Theory (Part 1)*, Wiley, 1968.

2. M.Schwartz and L.Shaw, *Signal Processing: Discrete Spectral Analysis, Detection and Estimation*, McGraw-Hill, 1975.

3. J.L.Melsa and D.L.Cohn, *Decision and Estimation Theory*, McGraw-Hill, 1978.

4. A.Papoulis, *Probability, Random Variables and Stochastic Processes*, McGraw-Hill, 3rd Ed, 1991.

5. M.B.Priestley, *Spectral Analysis and Time Series*, Academic Press, 1981.

6. E.L.Lehmann, *Theory of Point Estimation*, Wiley, 1982.

7. R.A.Horn and C.R.Johnson, *Matrix Analysis*, Cambridge University Press, 1985.

8. A.V.Oppenheim and R.W.Schafer, *Discrete Time Signal Processing*, Prentice-Hall, 1989.

9. L.R.Rabiner and R.W.Schafer, *Digital Processing of Speech Signals*, Prentice-Hall, 1978.

10. J.D.Markel and A.H.Gray, *Linear Prediction of Speech*, Springer-Verlag, 1975.

11. E.A.Robinson, *Statistical Communication and Detection with Special Reference to Radar and Seismic Signals*, Griffin, 1967.

12. J.Makhoul, "Linear prediction: A tutorial review," *Proc. IEEE*, vol. 63, pp. 561-581, 1975.

13. L.C.Wood and S.Treitel, "Seismic signal processing," *Proc. IEEE*, vol. 63, pp. 649-661, 1975.

14. S.T.Alexander, *Adaptive Signal Processing*, Springer-Verlag, 1986.

15. T.W.Parsons, *Voice and Speech Processing*, McGraw-Hill, 1987.

16. S.J.Orfanidis, *Optimum Signal Processing*, McGraw-Hill, 1988.

17. N.Levinson, "The Wiener RMS error criterion in filter design and prediction," *J. Math. Phys.*, vol. 25, pp. 261-278, 1947.

18. P.Strobach, *Linear Prediction Theory*, Springer-Verlag, 1990

19. M.L.Honig and D.G.Messerschmitt, *Adaptive Filters: Structures, Algorithms and Applications*, Kluwer, 1984.

20. S.Haykin, *Adaptive Filter Theory*, 2nd Ed, Prentice-Hall, 1991.

21. J.Durbin, "Efficient estimation of parameters in moving average models," *Biometrika*, vol. 46, pp. 306-316, 1959.

22. K.H.Waters, *Reflection Seismology: A Tool for Energy Resource Exploration*, Wiley, 1978.

23. E.A.Robinson, "Multichannel z transforms and minimum delay," *Geophysics*, vol. 31, pp. 482-500, 1966.

24. P.M.Clarkson and J.K.Hammond, "Deconvolution," in *Applied Digital Signal Processing*, ISVR Short Course Notes, University of Southampton, 1984.

25. S.M.Simpson, "Recursive schemes for normal equations of Toeplitz form," Sci. Rept. No 7, ARPA Project Vela-Uniform, M.I.T., 1963.

26. S.Treitel and R.J.Wang, "The determination of digital Wiener filters from an ill-conditioned system of normal equations," *Geophsyical Prospecting*, vol. 24, pp. 317-327, 1976.

27. S.Senmoto and D.G.Childers, "Signal resolution via digital inverse filtering," *IEEE Trans. Aero., Electr. Systs.*, vol. 8, pp. 633-640, 1972.

28. R.V.Churchill and J.W.Brown, *Introduction to Complex Variables and Applications*, McGraw-Hill, 4th Ed, 1984.

chapter four

Introduction to Adaptive Signal Processing

4.1 Adaptive Signal Processing with the LMS Algorithm

4.1.1 General

Our starting point for the development of the adaptive filter is the fixed least-squares operator given by equations (3.2.12) of Section 3.2. The least-squares technique provides a powerful approach to digital filtering in situations where a *fixed*, finite length filter is applicable. Indeed, this approach has achieved widespread application in many areas. However, there are a number of situations where this solution may be unsatisfactory. The principal problem arises when the input data is statistically nonstationary. Although the normal equations can be formulated for nonstationary inputs, the calculation of the nonstationary correlation coefficients presents difficulties. As noted in Section 2.6, for stationary and ergodic inputs we can obtain estimates of the correlation coefficients by replacing ensemble averages by estimates obtained by time averaging. Various estimators were discussed in Section 2.6. For nonstationary inputs, however, such an approach is clearly unsound. For those random processes which might be classed as 'locally stationary' (that is whose properties change only slowly with time), it may be reasonable to estimate the autocorrelation with respect to a point in time as a short-time average. A different least-squares operator is then computed for each optimization interval. Many commonly occurring signals might be classed as locally stationary in this way. This technique of subdividing a piece of data into smaller intervals referred to as windows or gates, over which the signal is considered stationary, is widespread in speech and seismic processing among other applications [1],[2]. Attempts have even been made to optimize the length of these gates as a trade-off between estimation error in the autocorrelation (which decreases with interval length) and the error implicit in replacing a nonstationary autocorrelation with a time average (which obviously increases with interval length) [3]. However, the resulting formula is somewhat unwieldy to use and is only applicable to a limited class of processes. In any case, gating can only provide a partial answer to filtering continuously time-varying signals. A more complete solution would be provided by a continuously adaptive filter. In attempting to develop such an adaptive filter, we are seeking to find an algorithm which is optimal in the least-squares sense, but which is responsive to changes in the optimal solution arising as each new data point becomes available. We want a filter which will *iterate* to *track* changes in the optimal solution. As a prelude to detailed consideration of particular algorithms, we specify the general components of an adaptive filter; these comprise three essential elements:

a) The Structure of the Filter − The most widely used are Finite Impulse Response (FIR) filters, implemented in transversal form. These have the advantage of being relatively easy to analyze, and furthermore have a straightforward implementation in tapped delay line form. For the moment, we will restrict attention to such structures. However, any of the usual digital filter structures can be implemented in adaptive form, and we will examine some of these possibilities in later chapters.

b) The Overall System Configuration − As we shall see, many different system configurations are possible depending on the application. However, the essential features are as depicted in Figure 4.1.1. The system consists of the adaptive filter f acting on an input sequence $x(n)$ to produce an output $y(n)$. The filter is designed so that the output should approximate a desired or training signal $d(n)$. The error $e(n)$ between the desired signal and the output is used to control the filter coefficients, hence making the filter data adaptive. This relationship is depicted in Figure 4.1.2. Note that the adaptive nature of the solution is represented by the notation $f_n(j)$ that is, the set of filter impulse response coefficients $j=0,1,...,L-1$ at sample index n. In terms of the vector notation introduced in Section 3.2 the data vector $\underline{x}_n$ is unchanged:

$$\underline{x}_n = [x(n), x(n-1), ..., x(n-L+1)]^t ,$$

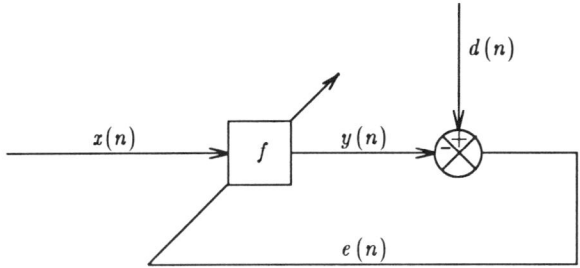

Figure 4.1.1 Basic adaptive filtering system. Inputs to the filter are $x(n)$ and $d(n)$. $y(n)$ is the filter output and $e(n)$ represents the difference between the desired input $d(n)$ and the actual output $y(n)$.

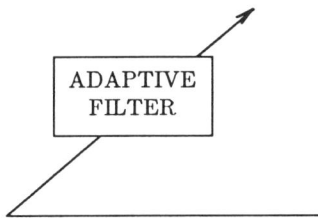

Figure 4.1.2 Block diagram representation of the adaptive filter. The arrow represents the forcing action of the data on the filter coefficients.

while the filter coefficient vector is modified to include the effect of sample index as

$$\underline{f}_n = [f_n(0), f_n(1), ..., f_n(L-1)]^t \ .$$

Note that the convolution summation is expressed as

$$y(n) = \sum_{i=0}^{L-1} f_n(i) x(n-i) \ ,$$

or, in vector form

$$y(n) = \underline{f}_n^t \underline{x}_n \ . \tag{4.1.1}$$

The input/output relation for the adaptive filter is therefore similar to that for the usual FIR filter, the only distinction is the subscript n, indicating the time-variant nature of the filter impulse response. As usual, the error is defined by

$$e(n)=d(n)-y(n) \ . \tag{4.1.2}$$

It is immediately apparent that the specification of the desired signal $d(n)$, is the key to the adaptive filter. This is entirely application dependent and much of the success of adaptive processing has been due to the ability of workers in the field to design this desired signal in novel and ingenious ways. Again, later chapters will examine such applications in detail.

c) Performance Criterion for Adaptation — The third element in the specification of the adaptive filter is the choice of the performance criterion by which the filter coefficients are updated. There are many possibilities although, as we have indicated, we shall concentrate in the main on one — least-squares minimization of the error signal. As we shall see later, this one criterion has given rise to a number of adaptation strategies. As far as this Chapter is concerned, we shall concentrate on introducing adaptive filters through one — the so-called Least Mean Squares (LMS) algorithm. Our reasons for making these apparently arbitrary selections are partly historical — the LMS algorithm was the first adaptive filter developed, and partly pragmatic — the least-squares method in general, and the LMS algorithm in particular, is simple both in concept and implementation.

The objective for this least-squares design is thus to solve the normal equations to find the optimal filter, but to continuously update this solution as each new data sample becomes available. Perhaps the most obvious way to compute such a time-varying solution is to recompute the correlations at each point, and solve the resulting normal equations. Leaving aside the difficulties implicit in the determination of these correlation coefficients, such a solution effected by the Levinson recursion at every single data sample involves an enormous computational burden, especially if the filter length L is large. Another possibility is to discard the idea of a direct solution to the normal equations and consider instead an **iterative solution**.

4.1.2 Iterative Solutions to Normal Equations

For stationary signals, iterative solutions have been applied to the derivation of the optimal least squares operator in place of the Levinson recursion [4]. Referring to Section 3.2, our objective is the minimization of

$$J = E\{e^2(n)\} \ . \tag{3.2.1}$$

Iterative techniques can be thought of as searching the function J (also known in this context as the 'performance surface') for the filter $\underline{f}^*$ which produces $J_{\min}$ (see Figure 4.1.3).

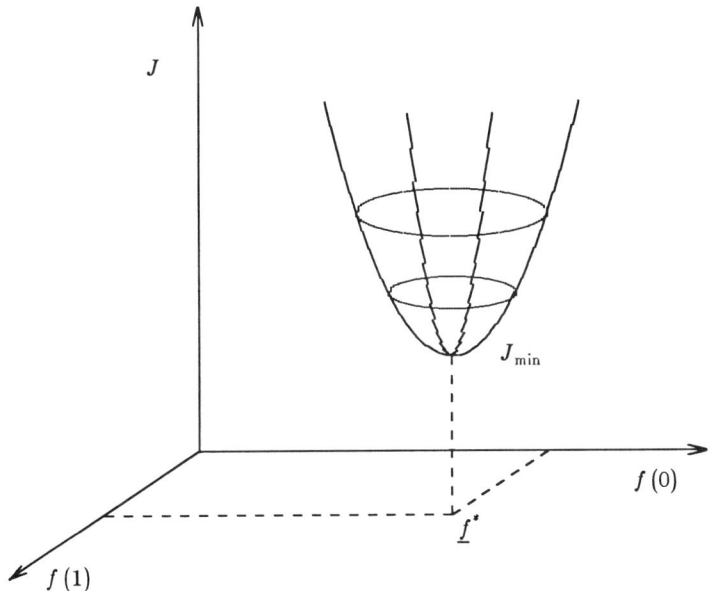

Figure 4.1.3 Quadratic performance surface J for a two-dimensional filter with coefficients $f(0)$, $f(1)$. Coordinates $f^*(0)$, $f^*(1)$ produce the minimum $J_{\min}$.

This iteration begins with an initial estimate or guess $\underline{f}_0$ as to the optimal filter. The filter is then successively updated producing estimates $\underline{f}_1, \underline{f}_2,...$ using an iteration formula designed to force the filter towards the optimal solution. Such an approach may have advantages even when the objective is to obtain the fixed least-squares solution. In particular, the matrix R may be singular (corresponding to equality in the positive semi-definite relation for

R) or close to singular. Iterative solutions usually avoid direct inversion of R and may produce stable solutions even when R is singular. Iterative methods generally involve greater overall computation compared to direct solutions but if the initial approximation is sufficiently close to the optimal value, then only a relatively small number of iterations may be required to obtain a satisfactory approximation to $\underline{f}^*$.

As we have already noted the performance index J can be thought of as an L-dimensional bowl-shaped (hyperparaboloid) surface. The important point is that such a surface has a single minimum value. As we have already indicated, this is one of the major attractions of least-squares minimization – there is no prospect of the minimization resulting in a solution which is only locally optimal as may occur with the minimization of many other functionals. The iterative process begins at some point on the surface with a value for J corresponding to the initial filter estimate $\underline{f}_0$, and subsequently attempts to iterate towards the minimum.

Recall from Section 3.2 that the least-squares filter is obtained from the solution of the normal equations (3.2.12) as

$$\underline{f}^* = R^{-1}\underline{g} \ . \tag{3.2.12}$$

A general iterative formula for the solution is given by

$$\underline{f}_{n+1} = \underline{f}_n - \alpha_n \underline{p}_n \ , \tag{4.1.3}$$

where $\underline{f}_n$ denotes the nth update of the filter coefficient vector, α_n is a **step-size** parameter, and where $\underline{p}_n$ is a vector which defines the **search direction** [5]. A fairly general class of algorithms are those which iterate on the basis of the **gradient of the mean-squared error**. These gradient search algorithms are characterized by search directions $\underline{p}_n$ of the form

$$\underline{p}_n = D_n . \nabla J_n \ , \tag{4.1.4}$$

where ∇J_n is the gradient of the performance index J with respect to the filter coefficient vector $\underline{f}_n$ and has elements

$$\nabla J_n = \frac{\partial J}{\partial \underline{f}_n} = \left[\frac{\partial J}{\partial f_n(0)}, \frac{\partial J}{\partial f_n(1)}, \dots, \frac{\partial J}{\partial f_n(L-1)} \right]^t \ ,$$

and where D_n is an $(L \times L)$ weighting matrix. A large number of possibilities for the selection of D_n and α_n exist, and many adaptive algorithms can be developed from particular choices for these parameters. For the moment we shall concentrate on one particular approach, one of the most straightforward, that of **steepest descent** [6].

The philosophy of steepest descent can be described as follows: At each iteration, move (in the space of filter weights) in a direction opposite to the gradient of the performance index J, and by a distance proportional to the magnitude of that gradient. In terms of the specification of equations (4.1.3) and (4.1.4) this gives a particularly simple form:

$$D_n = I ,$$

where I is the identity matrix, so that

$$\underline{p}_n = \nabla J_n = \frac{\partial E\{e^2(n)\}}{\partial \underline{f}_n} .$$

Also $\alpha_n = \alpha/2$, where α is a constant of proportion, and where the 2 is introduced for convenience.[1] The iteration equation (4.1.3) now becomes

$$\underline{f}_{n+1} = \underline{f}_n - \frac{\alpha}{2} \nabla J_n . \tag{4.1.5}$$

As in Section 3.2, if we differentiate J with respect to $\underline{f}$, then from equation (3.2.4) we have

$$\nabla J = \frac{\partial J}{\partial \underline{f}} = \frac{\partial E\{e^2(n)\}}{\partial \underline{f}} = 2(R\underline{f} - \underline{g}) . \tag{3.2.4a}$$

Thus using (4.1.5) and (3.2.4a), for filter $\underline{f}_n$ we may write

$$\underline{f}_{n+1} = \underline{f}_n - \alpha(R\underline{f}_n - \underline{g}) . \tag{4.1.6}$$

---

1. Some authors do not use the device of dividing by 2. This leads to the presence of a 2 in the steepest descent, the subsequent LMS update, and in all of the performance characterizations which follow. The reader should be aware of this possibility when consulting other texts or papers and should also be aware that conversion between one set of results and another can always be achieved by simply substituting $\alpha_1 = 2\alpha$, or conversely as required.

The term steepest descent arises from the fact that in the neighborhood of f_n the gradient is normal to lines of equal cost (that is, lines along which J is equal, as illustrated by the ellipses in Figure 4.1.3). Thus, the gradient direction is the line of steepest ascent in cost terms. We see that the algorithm has a natural tendency to take large steps when $\underline{f}$ is far from the optimal filter $\underline{f}^*$, and progressively smaller steps as $\underline{f}$ approaches $\underline{f}^*$ and the magnitude of ∇J decreases. Consideration of Figure 4.1.3 and the steepest descent equations suggests that provided the steps taken at each iteration are not too great (that is α is not too large), the process will, eventually converge to the optimal solution. In fact this intuitive view is correct, and in contrast to many iterative techniques convergence can be analytically established for the method of steepest descent [6] subject to a restriction on the step-size α. Formally, it can be shown that for $x(n)$, $d(n)$ zero-mean and jointly stationary, then

$$\lim_{n \to \infty} \{\underline{f}_n\} \to \underline{f}^* \ ,$$

provided

$$0 < \alpha < \frac{2}{\lambda_{\max}} \ , \tag{4.1.7}$$

where $\lambda_{\max}$ is the largest eigenvalue of the autocorrelation matrix R. We shall defer discussion of this result until the next section and for now merely note that while, subject to this condition, convergence can be assured, the algorithm may only converge slowly, particularly in the region of the minimum.

4.1.3 The LMS Adaptive Filter

So far we have considered only an iterative solution for the stationary (fixed) normal equations. To design a filter which is responsive to changes in the input signal environment we need an iterative structure that is dependent on the input data. We could construct a steepest descent scheme simply by replacing the fixed auto and cross-correlation matrices by their time-dependent equivalents, that is by replacing equation (4.1.6) by

$$\underline{f}_{n+1} = \underline{f}_n - \alpha (R_n \underline{f}_n - \underline{g}_n) \ , \tag{4.1.8}$$

where R_n, $\underline{g}_n$ are the autocorrelation matrix and cross-correlation vector, respectively, computed with reference to the n-th sample index of the input sequence. There are two main problems with this approach: Firstly, the calculation of R and g at each point is very expensive computationally, and secondly, as discussed in Section 2.6, it may not be possible to compute R_n and g_n if only a single realization of the process is available. These difficulties can be overcome, in part at least, by replacing the gradient ∇J_n by an estimate. One adaptive filtering method is based on estimating the gradient of the mean-squared error by the gradient of the instantaneous value of the squared error. That is

$$\nabla J_n = \frac{\partial J}{\partial \underline{f}_n} = \frac{\partial E\{e^2(n)\}}{\partial \underline{f}_n} \, , \tag{4.1.9}$$

is replaced by

$$\nabla \hat{J}_n = \frac{\partial \hat{J}}{\partial \underline{f}_n} = \frac{\partial e^2(n)}{\partial \underline{f}_n} \, . \tag{4.1.10}$$

The iterative update which results from this approximation is known as the **Least Mean Squares (LMS)** algorithm [7]. The filter coefficient iteration (update formula) for the LMS algorithm is derived by using the instantaneous estimate in the steepest descent update equation, so that equation (4.1.5) becomes:

$$\underline{f}_{n+1} = \underline{f}_n - \frac{\alpha}{2} \nabla \hat{J}_n \, . \tag{4.1.11}$$

Now,

$$\hat{J}_n = e^2(n) = (d(n) - \underline{f}_n^t \underline{x}_n)^2 \, ,$$

so that

$$\nabla \hat{J}_n = \frac{\partial \hat{J}}{\partial \underline{f}_n} = 2e(n) \frac{\partial e(n)}{\partial \underline{f}_n} \, , \tag{4.1.12}$$

or

$$\nabla \hat{J}_n = -2e(n)\underline{x}_n \, . \tag{4.1.13}$$

Hence equation (4.1.11) becomes

$$\underline{f}_{n+1} = \underline{f}_n + \alpha e(n)\underline{x}_n \tag{4.1.14}$$

This equation is known as the LMS adaptive filter update. It says simply that the filter impulse response at each iteration is equal to the previous filter impulse response plus a term proportional to the product of the error and the data vector. This filter represents by far the most widely used adaptive filter and, in fact, it provides the basis for most of the adaptive processing applications which have been reported in the literature.

A number of attractions of the LMS algorithm are immediately apparent from equation (4.1.14). Firstly, the update equation is computationally simple, requiring no matrix inversions or other expensive operations, and the total update requires approximately L multiplications and L additions per step (we shall be a little more precise on this subject in Chapter 6). Secondly, the algorithm requires no averaging, thus avoiding the difficulties, both analytical and computational, involved in computing correlations. The update equation also indicates as a consequence of the last two points discussed, the possibility of real-time implementation as a tapped delay line.

Intuitively, replacing the gradient estimate ∇J_n by the instantaneous value $\nabla \hat{J}_n$ corresponds to updating the filter at each point in a manner designed to keep the filter as close as possible to the instantaneous solution of the normal equations. On the other hand, by replacing the original performance index J by the instantaeous version $\hat{J}$ it is clear that we are no longer solving the same problem exactly. It is also apparent from the update that for random inputs $\underline{f}_n$ is random and is thus quite different from $\underline{f}^*$ which is a deterministic quantity. Consequently, we cannot in general expect the filter to converge exactly to the least-squares solution $\underline{f}^*$.

Notes on the LMS Filter

The LMS adaptive filter is specified by the update equation (4.1.14) together with the output and error equations of (4.1.1), (4.1.2). The update equation is nonlinear (because of the product $e(n)\underline{x}_n$), and recursive. By contrast, the filtering operation $y(n)=\underline{f}_n^t\underline{x}_n$ at each iteration is linear and non-recursive. Operation of the LMS filter requires selection of the constant α — known generally as the **adaptation constant**, the filter length L, and the initial filter coefficient vector $\underline{f}_0$. The selection of these parameters is application dependent, and we shall examine these questions in some detail in subsequent chapters. However, a few comments may be made immediately:

a) *Initialization* – Operation of the LMS filter requires that a set of values be selected for the initial filter impulse response $\underline{f}_0$. It should be noted that because of the unimodal nature of the performance index J, the initial value $\underline{f}_0$ will not affect the ultimate convergence (or otherwise) of the filter. This is certainly fortunate since in most circumstances there is little prior knowledge available to assist in the selection of $\underline{f}_0$. A common practice in adaptive filtering is to set $\underline{f}_0=\underline{0}$ This ensures that the filter output will initially be zero and will gradually build (hopefully) towards the optimal solution. On the other hand, selection of $\underline{f}_0$ close to the optimal solution is obviously advantageous if prior information exists, and as we shall see, there are some applications where this is possible.

b) *The Adaptation Constant* – The adaptation constant α plays a vital role in determining the behavior of the LMS algorithm. Empirically, increasing the magnitude of this constant increases the size of the steps taken at each iteration and thus increases the speed with which the adaptive filter approaches the optimal solution. However, increasing α also increases the propensity of the algorithm to respond to spurious events, that is, to enhance noise in the filter coefficients. This is due to the fact that the true gradient ∇J as used in steepest descent equation is replaced in the LMS algorithm by an instantaneous estimate $\nabla \hat{J}$ (see equation (4.1.11)). We might think of $\nabla \hat{J}$ as the sum of the true gradient ∇J and a 'noise term' representing the sample fluctuations. The larger the adaptation step-size, the more the noise from this term is incorporated into the filter coefficients. Even when $\underline{f}$ is in the region of $\underline{f}^*$, and ∇J approaches $\underline{0}$, $\nabla \hat{J}$ will generally fluctuate about this value. Once again a large α implies more noise in the coefficients and this is reflected in an increase in the mean-squared error associated with the solution over and above the minimum mean-squared error $J_{\min}$. Also, as in the case of the steepest descent algorithm, increasing the value of α beyond certain limits will cause instability in the algorithm (divergence of the iteration). These issues will be discussed in some detail in the next section which is devoted to the behavior of the LMS algorithm.

4.2 Performance of the LMS Adaptive Filter

4.2.1 Introduction

In spite of the apparent simplicity of the LMS filter as expressed by the update equation (4.1.14)

$$\underline{f}_{n+1}=\underline{f}_n+\alpha e(n)\underline{x}_n \ ,\tag{4.1.14}$$

analysis of the behavior of the filter is difficult. A considerable amount of work has been done, however, and results have been obtained for certain limited types of input signals. Essentially, the analysis of the filter performance can be sub-divided into two basic concerns:

i) The stability and convergence of the algorithm (towards the least-squares optimum solution).

ii) The mean-squared error of the filter.

The approach to the analysis is necessarily different according to whether the input processes are deterministic or random.

4.2.2 Random Inputs: Stability and Convergence

Although the adaptive filter is designed to track nonstationary inputs, the analysis of algorithm performance will focus on stationary zero-mean inputs $x(n)$, $d(n)$. This is an idealization used primarily because for such inputs the analysis becomes relatively tractable. From these stationary results we may make certain inferences about the tracking capability of the filter when subjected to nonstationary inputs.

The very simplest approach is to ignore the convergence issue entirely and assume that the filter behaves exactly like the solution to the Normal Equations for the infinite two-sided least-squares filter (the Wiener solution). This solution was derived in Section 3.2 and for inputs $x(n)$ and $d(n)$ is given by

$$F(e^{j\omega})==\frac{G(e^{j\omega})}{R(e^{j\omega})}=\frac{R_{xd}(e^{j\omega})}{R_{xx}(e^{j\omega})} \ ,\tag{4.2.1}$$

where $R_{xd}(e^{j\omega})$ is the cross-spectrum of the input and desired signals, and $R_{xx}(e^{j\omega})$ is the power spectrum of the input signal. This is a gross over-simplification of the LMS system because it completely neglects all adaptive aspects of the filter behavior, as well as causality and finite length. It often gives a surprisingly good approximation to the steady-state behavior of the filter however, and is a useful rule of thumb when trying to make predictions about the likely performance of an adaptive filtering system. This last point will be illustrated in later sections.

More generally, when the adaptive filter is subjected to random inputs it is important to realize that the filter vector is not a deterministic quantity derived from second order statistics, as it is in the steepest descent method, but is generated directly from nonlinear combinations of random vectors. Thus, the filter coefficients themselves are random. Consequently, we cannot expect the filter to converge precisely to the constant vector $\underline{f}^{*}$. The best we can hope for is that the filter will tend towards this quantity in the mean, that is:

$$\lim_{n \to \infty} E\{\underline{f}_n\} \to \underline{f}^{*} \ . \tag{4.2.2}$$

Even if this proves to be the case, the coefficients will have some variability about this solution. This itself is important to quantify, since a filter which satisfies (4.2.2) will be useless if the variance of the coefficients about that solution is too great.

Convergence of the LMS Filter in the Mean

Explicit proof of convergence for the LMS algorithm can be obtained by applying the so-called 'independence assumption' [8]. This condition corresponds to independence of $\underline{f}_n$ and $\underline{x}_n$. Due to the recursive nature of the update (4.1.14), however, $\underline{f}_n$ depends not only on $\underline{x}_{n-1}$ but also on $\underline{x}_{n-2}$, $\underline{x}_{n-3}, \ldots$ Consequently the independence assumption is tantamount to requiring that successive input vectors should be mutually uncorrelated. That is

$$E\{\underline{x}_n \underline{x}_m^t\} = 0 \qquad ; n \neq m \ . \tag{4.2.3}$$

This condition is actually stronger than requiring that the input be a white noise sequence, since if $x(n) = w(n)$ with

$$E\{x(n)x(m)\}=\sigma_w^2 \qquad n=m \; ,$$
$$=0 \qquad n \neq m \; , \qquad (4.2.4)$$

then the correlation matrix defined by equation $(4.2.3)$ with, for example, $m=n-1$ gives

$$E\{\underline{x}_n \underline{x}_{n-1}^t\}= \begin{bmatrix} 0 & 0 & 0 & 0 & 0 \\ \sigma_w^2 & 0 & 0 & . & . \\ 0 & \sigma_w^2 & . & . & . \\ . & 0 & . & 0 & . \\ 0 & . & 0 & \sigma_w^2 & 0 \end{bmatrix} ,$$

which does not satisfy $(4.2.3)$. Clearly, even white sequences cannot satisfy this 'independence assumption'.

Although this requirement for independent input vectors is highly unrealistic, it remains one of the few cases where convergence in the mean may be explicitly proved. The approach adopted for the proof begins with consideration of the filter coefficient vector, which, from the update equation $(4.1.14)$ is given by

$$\underline{f}_{n+1}=\underline{f}_n+\alpha(d(n)-\underline{f}_n^t\underline{x}_n)\underline{x}_n \; , \qquad (4.2.5)$$

or, rearranging slightly

$$\underline{f}_{n+1}=(I-\alpha\underline{x}_n\underline{x}_n^t)\underline{f}_n+\alpha d(n)\underline{x}_n \; . \qquad (4.2.6)$$

Taking the expectation of both sides of this equation and using the independence assumption gives

$$E\{\underline{f}_{n+1}\}=(I-\alpha R)E\{\underline{f}_n\}+\alpha \underline{g} \; . \qquad (4.2.7)$$

Using the normal equations $\underline{g}=R\underline{f}^*$ we can write

$$E\{\underline{f}_{n+1}\}=(I-\alpha R)E\{\underline{f}_n\}+\alpha R\underline{f}^* \; . \qquad (4.2.8)$$

Defining an error vector $\underline{u}_n$, as the difference between the expected value of the adaptive filter coefficients and the optimal solution:

$$\underline{u}_n=E\{\underline{f}_n\}-\underline{f}^* \; , \qquad (4.2.9)$$

and subtracting $\underline{f}^*$ from both sides of equation (4.2.8), yields

$$\underline{u}_{n+1}=(I-\alpha R)\underline{u}_n \ . \tag{4.2.10}$$

We now have a recursive equation describing the error in the LMS coefficients. Proving that the LMS algorithm converges in the mean to the optimal solution $\underline{f}^*$, reduces to proving that the error vector $\underline{u}_n$ converges to zero. Actually, conditions for this can be inferred directly from equation (4.2.10) by consideration of vector norms. However, it is more instructive, and necessary for what follows, to first **decouple** the update equations. This is done using a decomposition of the matrix R as

$$R=Q\Lambda Q^t \ , \tag{4.2.11}$$

where Q is referred to as the **modal matrix** because it is used to rotate the error vector, $\underline{u}_n$, into uncoupled 'modes'. The columns of Q are the eigenvectors of R, that is

$$Q=[\underline{q}_1,\underline{q}_2,...,\underline{q}_L] \ ,$$

where $\underline{q}_i$ is an eigenvector corresponding to the eigenvalue λ_i, and where for real input signals $x(n)$, real eigenvectors can be found [9]. The matrix Q satisfies

$$Q^tQ=I \ , \tag{4.2.12}$$

that is Q is **orthonormal**. Λ, the so-called **spectral matrix**, is a diagonal matrix whose elements consist of the eigenvalues of R, that is

$$\Lambda= \begin{bmatrix} \lambda_1 & 0 & . & . & 0 \\ 0 & \lambda_2 & . & . & . \\ . & 0 & . & . & . \\ . & . & . & . & 0 \\ 0 & 0 & . & 0 & \lambda_L \end{bmatrix} \ . \tag{4.2.13}$$

The factorization (4.2.11) is called a **unitary similarity transform** and it can be shown that any correlation matrix R may be diagonalized in this way [9] (see Appendix 4A for further detail on diagonalization).

Returning to equation (4.2.10) and substituting (4.2.11) we have:

$$\underline{u}_{n+1} = (I - \alpha Q \Lambda Q^t)\underline{u}_n \ . \tag{4.2.14}$$

The matrix Q is now used to define a 'rotated' error vector $\underline{u}'$

$$\underline{u}' = Q^t \underline{u} \ . \tag{4.2.15}$$

The purpose of this rotation becomes clear if we multiply both sides of (4.2.14) by Q^t and use the fact that $Q^t Q = I$

$$\underline{u}'_{n+1} = (I - \alpha \Lambda)\underline{u}'_n \ . \tag{4.2.16}$$

In view of the diagonal nature of Λ we can write separate, decoupled equations for each element of $\underline{u}'$, as

$$u'_{n+1}(j) = (1 - \alpha \lambda_j) u'_n(j) \ , \tag{4.2.17}$$

where $u'_{n+1}(j)$ is the j-th element of the rotated error vector $\underline{u}'_{n+1}$. This is a first order scalar difference equation which, using repeated substitution, can be expressed as

$$u'_n(j) = (1 - \alpha \lambda_j)^n u'_0(j) \ . \tag{4.2.18}$$

Consequently,

$$\lim_{n \to \infty} \{u'_n(j)\} \to 0 \ ,$$

provided

$$|1 - \alpha \lambda_j| < 1 \ .$$

Thus the error vector $\underline{u}'$ converges to zero, and hence in the mean the filter vector converges to $\underline{f}^*$, provided this condition holds for all j.

The eigenvalues λ_j of R are all real and positive since R is symmetric and positive definite (actually positive semi-definite which allows for the possibility of $\lambda_j = 0$). Consequently the condition for convergence is:

$$0 < \alpha < \frac{2}{\lambda_j} \qquad j = 0, 1, ..., L-1 \ . \tag{4.2.19}$$

Therefore the condition is satisfied for all modes, and the filter is convergent in the mean provided α satisfies

$$0<\alpha<\frac{2}{\lambda_{\text{max}}} \, , \tag{4.2.20}$$

where λ_{max} is the largest eigenvalue of the correlation matrix R. This result mirrors that for the steepest descent algorithm (see equation (4.1.7)).

Most authors would accept that the upper bound imposed on α by equation (4.2.20) is far too high to ensure stability. Also, it is not usually convenient to estimate the size of the maximum eigenvalue prior to executing the adaptive filter. One way to overcome this problem is to use the fact that

$$tr\{R\}=\sum_{i=1}^{L} \lambda_i \, , \tag{4.2.21}$$

where $tr\{\ \}$ denotes the trace of the matrix [9]. Now, since the eigenvalues of the positive semidefinite correlation matrix satisfy $\lambda_i \geq 0$, we have

$$\sum_{i=1}^{L} \lambda_i \geq \lambda_{\text{max}} \, ,$$

so that

$$\frac{2}{\sum_{i=1}^{L} \lambda_i} \leq \frac{2}{\lambda_{\text{max}}} \, .$$

Hence, combining this result and (4.2.20) we have

$$0<\alpha<\frac{2}{\sum_{i=1}^{L} \lambda_i} \, , \tag{4.2.22}$$

which is a stronger condition than (4.2.20). Also, in view of the Toeplitz nature of R

$$\sum_{i=1}^{L} \lambda_i = tr\{R\} = Lr(0) = LE\{x^2(n)\} \, ,$$

so that equation (4.2.22) becomes

$$0<\alpha<\frac{2}{LE\{x^2(n)\}} \ ,$$

or

$$0<\alpha<\frac{2}{L\ \{power\ of\ the\ input\}} \ . \qquad (4.2.23)$$

This is a stability condition which is stronger than (4.2.20) and which is relatively easy to compute. Notice, however, that this limit is quite different to (4.2.20), having a dependence on the filter length L.

A simple experiment is sufficient to illustrate the merits of the two stability limits. A zero-mean, Gaussian white noise sequence with variance σ_w^2, is supplied as both input and desired signal to the adaptive filter, as shown in Figure 4.2.1. This is, of course,

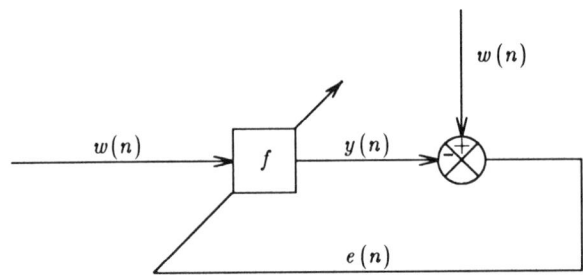

Figure 4.2.1 Adaptive filter setup for stability trials. $w(n)$ is a zero-mean iid Gaussian sequence with $\sigma_w^2=0.04$.

a trivial exercise in adaptive filtering. The solution which minimizes the mean-squared error is $f_n(0)=1$, $f_n(i)=0$ for $i\neq0$, which gives a zero error. Our purpose here is to examine the conditions under which the filter will converge to this solution. The experiment is repeated many times with different values for the adaptation constant. The first run uses a very small value for α. On each successive repetition of the experiment the adaptation constant is slowly increased until the point at which instability occurs. The values of α for which this occurs, together with those predicted by equations (4.2.20) and (4.2.23) are shown in Table 4.2.1 for the case of $\sigma_w^2=0.04$. Several conclusions can be drawn from the results:

L	*Eq. (4.2.20)*	*Eq. (4.2.23)*	*Observed*
5	50	14.19	13.22
10	50	7.09	6.60
20	50	3.54	3.54
50	50	1.41	1.5
100	50	0.71	0.82

Table **4.2.1** Comparison of LMS stability criteria. 'Observed' instability is defined as occurring whenever an error with magnitude in excess of 1×10^{10} occurs during the 10,000 sample trial.

i) The stability condition based on the input signal power (equation (4.2.23)) is much closer to the instability point observed in simulations than that implied by equation (4.2.20).

ii) The observed instability point has a clear dependence on filter length, L.

iii) The limit imposed by (4.2.20) actually lies well outside the true stability limit in all cases examined.

These results demonstrate that it is appropriate to employ equation (4.2.23) as a stability criterion rather than (4.2.20). The failure of the latter to predict the stability conditions with any accuracy can be explained as follows: Equation (4.2.20) gives a condition for the convergence *in the mean* for the filter coefficients. This condition is not sufficient for stability, however. We require additionally that the system have finite variance about this solution. Thus, stability depends on convergence to a finite mean-squared error and this is a quite different requirement from mean convergence. In fact for Gaussian inputs, by using extensions of the independence assumption, (4.2.23) can be derived directly from consideration of mean-squared error [10] though the algebra is rather involved. Note that, for practical purposes the overall stability is the important thing and thus the stronger condition (4.2.23) is appropriate.

The Eigenvalue Disparity Problem

For stationary inputs, we have seen that subject to the independence assumption, the LMS filter will ultimately converge, in the mean, to the least-squares optimum. However, this convergence does not usually take place *uniformly*. This can be a major practical problem for the LMS algorithm as we shall see. To understand this phenomenon we need only return to equation (4.2.17):

$$u_{n+1}^{'}(j)=(1-\alpha\lambda_j)u_n^{'}(j) .\qquad(4.2.17)$$

We see that all of the coefficients of the expected error vector will decrease in magnitude, provided

$$0<\alpha<2/\lambda_{\,\text{max}} .\qquad(4.2.20)$$

However, the rate of decay for each element $u_n^{'}(j)$ depends on the magnitude of the term $|(1-\alpha\lambda_j)|$ and since, in general, these eigenvalues will be distinct, some modes will converge more quickly than others. Thus, the LMS algorithm generally converges towards the optimum solution in a nonuniform manner. The exception to this is if all of the eigenvalues of the correlation matrix are equal, as is the case if the input signal is white. The phenomenon of nonuniform convergence is known for obvious reasons as the **eigenvalue disparity** problem. It is one of the two major disadvantages of the LMS algorithm (the other being the relatively slow convergence of steepest descent based algorithms). Its practical implications are difficult to quantify precisely. There are, however, some guidelines which are useful:

a) Eigenvalue Ratio and Spectrum — The ratio of the maximum to the minimum eigenvalue is bounded by ratio of the largest to the smallest component in the power spectrum [10]. That is,

$$\frac{\lambda_{\text{max}}}{\lambda_{\text{min}}}\leq\frac{S_{\text{max}}}{S_{\text{min}}} ,\qquad(4.2.24)$$

where S_{max}, S_{min} are the maximum and minimum components in the power spectrum of $x(n)$ (see also Problem 4.7). It can also be shown that the eigenvalue ratio asymptotically approaches this ratio as $L\rightarrow\infty$. Practically, for large L the bound may be taken as an equality. In some examples, the inequality is replaced by equality for all L. This can be illustrated by extreme cases; one such example is white noise which has spectral power equally distributed at all frequencies, and for which all the eigenvalues of R are equal. At the other extreme when the input is a sinusoid, the ratio of the peak spectral component to the smallest is infinite and the ratio of maximum to minimum eigenvalue is also infinite.

b) Eigenvalue Ratio and Filter Length — The ratio of the largest to smallest eigenvalues of R is a monotonically non-decreasing function of the filter length. This result tells us that we may *never* improve the eigenvalue disparity (and therefore the uniformity of the convergence) by increasing the filter length. This result is clearly basic to LMS performance but appears to have been widely overlooked. The result follows very simply as a consequence of the separation theorem for real symmetric matrices [9, page 104], (see [11] for more detail).

Time-Constants for Convergence

The mean convergence of the LMS algorithm is often quantified by means of **time-constants**. In particular, the time-constant for adaptation for an uncoupled mean coefficient error component $u'_n(j)$ is defined as the time required for that component to decay to $1/e$ of its initial value. Starting with (4.2.18):

$$u'_n(j) = (1 - \alpha\lambda_j)^n u'_0(j) , \tag{4.2.18}$$

we determine the value of $n = t_j$, say, for which

$$u'_{t_j}(j) = (1 - \alpha\lambda_j)^{t_j} u'_0(j) = \frac{u'_0(j)}{e} . \tag{4.2.25}$$

Taking logarithms we have

$$t_j \ln(1 - \alpha\lambda_j) = -1 . \tag{4.2.26}$$

For $\alpha\lambda_j \ll 1$, we have

$$\ln(1 - \alpha\lambda_j) \approx -\alpha\lambda_j ,$$

so that

$$t_j \approx \frac{1}{\alpha\lambda_j} . \tag{4.2.27}$$

Of course, the time-constants associated with the different decoupled components will vary if the eigenvalues of R are distinct, but in view of the comments of the preceding section this is hardly unexpected. Overall convergence is clearly limited by the slowest mode of convergence which in turn stems from the smallest eigenvalue:

$$t=\frac{1}{\alpha\lambda_{\min}} \ . \tag{4.2.28}$$

Example: Time-Delay Estimation

Consider the LMS adaptive filter as depicted in Figure 4.2.2 with $L=2$ and with inputs:

$$x(n)=w(n)+aw(n-1) \ ,$$
$$d(n)=x(n-1) \ , \tag{4.2.29}$$

where $E\{w(n)w(m)\}=\delta(n-m)$.

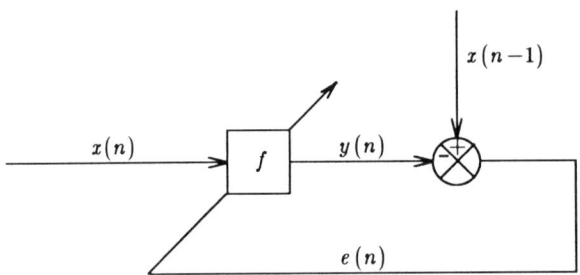

Figure 4.2.2 Time-delay estimation using an adaptive filter.

$x(n)$ is a first order Moving Average (MA) process. This is actually a simple time-delay estimation problem (the topic is discussed in detail in Section 5.3). For now, this example will be used to illustrate some of the aspects of LMS behavior discussed above.

a) Least-Squares Solution – Firstly, the least squares solution to this problem is obtained from $R\underline{f}^{*}=\underline{g}$. In this case, evaluating $r(i)$ and $g(i)$ using

$$r(i)=E\{x(n)x(n-i)\}$$

and
$$g(i)=E\{d(n)x(n-i)\}$$

gives

$$\begin{bmatrix} (1+a^2) & a \\ a & (1+a^2) \end{bmatrix} \begin{bmatrix} f(0)^* \\ f(1)^* \end{bmatrix} = \begin{bmatrix} a \\ (1+a^2) \end{bmatrix} , \qquad (4.2.30)$$

and hence
$$\underline{f}^* = [0,1]^t . \qquad (4.2.31)$$

That is, the filter acts as a unit delay. This result is intuitively reasonable, the filter delays the input $x(n)$ and thus cancels $d(n)=x(n-1)$, leaving $e(n)=0$, obviously the minimum mean-squared error solution.

b) *Stability Limit* — Next consider the LMS algorithm as characterized by the update equation (4.1.14). For this problem the overall convergence condition (4.2.23)

$$0<\alpha<\frac{2}{Lr(0)} , \qquad (4.2.23)$$

gives simply

$$0<\alpha<\frac{1}{(1+a^2)} . \qquad (4.2.32)$$

c) *Non-Uniform Convergence* — Each mode of the adaptive filter converges according to equation (4.2.17). To interpret this we need the eigenvalues of R. These are given by:

$$\det|R-\lambda I|=0 ,$$

which yields

$$\lambda^2-2\lambda(1+a^2)+(a^4+a^2+1)=0 ,$$

from which

$$\lambda_1=(1+a^2)+a , \quad \lambda_2=(1+a^2)-a . \qquad (4.2.33)$$

From equation (4.2.17) the rotated filter error vector components are given by:

$$u'_{n+1}(0) = [1 - \alpha((1+a^2)+a)]u'_n(0) ,$$

$$u'_{n+1}(1) = [1 - \alpha((1+a^2)-a)]u'_n(1) . \qquad (4.2.34)$$

The rate of convergence of the two modes is accordingly governed by the magnitudes of the two eigenvalues and can easily be determined once a is specified. Because of the small order of the problem $(L=2)$ it is possible to recover the original (unrotated) error vector $\underline{u}_n$. The algebra is straightforward though somewhat tedious. We will sketch the process and leave the reader to fill in the details. The result is obtained by firstly finding the matrix Q from the eigenvectors of R, that is

$$Q = [\underline{x}_1, \underline{x}_2] ,$$

where

$$R\underline{x}_i = \lambda_i \underline{x}_i \quad ; i=1,2 .$$

An orthonormal Q can be found as:

$$Q = \begin{bmatrix} \dfrac{1}{\sqrt{2}} & \dfrac{1}{\sqrt{2}} \\ \dfrac{1}{\sqrt{2}} & \dfrac{-1}{\sqrt{2}} \end{bmatrix} .$$

Using equations (4.2.15) and (4.2.16), assuming $\underline{f}_0 = \underline{0}$, and applying repeated substitution finally yields:

$$u_n(0) = (1-\alpha\lambda_1)^n(-1/2) + (1-\alpha\lambda_2)^n(1/2) ,$$

$$u_n(1) = (1-\alpha\lambda_1)^n(-1/2) + (1-\alpha\lambda_2)^n(-1/2) , \qquad (4.2.35)$$

where λ_1, λ_2 are given by (4.2.33). Some further simplification is possible, but the general result is clear enough. The point is that while the initial error, $u_0(0)$, is zero, the errors at later iterations, $u_1(0), u_2(0),...$ are not. In other words, as a result of the coupling between modes, the error associated with individual components can actually *increase* during the adaptation, even though it ultimately converges to zero. Further illustrations of the practical relevance of this effect will be given in later Sections.

4.2.3 Random Inputs: Steady-State Mean-Squared Error

In the previous section, we described the basic stability conditions and convergence behavior of the LMS algorithm for stationary random inputs. As we have already stated, however, consideration of convergence towards $\underline{f}^*$ is of limited value without corresponding results for the behavior of the system in the region of that steady-state solution. This analysis is essentially concerned with the mean-squared error J for the adaptive filter and, in particular, for stationary inputs the analysis focuses on the steady-state value for J, which we denote J_∞.

We recall that the error energy of the fixed least-squares filter is given by equation (3.2.14a):

$$J_{\min} = \sigma_d^2 - \underline{f}^{*t}\underline{g} \; . \tag{3.2.14a}$$

Since the LMS filter tries to emulate the optimal filter, $J_{\min}$ could be considered as an ideal which the adaptive filter strives to attain. In general, due to the random fluctuations of the filter, the steady-state error associated with the adaptive filter will be higher than this even if the adaptive filter converges in the mean. These random fluctuations are caused by the fact that even if the behavior of $\underline{f}$ is ideal, the error $e(n)$ is still generally non-zero and acts as a perturbing influence on the filter coefficients. The steady-state mean-squared error for the adaptive filter can be qualitatively described via a sum

$$J_\infty = J_{\min} + excess \; mean-squared \; error \; , \tag{4.2.36}$$

where the excess mean-squared error is due to adaptation effects.

As we have seen, difficulties in the analysis of $E\{\underline{f}_n\}$ are sufficient to necessitate the use of unrealistic 'independence' assumptions. The difficulties associated with the analysis of the mean-squared error are, however, even greater. The simplest approach[2] is that of Widrow [12], which in spite of its dependence

---

2. The term 'simplest approach' is certainly a relative one. Deriving the desired result takes the next four pages. The reader who is more interested in results than derivations may wish, on a first reading, to skip to equation (4.2.52).

on some statistical assumptions that cannot be strictly justified, produces results which are consistent with those observed in simulations, at least for small values for α.

Beginning with the LMS update of equation (4.1.14), we may define a coefficient error vector $\underline{v}_n = \underline{f}_n - \underline{f}^*$. Note the distinction between this error vector and $\underline{u}_n$ of Section 4.2.2 which was defined as $\underline{u}_n = E\{\underline{f}_n\} - \underline{f}^*$. The definition for $\underline{v}_n$ is analogous to that used for the fixed least-squares filter in Section 3.2. Using $\underline{v}_n$, the update equation may be expressed as

$$\underline{v}_{n+1} = \underline{v}_n + \alpha e(n)\underline{x}_n .\tag{4.2.37}$$

The raw gradient estimate $\nabla \hat{J}_n = \nabla e^2(n) = -2e(n)\underline{x}_n$ can be written as the sum of two components; the true gradient and a noise term

$$-2e(n)\underline{x}_n = \nabla J_n + 2\underline{N}_n ,\tag{4.2.38}$$

where ∇J_n is the gradient of the mean-squared error with respect to the filter vector $\underline{f}_n$, $\underline{N}_n$ is a noise vector arising as a consequence of the instantaneous nature of the gradient estimate, and the 2 is included for convenience. As a consequence of the independence assumption, the instantaneous gradient estimate is unbiased and we may write

$$E\{\underline{N}_n\} = \underline{0} .\tag{4.2.39}$$

We recall from Section 3.2 that the true gradient is

$$\nabla J_n = 2(R\underline{f}_n - \underline{g}) ,\tag{3.2.4a}$$

so that, combining (3.2.4a) and (4.2.38) gives

$$e(n)\underline{x}_n = -R\underline{v}_n - \underline{N}_n ,\tag{4.2.40}$$

where, additionally, we have used $\underline{g} = R\underline{f}^*$. Substituting for $e(n)\underline{x}_n$ into the update equation (4.2.37) yields

$$\underline{v}_{n+1} = (I - \alpha R)\underline{v}_n - \alpha \underline{N}_n .\tag{4.2.41}$$

Note that, applying the expectation operation to both sides of (4.2.41) and recognizing that $E\{\underline{v}_n\} = \underline{u}_n$ shows that (4.2.41) reduces to equation (4.2.10). At this point, however, our interest is in the raw (noisy) version (4.2.41) rather than the expected value. Using the rotation defined above we may decouple the equations (4.2.41) as

$$\underline{v}'_{n+1} = (I - \alpha\Lambda)\underline{v}'_n - \alpha\underline{N}'_n \ , \qquad (4.2.42)$$

where

$$\underline{N}'_n = Q^t \underline{N}_n \ ,$$

and

$$\underline{v}'_n = Q^t \underline{v}_n \ .$$

Using the decoupled form to isolate the *j*-th element we have

$$v'_{n+1}(j) = (1 - \alpha\lambda_j)v'_n(j) - \alpha N'_n(j) \ .$$

Squaring both sides and taking expectations gives

$$\gamma_{n+1}(j) = (1 - \alpha\lambda_j)^2 \gamma_n(j) + \alpha^2 E\{N'^2_n(j)\} - 2\alpha(1 - \alpha\lambda_j)E\{v'_n(j)N'_n(j)\}$$

where now $\gamma_n(j) = E\{v'^2_n(j)\}$. Next, we assume that

$$E\{v'_n(j)N'_n(j)\} = 0 \ , \qquad (4.2.43)$$

so that

$$\gamma_{n+1}(j) = (1 - \alpha\lambda_j)^2 \gamma_n(j) + \alpha^2 E\{N'^2_n(j)\} \ . \qquad (4.2.44)$$

This equation gives an approximate description of how the mean-squared error is propagated in the rotated error vector. In itself, this is not particularly helpful without some knowledge of $\underline{N}_n$. Since our interest is in the steady-state solution we may assume that

$$\lim_{n \to \infty} \{E\{\underline{f}_n\}\} \to \underline{f}^* \ ,$$

and hence that $\nabla J_n = 0$. From (4.2.40) we may then write

$$e(n)\underline{x}_n = -\underline{N}_n \ ,$$

so that

$$E\{\underline{N}_n \underline{N}^t_n\} = E\{e^2(n)\underline{x}_n \underline{x}^t_n\} \ ,$$

$$\approx E\{e^2(n)\}E\{\underline{x}_n\underline{x}_n^t\} \ ,$$

or

$$E\{\underline{N}_n\underline{N}_n^t\}\approx J_{\min}R \ , \tag{4.2.45}$$

where the result depends on the further assumption of independence of the error and the data correlation. Using the rotated form for $\underline{N}$, we may write

$$E\{\underline{N}_n'\underline{N}_n^{'t}\}=E\{Q^t\underline{N}_n\underline{N}_n^tQ\} \ ,$$

$$\approx J_{\min}Q^tRQ \ ,$$

$$\approx J_{\min}\Lambda \ .$$

The matrix $E\{\underline{N}_n'\underline{N}_n^{'t}\}$ is therefore diagonal with j-th element:

$$E\{N_n^{'2}(j)\}=J_{\min}\lambda_j \ . \tag{4.2.46}$$

Now that we have some useful information about $\underline{N}$, we may return to the equation describing the mean-squared error propagation. Accordingly, combining this result with (4.2.44) and assuming the expression for the mean-squared value of $N_n'(j)$ is valid for all n we have

$$\gamma_{n+1}(j)=(1-\alpha\lambda_j)^2\gamma_n(j)+\alpha^2J_{\min}\lambda_j \ . \tag{4.2.47}$$

Applying repeated substitution yields

$$\gamma_1(j)=(1-\alpha\lambda_j)^2\gamma_0(j)+\alpha^2J_{\min}\lambda_j \ ,$$

$$\gamma_2(j)=(1-\alpha\lambda_j)^4\gamma_0(j)+(1-\alpha\lambda_j)^2\alpha^2J_{\min}\lambda_j+\alpha^2J_{\min}\lambda_j \ ,$$

and so on. Finally, we observe

$$\gamma_n(j)=(1-\alpha\lambda_j)^{2n}\gamma_0(j)+\alpha^2\sum_{i=0}^{n/2}(1-\alpha\lambda_j)^{2i}J_{\min}\lambda_j \ . \tag{4.2.48}$$

This equation gives an expression for the propagation of the mean-squared error for the j-th filter mode. In view of the assumption that the steady-state value for $E\{N_n^{'2}(j)\}$ applies throughout, we anticipate equation (4.2.48) being most useful for large n. In particular, if

$$0 < \alpha < \frac{2}{\lambda_j} \; ,$$

then the term in (4.2.48) due to $\gamma_0(j)$ decays to zero, and in the limit as n increases

$$\lim_{n \to \infty} \{\gamma_n(j)\} = \alpha^2 J_{\min} \lambda_j \left[\frac{1}{1 - (1 - \alpha\lambda_j)^2} \right] \; ,$$

or

$$\lim_{n \to \infty} \{\gamma_n(j)\} = \frac{\alpha J_{\min}}{2 - \alpha\lambda_j} \; . \tag{4.2.49}$$

Equation (4.2.49) gives an expression for the steady-state mean-squared error for the j-th orthogonal coefficient error. This is itself a useful relation. Our main objective, however, is the overall mean-squared error. In Section 3.2, for a fixed filter $\underline{f}$, the excess mean-squared error was shown to be

$$J = J_{\min} + \underline{v}^t R \underline{v} \; , \tag{3.2.17}$$

where $\underline{v} = \underline{f} - \underline{f}^*$. By similar reasoning, and subject to the usual independence assumption, for the adaptive filter error vector $\underline{v}_n$ we have (see also Problem 4.11)

$$J_n = J_{\min} + E\{\underline{v}_n^t R \underline{v}_n\} \; . \tag{4.2.50}$$

Using the now familiar decomposition of R we may write

$$J_n = J_{\min} + E\{\underline{v}_n^t Q \Lambda Q^t \underline{v}_n\} \; ,$$

and in terms of the rotated error vector $\underline{v}_n'$ we have

$$J_n = J_{\min} + E\{\underline{v}_n'^t \Lambda \underline{v}_n'\} \; .$$

Therefore, the excess mean-squared error over and above $J_{\min}$ is given by

$$excess \; mse = J_n - J_{\min} = E\{\underline{v}_n'^t \Lambda \underline{v}_n'\} \; .$$

Expanding the right-hand side of this equation gives

$$excess\ mse = \sum_{j=0}^{L-1} \lambda_j E\{v_n'^2(j)\} = \sum_{j=0}^{L-1} \lambda_j \gamma_n(j) \ .$$

Hence using (4.2.49), as $n \to \infty$ we have

$$steady-state\ excess\ mse = J_\infty - J_{min} = \sum_{j=0}^{L-1} \frac{\alpha J_{min} \lambda_j}{2 - \alpha \lambda_j} \ . \qquad (4.2.51)$$

which we may expand as

$$J_\infty = J_{min} + \sum_{j=0}^{L-1} \frac{\alpha}{2} J_{min} \lambda_j (1 + \frac{\alpha}{2} \lambda_j + \frac{\alpha^2}{4} \lambda_j^2 + ...) \ .$$

For $\alpha \ll 2/\lambda_j$, we may approximate this by

$$J_\infty \approx J_{min} + \sum_{j=0}^{L-1} \frac{\alpha}{2} J_{min} \lambda_j \ ,$$

or

$$J_\infty \approx J_{min} (1 + \frac{\alpha}{2} tr\{R\}) \ , \qquad (4.2.52a)$$

or, finally

$$J_\infty \approx J_{min} (1 + \frac{\alpha}{2} L\ power\ of\ the\ input) \ . \qquad (4.2.52b)$$

As we have seen, this analysis has been weakened by several assumptions and approximations, and the reader might be tempted to wonder if the result accurately reflects the practical situation. We can try to address this question by constructing a simple simulation: Consider the setup shown in Figure 4.2.3. The adaptive filter is supplied with inputs $w_1(n)$, $w_2(n)$ both zero-mean white noise sequences with

$$E\{w_1(n)w_2(m)\} = 0 \quad ;\ for\ all\ n,m \ ,$$

and

$$E\{w_1^2(n)\} = E\{w_2^2(n)\} = \sigma_w^2 \ .$$

With this setup we may readily observe that $f^* = 0$ and hence that $J_{min} = \sigma_w^2$. Figure 4.2.4 shows the value of J_∞ obtained from

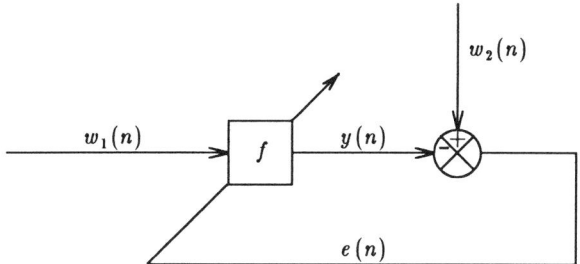

Figure 4.2.3 Adaptive Filter setup for mean-squared error trials. $w_1(n)$, $w_2(n)$ are zero-mean iid Gaussian unit variance sequences.

equation (4.2.52), denoted J_{theory}, and that obtained by simulation, $J_{\exp}$, say.

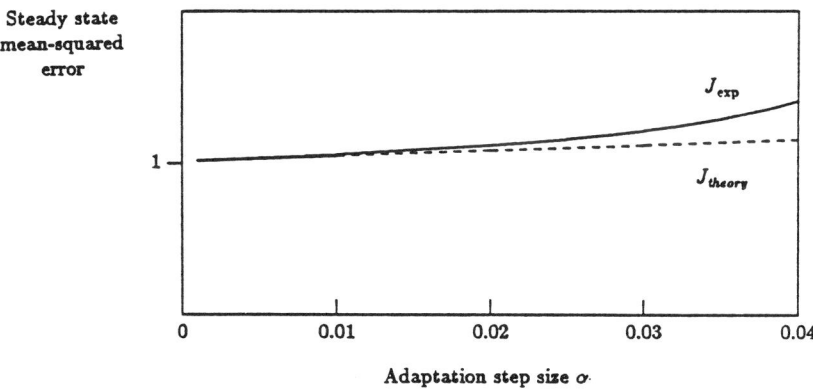

Figure 4.2.4 Mean-squared error with the LMS algorithm: theory versus simulation. J_{theory} is the steady-state mean-squared error as given by equation (4.2.52). $J_{\exp}$ was obtained by averaging the steady-state squared errors (using a time-average).

Note that strictly $J_{\exp}$ should be obtained by repeating the trial many times to form an 'ensemble' and averaging $e^2(n)$ across this ensemble (with a large value for n). In practice, after the filter reaches steady-state the resulting $e^2(n)$ can be assumed to be a stationary, ergodic sequence and $J_{\exp}$ can be approximated using a time-average over interval M, beginning after many iterations $n = N$, say

$$J_{\exp} \approx \frac{1}{M} \sum_{n=N}^{N+M-1} e^2(n) \; . \qquad (4.2.53)$$

The results displayed in Figure 4.2.4 were obtained using $N=5192$ and $M=3000$. Examining the figures, it is clear that theory and simulation are in good agreement for small values of α. As α increases the prediction of equation (4.2.52) becomes less accurate. Finally, we note that the simulation produces an unstable result at a lower value for α than predicted by the theory (though this is not shown in Figure 4.2.4). This is hardly unexpected in view of the discussion in Section 4.2.2 above. Overall the point is that for small values of α, equation (4.2.52) provides a good indication of mean-squared error and that in practice these small α values are the ones which are invariably employed.

4.2.4 Deterministic Signals: Transfer Function Analysis

If the LMS adaptive filter is subjected to deterministic input signals, the analysis of the previous sections cannot be applied. An alternative approach based on the use of the z transform was introduced by Glover [13] who considered the behavior of the LMS filter when it is subjected to sinusoidal inputs. In particular, Glover showed that for such inputs the entire nonlinear LMS structure can be approximated by a simple linear transfer function. More recently it has been shown that this transfer function description can be applied to the analysis of the adaptive filter for a broad range of deterministic inputs [14]. A particular attraction of this approach is that in many instances the results are exact. This is in stark contrast to the random analysis given above, where a variety of unlikely approximations are involved.

We begin with the filter coefficient update of equation (4.1.14)

$$\underline{f}_{n+1} = \underline{f}_n + \alpha e(n)\underline{x}_n \; , \qquad (4.1.14)$$

where $x(n)$ is now an arbitrary deterministic input. Let us assume that the initial filter coefficient vector satisfies

$$\underline{f}_0 = \underline{0} \; . \qquad (4.2.54)$$

As discussed in Section 4.1, this represents no loss of generality because of the unimodal nature of the performance surface. Now, applying repeated substitution to (4.1.14) we may write

$$\underline{f}_n = \alpha \sum_{i=0}^{n-1} e(i)\underline{x}_i \ .$$
(4.2.55)

The filter output is

$$y(n) = \underline{f}_n^t \underline{x}_n \ .$$
(4.2.56)

Hence, substituting (4.2.55) we have

$$y(n) = \alpha \sum_{i=0}^{n-1} e(i)\underline{x}_i^t \underline{x}_n \ ,$$

$$= \alpha L \sum_{i=0}^{n-1} e(i) r_{i,n} \ ,$$
(4.2.57)

where we have defined

$$r_{i,n} = \frac{1}{L}\underline{x}_i^t \underline{x}_n = \frac{1}{L}\sum_{j=0}^{L-1} x(i-j)x(n-j) \ .$$
(4.2.58)

Also,

$$e(n) = d(n) - y(n) \ ,$$
(4.2.59)

so that equation (4.2.57) can be written as

$$e(n) + \alpha L \sum_{i=0}^{n-1} e(i) r_{i,n} = d(n) \ .$$
(4.2.60)

Equation (4.2.60) is a relation between the desired input and the output error in terms of a purely recursive difference equation of n-th order. In general, the coefficients of this difference equation are time-varying. However, as $L \rightarrow \infty$, providing the signal has finite energy

$$r_{n,i} \approx r(n-i) \ .$$
(4.2.61)

Consequently, for sufficiently long filters we can write

$$e(n) + \alpha L \sum_{i=0}^{n-1} r(n-i)e(i) = d(n) \ .$$
(4.2.62)

Recognizing that the term within the summation is the convolution of the sequence $r(n)$ with $e(n)$ this equation may be z transformed to give[3]

$$E(z)+\alpha LR(z)E(z)=D(z) , \qquad (4.2.63)$$

where $R(z)=r(1)z^{-1}+r(2)z^{-2}+....$, and where we have used the one-sided z transform [15] defined for a signal $s(n)$ by

$$S(z)=\sum_{n=0}^{\infty} s(n)z^{-n} .$$

Equation (4.2.63) may be rewritten as

$$\frac{E(z)}{D(z)}=\frac{1}{1+\alpha LR(z)}=H(z) . \qquad (4.2.64)$$

The significance of this result is that the nonlinear adaptive filter has been replaced by a linear transfer function $H(z)$ (linear, that is, once $x(n)$ is specified). This transfer function relates the desired input $d(n)$ to the output error $e(n)$ (see Figure 4.2.5).

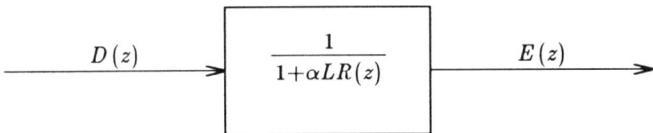

$$D(z) \longrightarrow \boxed{\frac{1}{1+\alpha LR(z)}} \longrightarrow E(z)$$

Figure 4.2.5 Equivalent transfer function descriptor for the LMS filter. $D(z)$, $E(z)$ are one-sided transforms of $d(n)$, $e(n)$ respectively, and $R(z)=r(1)z^{-1}+r(2)z^{-2}+....$

By utilizing the simple relationship between $d(n)$, $e(n)$ and $y(n)$, a similar transfer function can be found relating $d(n)$ and $y(n)$ (see Problem 4.12) as

$$\frac{Y(z)}{D(z)}=\frac{\alpha LR(z)}{1+\alpha LR(z)} . \qquad (4.2.65)$$

---

3. This can be verified by writing out (4.2.62) for $n=0,1,2,...$ etc. and then applying the transform to each term.

Special Case: Periodic Inputs

In general, the transfer function of equation (4.2.64) is an approximate relation whose accuracy depends on the assumption that $r_{n,i} \approx r(n-i)$ used in the derivation. For some inputs, however, the results can be exact. One such class of inputs is periodic signals. In particular, if the filter length is an integer multiple of the period of the input, then $r_{n,i} = r(n-i)$ and the transfer function (4.2.64) is a precise description of the adaptive system behavior. Consider a signal which is periodic with period N (an integer number of samples). For this case, $r(i)$ is periodic with maximum period N. $R(z)$ becomes

$$R(z) = r(1)z^{-1} + ... + r(N-1)z^{-(N-1)} + r(0)z^{-N} + r(1)z^{-N-1} + ... ,$$

$$= r(1)z^{-1}[1 + z^{-N} + ...] + r(2)z^{-2}[1 + z^{-N} + ...] + ...$$

$$+ r(0)z^{-N}[1 + z^{-N} + ...] ,$$

or

$$R(z) = \frac{\hat{R}(z)}{1 - z^{-N}} ,$$

where $\hat{R}(z) = r(1)z^{-1} + r(2)z^{-2} + ... + r(0)z^{-N}$. Substituting into equation (4.2.64) and rearranging gives

$$\frac{E(z)}{D(z)} = \frac{1 - z^{-N}}{1 + \alpha L \hat{R}(z) - z^{-N}} . \tag{4.2.66}$$

This transfer function has zeros equi-spaced around the unit circle of the z-plane, and poles whose locations are determined by the zeros of

$$1 + \alpha L \hat{R}(z) - z^{-N} .$$

It is difficult to obtain the zeros of this function explicitly for a general periodic signal. However, simple results can be obtained for some special cases such as sinusoidal signals, as discussed below. Other examples are considered in the problems at the end of the chapter.

Example: Sinusoidal Inputs [13]

We next consider inputs of the form

$$x(n) = A\cos(\omega_0 n T + \phi) , \qquad (4.2.67)$$

where $\omega_0 = 2\pi f_0$, and T is the sample interval. Equation (4.2.58) gives

$$
\begin{aligned}
r_{n,n-i} &= \frac{1}{L} x_n^t x_{n-i} \\
&= \frac{1}{L} \sum_{j=0}^{L-1} \left[A^2 \cos(\omega_0(n-j)T+\phi)\cos(\omega_0(n-i-j)T+\phi) \right] , \\
&= \frac{A^2}{2L} \sum_{j=0}^{L-1} \left[\cos(\omega_0(2n-2j-i)T+2\phi)+\cos(\omega_0 i T) \right] .
\end{aligned}
$$

Now, if we assume that the filter length L is an integer multiple of the period of the sinusoid (that is $L = l/f_0 T$, where l is any integer), then the first term of this equation is zero. Hence

$$r(i) = \frac{A^2}{2}\cos(\omega_0 i T) . \qquad (4.2.68)$$

We see that $r_{n,n-i} = r(i)$ and the transfer function (4.2.64) holds exactly. For this particular example it is more straightforward to obtain the desired transfer function directly by substituting (4.2.68) into equation (4.2.64), rather than using the periodic form (4.2.66). We require $R(z) = r(1)z^{-1} + r(2)z^{-2} + ...$, so that

$$R(z) = \frac{A^2}{2} \left[\sum_{i=0}^{\infty} (\cos(\omega_0 i T)z^{-i}) - 1 \right] ,$$

which is easily obtained as

$$R(z) = \frac{A^2}{2} \left\{ \left[\frac{1-\cos(\omega_0 T)z^{-1}}{1-2\cos(\omega_0 T)z^{-1}+z^{-2}} \right] - 1 \right\} . \qquad (4.2.69)$$

Rearranging equation (4.2.69) and substituting into (4.2.64) gives the desired transfer function

$$\frac{E(z)}{D(z)} = \frac{1-2\cos(\omega_0 T)z^{-1}+z^{-2}}{[1-\frac{\alpha L A^2}{2}]z^{-2}+[\frac{\alpha L A^2}{2}-2]\cos(\omega_0 T)z^{-1}+1} . \qquad (4.2.70)$$

This expression shows that for this particular input, the relationship between the desired signal $d(n)$ and the error $e(n)$ has been replaced by a simple second order transfer function. The zeros of this transfer function lie on the unit circle at $z=e^{j\omega_0 T}$ and $z=e^{-j\omega_0 T}$. The poles are obtained from the zeros of the denominator, and for small α are located approximately[4] at

$$z_p \approx \left(1-\frac{\alpha L A^2}{4}\right)e^{\pm j\omega_0 T} . \qquad (4.2.71)$$

This corresponds to a notch filter with the notch centered on the position of the zero, $\omega_0 T$ (see Figure 4.2.6).

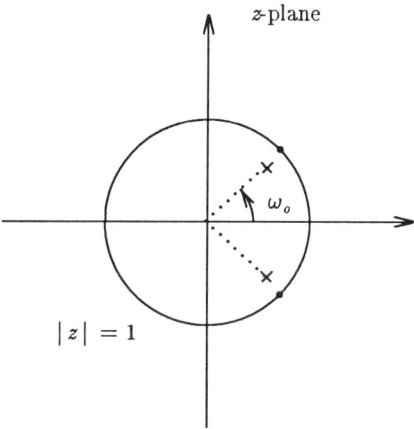

Figure 4.2.6 Pole-zero plot for the sinusoidal transfer function of equation (4.2.70). Result is a notch filter at ω_0. Note that pole locations are based on the approximation (4.2.71) and are only valid for small α.

The bandwidth of the notch is controlled by the poles, which lie on the same radial lines as the zeros. As the poles approach the unit circle the notch becomes tighter until at $|z_p|=1$ the notch disappears. Conversely as the poles move away from $|z_p|=1$, the notch widens. Consequently the notch width depends on α, L and

---

4. In fact the pole behavior is more complex than (4.2.71) suggests because as α increases, neglecting terms of α^2 is no longer reasonable. This is particularly marked in relation to the instability point for the filter. Use of the approximation (4.2.71) to find the condition $|z_p|=1$ produces a result which is incorrect by a factor of 2 (see Problem 4.18).

A^2, with the width of the notch increasing with an increase in these parameters. In fact this is entirely reasonable, though we shall postpone further interpretation until Section 5.1 when the sinusoidal input will reappear in an application example.

Note that the poles of the transfer function can also be used to derive simple stability limits for the filter. Since the adaptive filter is equivalent to the linear system (4.2.70), its stability is governed by similar rules, that is for the system to be stable the poles must satisfy $|z_p| < 1$. From equation (4.2.70) it is easy to verify that this occurs when

$$0 < \alpha < \frac{4}{LA^2} \ , \tag{4.2.72}$$

and this gives the required LMS stability limit. Notice that in spite of the fact that this result is for a deterministic input, the limit shows the same dependence on α, L and the power of the input as did the stability condition for stationary random inputs (see equation (4.2.23)).

A question which usually arises is as follows: If the adaptive filter with this form of input gives rise to a transfer function, why not simply use this transfer function in the first place? In other words why use an adaptive filter at all? The answer is that the sinusoid considered is an *idealized* input. If such an input were to be actually used (and assuming ω_0 could be accurately estimated) there would indeed be no point in constructing an adaptive filter. In practice, it is far more realistic to expect sinusoidal inputs whose amplitudes, phase and frequencies 'drift' slowly in time. In such cases we hope the adaptive filter will track the changes in the input and *at each instant* will approximate the behavior indicated by equation (4.2.70).

Appendix 4A — Diagonalization of the Correlation Matrix

Consider a correlation matrix R with eigenvalues λ_i ; $i = 1, 2, ..., L$. Define a matrix Q with

$$Q = [\underline{q}_1, \underline{q}_2, ..., \underline{q}_L] \ , \tag{A1}$$

where $\underline{q}_1$, $\underline{q}_2$,..., $\underline{q}_L$ form a set of real eigenvectors satisfying

$$\underline{q}_i^t \underline{q}_j = \delta(i-j) , \tag{A2}$$

that is the q_i's are orthonormal.[5] Q represents the **modal matrix** for R and, as a consequence of (A2)

$$Q^t Q = I . \tag{A3}$$

We define a **spectral matrix** Λ with

$$\Lambda = \begin{bmatrix} \lambda_1 & 0 & 0 & 0 \\ 0 & \lambda_2 & . & . \\ 0 & 0 & . & . \\ . & . & . & \lambda_L \end{bmatrix} . \tag{A4}$$

We may write

$$Q\Lambda = [\lambda_1 \underline{q}_1, \lambda_2 \underline{q}_2, \ldots, \lambda_L \underline{q}_L] ,$$

and

$$RQ = [R\underline{q}_1, R\underline{q}_2, ..., R\underline{q}_L] .$$

Now,

$$R\underline{q}_i = \lambda_i \underline{q}_i , \tag{A5}$$

is the eigenvalue-eigenvector relation. Hence

$$RQ = Q\Lambda ,$$

and so postmultiplying by Q^t we have

$$R = Q\Lambda Q^t , \tag{A6}$$

because of the orthogonality of Q.

---

5. We may identify such an orthonormal set for any correlation matrix R, even if R has repeated eigenvalues [20].

Problems

4.1 For each of the following, for a filter of length L, use equation (4.2.23) to find a stability limit for the LMS adaptive algorithm

    i) $x(n)=w(n)+a_1 w(n-1)+a_2 w(n-2)$,

    ii) $x(n)=ax(n-1)+w(n)$,

where a, a_1 and a_2 are constants, and where $w(n)$ is a zero-mean unit variance iid sequence.

4.2 Taking $L=2$, find stability limits for the LMS algorithm as given by both (4.2.20) and (4.2.23) for both *i)* and *ii)* in **4.1** above.

4.3 If an input signal $x(n)$ has

$$r_{xx}(0)=2, \quad r_{xx}(1)=1, \text{ and } r_{xx}(i)=0 \text{ for } i\geq 2 ,$$

then for a filter length $L=3$:

a) Quantify the stability limits for the LMS filter using both equations (4.2.20) and (4.2.23).

b) Find time-constants t_1, t_2, t_3 for this system.

4.4 An adaptive algorithm based on the method of steepest descent has an update equation given by

$$\underline{f}_{n+1}=\underline{f}_n-\alpha(R\underline{f}_n-\underline{g}) . \qquad (4.1.6)$$

If $\underline{v}_n$ is defined by

$$\underline{v}_n=(\underline{f}_n-\underline{f}^*)$$

where $\underline{f}^*=R^{-1}\underline{g}$ is the fixed least-squares solution, and given that $\underline{f}_0=\underline{0}$, show that

$$\underline{v}_n=-(I-\alpha R)^n \underline{f}^* .$$

If $x(n)$ is zero-mean white noise with variance $\sigma_w^2 = 0.5$ and $d(n) = x(n) + 0.5x(n-1)$, find an explicit form for $\underline{v}_n$.

4.5 Show that the LMS update may be written

$$\underline{v}_{n+1} = (I - \alpha \underline{x}_n \underline{x}_n^t)\underline{v}_n + e_0(n)\underline{x}_n ,$$

where $e_0(n)$ is the error associated with the optimal filter $\underline{f}^*$.

4.6 A useful way to bound the eigenvalues of a matrix is through the Gerschgorin circles theorem [16]. The theorem states that every eigenvalue of a matrix A with elements $a(i,j)$ lies in at least one of the circular discs with centers $a(i,i)$ and radii

$$\rho_i = \sum_{j=1}^{N} |a(i,j)| \quad ; j \neq i .$$

For an $(L \times L)$ correlation matrix R, the eigenvalues are real and the i-th eigenvalue is restricted to the interval on the real axis centered on $r(0)$ with length

$$\rho_i = \sum_{j=1}^{L} |r(i-j)| \quad ; j \neq i .$$

This theorem is helpful in bounding eigenvalue behavior in the adaptive filter when the correlation matrix is 'sparse'. Consider the following problem: A zero-mean iid sequence $w(n)$ is corrupted by a single reflection of magnitude a at time n_0. The resulting signal is input to an LMS filter. Use the Gerschgorin circles theorem to bound the eigenvalues, and hence the eigenvalue disparity in the LMS system.

4.7 For a stationary zero-mean signal $x(n)$, with power spectral density $S(e^{j\omega})$ and correlation matrix R, show that

$$\frac{\lambda_{\max}}{\lambda_{\min}} \leq \frac{S_{\max}}{S_{\min}} ,$$

where $S_{\max}$ and $S_{\min}$ are the largest and smallest components respectively, of $S(e^{j\omega})$, and where $\lambda_{\max}$, $\lambda_{\min}$ are the largest and smallest eigenvalues respectively of R.

Hint: use the relation:

$$\lambda_i = \frac{q_i^t R q_i}{q_i^t q_i} ,$$

where q_i is the i-th eigenvector of R.

4.8 Consider an input to an LMS filter in the form $x(n)=h(n)*w(n)$, where $w(n)$ is a zero-mean iid sequence. For each of the following cases, use the relation

$$\frac{\lambda_{\max}}{\lambda_{\min}} \le \frac{S_{\max}}{S_{\min}} ,$$

derived in the previous problem, to bound the eigenvalue disparity

i) $h(n)=\delta(n)+c\delta(n-1)$; c constant,

ii) $h(n)=a^n u(n)$; $|a|<1$,

where $u(n)$ is the unit step sequence.

4.9 Use the similarity transform $R=Q\Lambda Q^t$ to show that for the steepest descent algorithm with update

$$\underline{f}_{n+1}=\underline{f}_n-\alpha(R\underline{f}_n-\underline{g}) ,$$

then

$$\lim_{n\to\infty} \{\underline{f}_n\}\to\underline{f}^* .$$

4.10 An adaptive filtering problem is called 'homogeneous' if

$$d(n)=\underline{f}^{*t}\underline{x}_n ,$$

otherwise it is 'inhomogeneous' [17]. Assuming that signals $w(n)$, $v(n)$ are zero-mean iid, have variances σ_w^2 and σ_v^2, respectively, and are mutually uncorrelated, then for each of the following cases state whether the problem is homogeneous or inhomogeneous and find $J_{\min}$:

$i)\ x(n)=w(n),\ \ d(n)=\sum_{i=0}^{N-1} h(i)w(n-i)\ \ \ ;\ N\leq L.$

$ii)\ x(n)=w(n),\ \ d(n)=\sum_{i=0}^{N-1} h(i)w(n-i)\ \ ;\ N>L.$

$iii)\ x(n)=w(n),\ \ d(n)=x(n)+v(n).$

$iv)\ x(n)=w(n)+v(n),\ \ d(n)=w(n).$

4.11 Using the independence assumption, show that for the LMS algorithm with stationary random inputs, at iteration n the mean-squared error may be written as

$$J_n = J_{\min} + E\{\underline{v}_n^t R \underline{v}_n\}\ ,$$

where $\underline{v}_n = \underline{f}_n - \underline{f}^*.$

4.12 Using the procedure of Section 4.2.4, and assuming that

$$r_{i,n} = \frac{1}{L}\underline{x}_i^t \underline{x}_n = r(i-n)\ ,$$

show that for a deterministic input $x(n)$

$$\frac{Y(z)}{D(z)} = \frac{\alpha L R(z)}{1+\alpha L R(z)}\ ,$$

where $R(z) = r(1)z^{-1} + r(2)z^{-2} + \cdots$

4.13 The normalized LMS algorithm (see also Section 6.2.1) is given by:

$$\underline{f}_{n+1} = \underline{f}_n + \alpha \frac{e(n)\underline{x}_n}{\underline{x}_n^t \underline{x}_n}\ . \tag{6.2.3}$$

a) If $\underline{x}_n$ is a deterministic input signal, find an equivalent transfer function relating $E(z)$ to $D(z)$ for this algorithm. (You may assume that $r_{i,j} = \frac{1}{L}\underline{x}_i^t\underline{x}_j = r(i-j)$, and that $r(0)\neq0$.)

b) If $x(n)=A\cos(\omega_0 nT)$, find a specific form for the transfer function.

4.14 a) In the diagram, show that

$$A(n)=\sqrt{f_0^2+f_1^2}, \quad \phi(n)=\tan^{-1}\left[\frac{f_1}{f_0}\right],$$

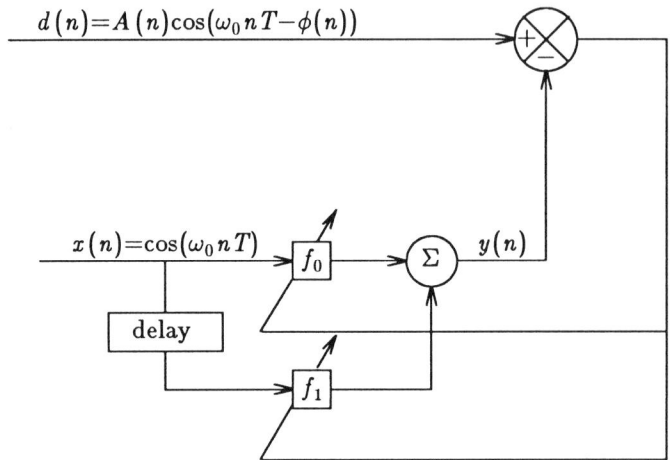

implies that $e(n)\to 0$.

b) For the same diagram, assuming the coefficients f_0 and f_1 are updated using the LMS update equation show that

$$e(n)+\alpha\sum_{i=0}^{n-1} e(i)\cos(\omega_0(n-i)T)=d(n).$$

Note: $T=1/12000$, $\omega_0=2\pi\times 1kHz$, the delay $=3$ samples, and f_0, f_1 are scalar coefficients.

4.15 Assume that the LMS adaptive filter has an input of the form

$$x(n)=1 \quad ; k=0,(N-1),(2N-1),...$$
$$=0 \quad otherwise.$$

a) Use equation (4.2.66) to find an equivalent transfer function relating $e(n)$ and $d(n)$ [18].

b) Plot the poles and zeros in the z-plane.

c) Obtain a stability limit for the adaptation constant α.

4.16 Repeat **4.15** if the input, $x(n)$, is the square wave [19] given by

$$x(n)=\{1,1,-1,-1,1,1,-1,-1,...\} .$$

4.17 Consider an input

$$x(n)=A_0\cos(\omega_0 nT)+A_1\cos(\omega_1 nT) .$$

Use the analysis of Section 4.2.4 to show that the transfer function approximation for $x(n)$ is a sum of two notch filters at frequencies ω_0, ω_1. State carefully the assumptions involved in your derivation.

4.18 Consider the sinusoidal transfer function relation of equation (4.2.70).

a) Show that the system has poles located at $|z_p|=1$ when
$$\alpha=\frac{4}{LA^2} .$$

b) Use $\alpha=\frac{4}{LA^2}-\delta$, where $\delta>0$, to show that the system is stable when

$$0<\alpha<\frac{4}{LA^2} .$$

c) Find the stability condition from the approximate relation (4.2.71). Are the two the same?

4.19 a) For a zero-mean stationary input, and assuming

$$E\{e(n)\underline{x}_i^t\underline{x}_n\}=E\{e(n)\}E\{\underline{x}_i^t\underline{x}_n\} ,$$

show that for the LMS update

$$\underline{f}_{n+1}=\underline{f}_n+\alpha e(n)\underline{x}_n , \qquad (4.1.14)$$

an equivalent transfer function relating the mean output and desired signals is

$$\overline{Y}(z) = \left[\frac{\alpha LR(z)}{1+\alpha LR(z)} \right] \overline{D}(z) ,$$

where $\overline{Y}(z) = \sum\limits_{n=0}^{\infty} E\{y(n)\}z^{-n}$, with $\overline{D}(z) = \sum\limits_{n=0}^{\infty} E\{d(n)\}z^{-n}$.

b) Use the result in a) to find $\overline{y}(\infty)$ if

$$d(n) = v(n) + 1 ,$$

$$x(n) = w(n) + aw(n-1) ,$$

where $w(n)$, $v(n)$ are zero-mean iid Gaussian sequences, and a is a constant. Comment on the result.

4.20 Consider the LMS system with input $x(n) = c = constant$. Derive a transfer function approximation relating $E(z)$ to $D(z)$. What type of filter is this?

Computer Problem 1. – LMS Adaptive Filter

Part 1 – Implement the LMS adaptive filter given by the update equation:

$$\underline{f}_{n+1} = \underline{f}_n + \alpha e(n)\underline{x}_n ,$$

where $\qquad e(n) = d(n) - y(n),$

and $\qquad y(n) = \underline{f}_n^t \underline{x}_n,$

and $\underline{f}_n = [f_n(0), f_n(1), ..., f_n(L-1)]^t$,
  $\underline{x}_n = [x(n), x(n-1), ..., x(n-L+1)]^t$.

You may assume $\underline{f}_0 = \underline{0}$ throughout. The inputs to this program should be:

i) Data files for $d(n)$ and $x(n)$.

ii) Filter length, L.

iii) Adaptation constant, α.

Part 2 – Generate 5,000 samples of each of the following types of test data:

i) Uniformly distributed random numbers with range [-1,1].

ii) Gaussianly distributed random numbers with zero-mean and unit variance.

iii) Sinusoid with digital frequency $\pi/5$ and phase angle zero.

a) Test your adaptive filter using each set of data. Use $d(n)=x(n)$ in all cases. In each case there should be a range of values for α for which the filter adapts to force the error to zero. Try several different values for L to confirm your results.

b) Evaluate the stability limits for your filter with each class of data. To do this you may assume that if the filter remains stable after 5000 iterations it will be considered stable. Check for instability within the program by examining the magnitude of the error. Large errors obviously correspond to unstable solutions. To establish stability limits, you will need to conduct some form of search. One possibility is as follows: Begin with a small value for α and run the filter, if no instability results increase α by a small amount, reset the program and run again. This procedure is continued until a value of α is reached for which instability occurs. This procedure is computationally rather lengthy so you may wish to use a binary search instead. You should find the highest stable value for α for each class of data and for values of $L=5,10,20,50,100$.

Part 3 – Generate another 5,000 samples of uniformly distributed noise (uncorrelated with the first set). Repeat the stability trials but use $d(n)$ as one set of noise and $x(n)$ as the other. How do your results compare with those of Part 2? Can you explain this?

Computer Problem 2. − Single Input Adaptive Filter

A *single input* LMS adaptive filter can be obtained by using $d(n)=x(n)$ and delaying the input $x(n)$ (see also Chapter 5). Implement such a filter using the update equation:

$$\underline{f}_{n+1}=\underline{f}_n+\alpha e(n)\underline{x}_{n-\Delta} \, ,$$

where $e(n)=x(n)-y(n),$

and $y(n)=\underline{f}_n^t\underline{x}_{n-\Delta},$

where $\underline{f}_n=[f_n(0),f_n(1),....,f_n(L-1)]^t,$
$\underline{x}_n=[x(n),x(n-1),...,x(n-L+1)]^t,$

where Δ is an integer delay. You may assume $\underline{f}_0=\underline{0}$ throughout. The inputs to this program should be:

i) Data file $x(n)$.
ii) Delay value Δ.
iii) Filter length L.
iv) Adaptation constant α.

The outputs should be:

i) Data file of error values.
ii) Data file of filter output values.

Test the program using:

i) $x(n)$ as 5000 points of Gaussian white noise with unit variance. $\Delta=0$, filter length $L=32$, and $\alpha=0.01$. Plot the error and confirm that this decays to zero as n increases.
ii) $x(n)=\cos(2\pi500\times0.0001)$, $\Delta=20$, $L=20$, and $\alpha=0.01$. The output should adapt to become a unit amplitude sinusoid, and the error should decay to zero.

a) Using the data in *i)* evaluate the effect of varying the delay. Does the system continue to operate satisfactorily?

b) Using the data in part b) evaluate the impact of delay. In what way do your results differ from those for the random input? Can you account for this?

c) Use the sinusoidal input to confirm the instability point predicted by the equivalent transfer function (4.2.70).

d) Evaluate the performance of your filter as the frequency of the input sinusoid changes. What conclusions can you draw?

References

1. K.L.Peacock and S.Treitel, "Predictive deconvolution – theory and practice," *Geophysics*, vol. 34, pp. 155-169, 1969.

2. L.R.Rabiner and R.W.Schafer, *Digital Processing of Speech Signals*, Prentice-Hall, 1978.

3. R.J.Wang, "The determination of optimum gate lengths for time-varying Wiener filtering," *Geophysics*, vol. 34, pp. 683-695, 1969.

4. R.J.Wang and S.Treitel, "The determination of digital Wiener filters by means of gradient methods," *Geophysics*, vol. 38, pp. 310-326, 1973.

5. H.W.Sorenson, *Parameter Estimation: Principles and Problems*, Dekker, 1980.

6. B.Widrow, "Adaptive filters I: fundamentals," Stanford Electronics Labs., Stanford, CA, SEL-66-126, 1966.

7. B.Widrow and M.E.Hoff, "Adaptive switching circuits," in *IRE Wescon Conv. Rec.*, pt 4, pp. 96-104, 1960.

8. B.Widrow, J.M.McCool, M.G.Larimore, and C.R.Johnson Jr., "Stationary and nonstationary learning characteristics of the LMS adaptive filter," *Proc. IEEE*, vol. 64, pp. 1151-1162, 1976.

9. R.A.Horn and C.R.Johnson, *Matrix Analysis*, Cambridge University Press, 1985.

10. S.Haykin, *Adaptive Filter Theory*, Prentice-Hall, 1985.

11. M.H.Lu and P.M.Clarkson, "The performance of adaptive noise cancellation systems in reverberant rooms," *J. Acoust. Soc. Amer.*, to appear.

12. B.Widrow and S.D.Stearns, *Adaptive Signal Processing*, Prentice-Hall, 1985.

13. J.R.Glover, "Adaptive noise cancelling applied to sinusoidal interferences," *IEEE Trans. Acoust., Speech, Signal Processing*, vol. ASSP-25, pp. 484-491, 1977.

14. P.M.Clarkson and P.R.White, "Simplified analysis of the LMS adaptive filter using a transfer function approximation," *IEEE Trans. Acoust., Speech, Signal Processing*, vol. ASSP-35, pp. 987-993, 1987.

15. A.V.Oppenheim and R.W.Schafer, *Discrete-Time Signal Processing*, Prentice-Hall, 1989.

16. B.N.Partlett, *The Symmetric Eigenvalue Problem*, Prentice-Hall, 1980.

17. R.R.Bitmead and B.D.O.Anderson, "Performance of adaptive estimation algorithms in dependent random environments," *IEEE Trans. Automat. Contr.*, vol. AC-25, pp. 782-787, 1980.

18. S.J.Elliott and P.Darlington, "Adaptive cancellation of periodic, synchronously sampled interferences," *IEEE Trans. Acoust., Speech, Signal Processing*, vol. ASSP-33, pp. 715-717, 1985.

19. P.M.Clarkson, "Analysis of the LMS adaptive filter applied to periodic signals," in *Mathematics in Signal Processing* (T.Durrani, Ed.), pp. 593-605, Oxford University Press, 1987.

20. G.Stephenson, *An Introduction to Matrices, Sets and Groups for Science Students*, Dover, 1986.

chapter five

Applications of Adaptive Filtering

5.1 Adaptive Noise Cancellation

5.1.1 Introduction

One of the most useful applications of adaptive filtering is the method of Adaptive Noise Cancellation (ANC). Originally proposed by Widrow *et al.* [1] in a classic paper, the method has been the subject of numerous studies involving a wide range of applications. ANC is concerned with the enhancement of noise corrupted signals. Its great strength is that in contrast to other enhancement techniques no *a priori* knowledge of signal or noise is required for the method to be applied. On the other hand, these advantages are not obtained without some cost. In particular, noise cancelling is a variation of optimal filtering which is designed to use a secondary or **reference** input. This reference input should contain little or no signal but should contain a noise measurement which is correlated (in some unknown way) with the noise component of the main or **primary** input. The ANC method is based on the principle of adaptively filtering this reference input to produce a replica of the noise which can then be subtracted from the primary measurement (see Figure 5.1.1). The relationship between this diagram and the adaptive filter introduced in the previous Chapter may not be immediately obvious. A moments thought, however, will confirm that they are equivalent with the desired input replaced by the primary, and the system input replaced by the reference. It should also be noted that in this application it is the *error signal* which is the system output.

The most important step in the ANC process is obtaining a reference signal which satisfies the stated requirements. This is a problem which obviously depends on the particular application, and we shall see several examples as we proceed. One obvious application, which has received considerable attention, is that of enhancing a voiced communication channel. In this problem, a

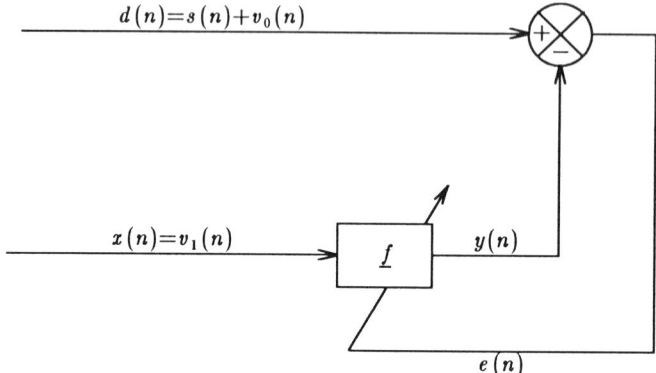

Figure 5.1.1 Adaptive noise cancelling system. $d(n)$ is the primary measurement containing signal $s(n)$ plus noise $v_0(n)$. The reference input $x(n)$ contains noise $v_1(n)$ which is correlated with $v_0(n)$ but not with $s(n)$.

speech signal $s(n)$ is measured in a noisy room or other acoustic enclosure. A typical setup for this problem is depicted in Figure 5.1.2.

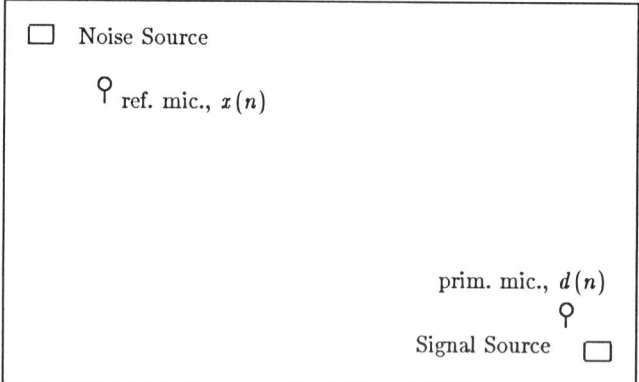

Figure 5.1.2 Symbolic representation of ANC in an acoustic enclosure. Noise propagates to primary microphone and corrupts signal measurements. ANC uses a second microphone close to the noise source to form a reference.

The primary measurement is derived from a source, typically a human speaker, located within the enclosure (see Figure 5.1.1). The speech signal is corrupted by noise $v_0(n)$ received from a noise source at some other location within the enclosure. A second

measurement is then obtained by placing another microphone close
to the noise source, thus ensuring that the reference contains
primarily noise. The enclosure in question might be, for example,
an aircraft cockpit where it is desired to improve the quality of the
speech transmitted to the control tower. Or the measurements
might be made in the interior of a car where the cancellation is
intended to facilitate the use of a 'hands-free' mobile telephone.

To obtain an intuitive feeling for why the ANC process should
work, we recall that the adaptive filter is based on approximately
minimizing the mean-squared error between the filter output and
the desired signal. Now, following Widrow *et al.* [1], we see from
Figure 5.1.1 that

$$e(n)=s(n)+v_0(n)-y(n) \ . \tag{5.1.1}$$

Squaring and taking expectations of both sides of (5.1.1) gives

$$E\{e^2(n)\}=E\{s^2(n)\}+E\{(v_0(n)-y(n))^2\}$$
$$+2E\{s(n)(v_0(n)-y(n))\} \ . \tag{5.1.2}$$

It is assumed throughout that $s(n)$ is uncorrelated with $v_0(n)$ and
$v_1(n)$. Hence, for a fixed filter, $s(n)$ is also uncorrelated with
$y(n)=f(n)*v_1(n)$, so that the final term in (5.1.2) is zero and we
have

$$E\{e^2(n)\}=E\{s^2(n)\}+E\{(v_0(n)-y(n))^2\} \ , \tag{5.1.3}$$

from which it is clear that $E\{e^2(n)\}$ is minimized when
$y(n)=v_0(n)$, and hence $e(n)=s(n)$, the desired result.

Another straightforward method of analysis, again following the
example of Widrow *et al.*, is to employ the infinite two-sided
Wiener approximation for the adaptive filter (see Section 4.2.2):

$$F(e^{j\omega})=\frac{R_{xd}(e^{j\omega})}{R_{xx}(e^{j\omega})} \ . \tag{5.1.4}$$

Given that in the noise cancelling system of Figure 5.1.1

$$d(n)=s(n)+v_0(n) \ ,$$
$$x(n)=v_1(n) \ , \tag{5.1.5}$$

then

$$F(e^{j\omega}) = \frac{R_{v_1 v_0}(e^{j\omega})}{R_{v_1 v_1}(e^{j\omega})} \ . \tag{5.1.6}$$

The noises v_1 and v_0 are, by assumption, correlated in some unknown manner. For our purposes it is helpful to think of this in terms of a transmission path relating v_0 and v_1 which can be modelled by a linear system, $h(n)$, say. That is

$$v_0(n) = h(n) * v_1(n) \ . \tag{5.1.7}$$

In the acoustic noise cancellation problem discussed above, for example, $h(n)$ would correspond to the impulse response function of the acoustic enclosure for the positions of the primary and reference microphones. Consequently, $h(n)$ generally would include a bulk delay corresponding to the transmission time for sound to travel between the two locations, together with reflections from the surfaces of the enclosure. Obviously, this impulse response depends on the particular enclosure, as well as the microphone locations. There is, however, no suggestion that we need know anything about $h(n)$ for the successful operation of the ANC system, it has been introduced here purely to facilitate our understanding of the process.

In Section 2.3, we obtained expressions relating the input and output spectra for a system $h(n)$ with stationary input (see equations (2.3.10) and (2.3.12)). Using equation (2.3.12), with input $v_1(n)$ and output $v_0(n)$, the cross-spectrum of the input with the output is

$$R_{v_1 v_0}(e^{j\omega}) = H(e^{j\omega}) R_{v_1 v_1}(e^{j\omega}) \ . \tag{5.1.8}$$

The corresponding relation for the input and output power spectra, obtained using equation (2.3.10) is

$$R_{v_0 v_0}(e^{j\omega}) = |H(e^{j\omega})|^2 R_{v_1 v_1}(e^{j\omega}) \ . \tag{5.1.9}$$

Substituting equation (5.1.8) into (5.1.6) gives

$$F(e^{j\omega}) = H(e^{j\omega}) \ . \tag{5.1.10}$$

The conclusion is that the adaptive filter models the unknown transmission path, $h(n)$.

In the time domain the filter output is

$$y(n) = f(n) * v_1(n) = h(n) * v_1(n) , \qquad (5.1.11)$$

which, from (5.1.7) is

$$y(n) = v_0(n) . \qquad (5.1.12)$$

Hence, finally

$$e(n) = s(n) + v_0(n) - v_0(n) ,$$

and perfect cancellation is the result.

In practice, the requirement for a perfect reference measurement is rarely met. Returning to Figure 5.1.2, it is obvious that if the noise can be transmitted from the reference to primary positions, then equally the signal can 'leak' in the opposite direction. Such leakage is clearly detrimental to performance since the system can now act to cancel the signal. The analysis used above to describe the ANC system can be extended to include this effect. For the sake of brevity we leave the development of these results to the Problems section. The result is of sufficient importance, however, that we will state it here: Assuming signal leakage can be modelled through a linear transmission path, $g(n)$, the ANC model of (5.1.5) becomes

$$d(n) = s(n) + v_0(n)$$
$$x(n) = g(n) * s(n) + v_1(n) . \qquad (5.1.13)$$

Now, we define the **signal-to-noise density ratio** of a measurement, $y(n)$

$$y(n) = s(n) + v(n) , \qquad (5.1.14)$$

by

$$\rho_y(e^{j\omega}) = \frac{R_{ss}(e^{j\omega})}{R_{vv}(e^{j\omega})} . \qquad (5.1.15)$$

That is, $\rho_y(e^{j\omega})$ is the ratio of the signal and noise power spectra and is thus effectively the Signal-to-Noise Ratio (SNR) as a function of frequency. In terms of this definition the performance of the ANC system with inputs (5.1.13) is derived in Problem 5.2 as

$$\rho_{out}\left(e^{j\omega}\right) = \frac{1}{\rho_{ref}\left(e^{j\omega}\right)} \; . \tag{5.1.16}$$

This equation, which is known as the **signal-to-noise inversion principle** says simply that the SNR of the ANC system output (that is the error) is the inverse of the SNR of the reference. The result is intuitively reasonable, the effect of signal leakage is to degrade the ANC process, the more leakage the better the reference SNR and the worse the result.

All of the foregoing is based on the idealized approximation of the adaptive filter as a fixed, infinite, two-sided Wiener filter. As such it completely neglects the issues of stability, convergence and steady-state error as discussed in Section 4.2, as well as the limitations imposed on the filter due to causality and finite length. In spite of these deficiencies, simulations often agree closely in steady-state with the behavior expected from the Wiener approximation. As such it is a simple but useful way to evaluate the potential for ANC in various applications.

5.1.2 Examples of ANC Systems

1. Own-Ship Noise Cancellation in a Towed-Array System

Consider the system depicted in Figure 5.1.3. A vessel tows an array of hydrophones a distance D behind the ship, with the objective of making broadband ambient noise measurements. The signals received at the hydrophones are summed and sampled in the manner of a conventional beamformer to give a single output $d(n)$, say.[1] The situation is complicated by the presence at the hydrophones of noise from the towing vessel which degrades the desired signal.

This problem can be made amenable to the techniques of adaptive noise cancellation by placing a separate hydrophone close to the towing vessel (or by steering the array directly at the ship). This second measurement constitutes a reference $x(n)$, which can

---

1. Beamforming is described in Section 8.2

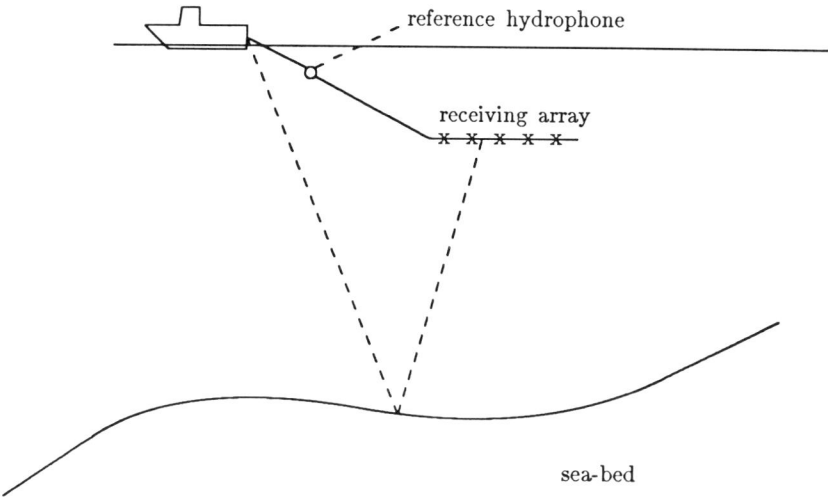

Figure 5.1.3 'Own-Ship' noise cancellation with a towed array. A ship tows an array of hydrophones. Received signals are corrupted by noise from the tow-ship. An ANC system constructed by using a reference derived from an extra hydrophone close to the tow-ship.

be used in conjunction with $d(n)$ to effect noise cancellation. The interference $i(n)$, say, measured at the reference position, will not be an exact replica of that at the hydrophones. There will in general be a delay, together with a transmission loss, due to the physical distance between them. Furthermore, the ship noise measured at the hydrophones will probably have arrived along multiple paths, via surface and bottom reflections. However, assuming all of these effects can be modelled by a linear system, then we can write

$$d(n) = s(n) + h(n) * i(n)$$
$$x(n) = i(n) , \tag{5.1.17}$$

where $h(n)$ is a system representing the effects of the transmission path on the interference. If $h(n)$ is linear, then the noise cancelling system can be employed.

Generally, the noise spectrum for this 'own-ship' problem contains both narrowband and broadband components. In many instances, however, the spectrum is dominated by a number of sinusoidal components with slowly varying amplitude and phase (see also Section 5.2). The effect of a linear system on a sinusoid can only be to change the amplitude and phase of the sinusoid, because the linear system cannot generate new frequencies. A model for the problem is thus given by

$$d(n)=s(n)+\sum_{i=0}^{M-1} B_i\sin(\omega_i nT+\theta_i) ,$$

$$x(n)=\sum_{i=0}^{M-1} A_i\sin(\omega_i nT+\phi_i) . \qquad (5.1.18)$$

Hence, the amplitudes and phases of the interference are different at the primary and reference locations but the frequencies are the same. In the simplest case, the interference might consist of a single tone $(M=1)$, so that

$$d(n)=s(n)+B_0\sin(\omega_0 nT+\theta) ,$$

$$x(n)=A_0\sin(\omega_0 nT+\phi) , \qquad (5.1.19)$$

We recall from Section 4.2.4, that the performance of the LMS adaptive filter subjected to a sinusoidal input has a relatively simple representation as a transfer function, given for small α by[2]

$$\frac{E(z)}{D(z)} \approx \frac{\left(1-e^{j\omega_0 T}z^{-1}\right)\left(1-e^{-j\omega_0 T}z^{-1}\right)}{\left(1-[1-\dfrac{\alpha LA_0^2}{4}]e^{j\omega_0 T}z^{-1}\right)\left(1-[1-\dfrac{\alpha LA_0^2}{4}]e^{-j\omega_0 T}z^{-1}\right)} ,$$

which is a notch filter, centered on the interference frequency $\omega_0 T$ (see Section 4.2.4, equations (4.2.70), (4.2.71)). The overall effect of applying the adaptive noise canceller to a system with a single corrupting sinusoid is to form a notch at the interference frequency whose bandwidth is controlled by α, L and A_0^2. Notching out the interference is precisely the solution we require. To take this a stage further, the notch width gradually increases as α, L and A_0^2 increase. Conceptually, assuming L and A_0^2 are fixed, when α is

---

2. As we indicated in Section 4.2, the approximation of the behavior of the transfer function as a simple notch as in equation (4.2.71) can only be sustained for small α. In other cases, the transfer function continues to act as a notch, but the pole positions become more intricate.

small the filter removes the interference without impacting the spectrum at adjacent frequencies. As α increases, as is necessary if the parameters of the sinusoid are likely to change significantly, the interference is still attenuated but the broader notch also removes part of the signal spectrum. We can conclude, therefore, that the adaptation rate should generally be as small as possible while retaining the ability to provide adequate tracking of drifting sinusoidal parameters.

2. ANC Without An External Reference Measurement

For obvious reasons there are many situations in which the application of ANC would potentially be useful but where it is impractical to obtain a separate reference measurement of the noise. In some situations, principally when the interference is periodic or at least 'pseudo-periodic' and the signal is broadband (as in the own-ship noise example above), it is possible to construct an ANC system without an external reference. In these examples the technique employed is to construct a reference measurement internally using a delayed version of the primary input (see Figure 5.1.4).

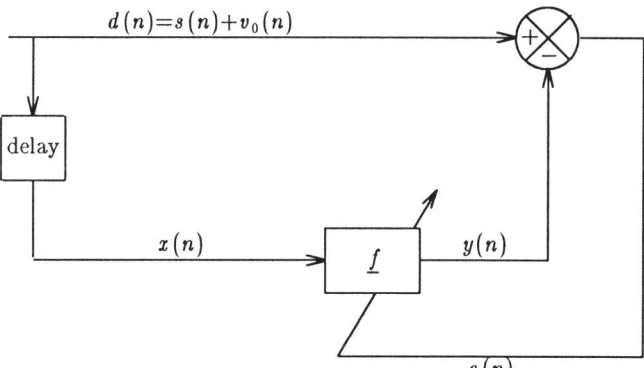

Figure 5.1.4 ANC with reference generated by delay. Pseudo-periodic interference $v_0(n)$ corrupts broadband signal components. Periodic interference components remain correlated in the reference after delay Δ, signal components decorrelate.

The approach is to use a delay Δ which is sufficient to decorrelate the signal component, that is

$$E\{s(n)s(n-\Delta)\}\approx 0 \ . \tag{5.1.17}$$

The delay will not decorrelate the periodic interference components. The extent to which this system is effective depends principally on how well the assumption of zero signal correlation between primary and reference measurements holds up. It should be noted, however, that the system performance will always be inferior to that of the ANC system with an ideal reference due to the action of the filter on the signal component of the input.

3. A Cautionary Note

A cautionary note concerning ANC is appropriate. The example considered is that of ANC for noise reduction in the cockpit of a fighter aircraft. In such circumstances noise levels in excess of 90 dB are typically observed across the whole speech spectrum. Noise cancelling microphones can be used to improve the situation. However, with increasing use of vocoders and automatic speech recognition, whose performance is badly degraded by noise, there is a pressing need for further noise reduction. Early simulations of ANC systems [2],[3] for acoustic noise reduction typically used a setup similar to that shown in Figure 5.1.2. A single loudspeaker is used to generate the noise, typically using a broadband noise generator. The reference microphone is placed close to the speaker. The speech input, which is corrupted by the noise field, is recorded by a primary microphone remote from the noise. Early investigators found that the major problem was that the filter length, L, needed to be large in order to effectively model the noise transmission path. That is, to model the impulse response of the room between reference and primary locations. This transmission path includes a bulk delay (the speed of sound in air is approximately 330 m/s. Consequently, at a separation of 2 m the time delay is 0.006 s which at a sample rate of, say, 10 kHz gives a 60 sample delay). Additionally, the filter must be sufficiently long to account for at least the bulk of any secondary reflections, that is for the reverberation of the room. These can potentially last for up to several seconds. Even for relatively small rooms, the response has a significant length. For example, Figure 5.1.5 shows a synthetic room response for a rectangular room. The room in

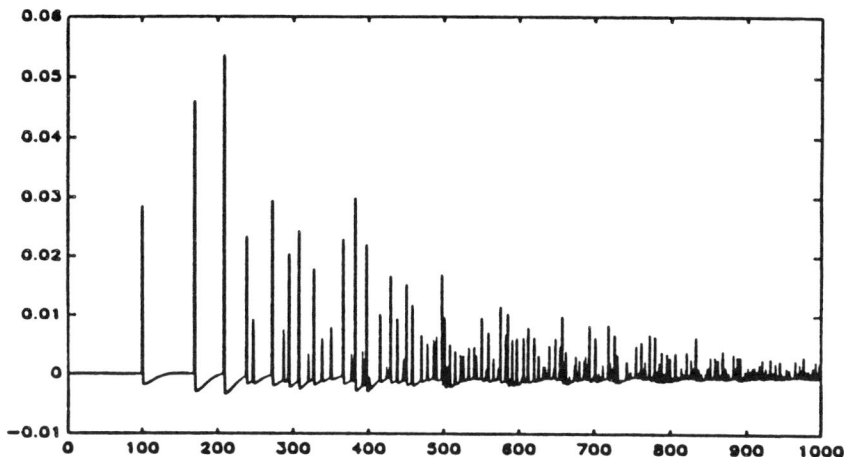

Figure 5.1.5 A typical room impulse response function. Response was generated using the image method [4], dimensions of the room were 10', 10', 10' and reflection coefficient of 0.8 was used for the surfaces of the enclosure.

question has dimensions 10' by 10' by 10', a reflection coefficient for each surface of 0.8 and the source and receiver were arbitrarily positioned.[3] The response easily exceeds 1000 points in length and it is clear that a filter of significant length would be required even to approximately model this transmission. However, given a sufficiently long filter (for example, $L=1500$ in the study by Boll and Pulispher [2]) and adequate adaptation time, noise attenuation of up to 20 dB and corresponding intelligibility increases have been observed.

Later investigators attempted to apply the principal to the fighter cockpit problem [5]. In this case physical separation of the primary and reference is replaced by an acoustic barrier – the pilot's oxygen mask and helmet. The primary microphone is placed inside the mask, close to the pilot's mouth. The reference is

---

3. The simulation for this room response was produced using a Fortran implementation of an image-based model which is available in reference [4]. The response has been high-pass filtered at 50 Hz, to remove low-frequency regions which the image model cannot accurately model. The program contains a number of simplifying assumptions about the room, but these are of secondary importance here.

placed at some point on the outside of the mask. The mask typically provides 15-30 dB of acoustic attenuation so that, in principal, the conditions of low SNR reference and high SNR primary are maintained. The initial work used omnidirectional microphones and reported improvements on the order of 10 dB. However, a subsequent investigation by Darlington *et al.* [6],[7], using more realistic simulations based on positioning several loudspeakers at various locations within the enclosure, produced far less impressive results. It was found that almost all improvements were limited to frequencies lower than 1 kHz, for which the use of directional microphones can create similar reduction. The same authors showed that the cause of the poor performance was simply that the noise transmission could not be modelled as a *single* linear transmission path. This was established by measuring the **coherence** between the reference and primary locations. The coherence was found to approximate that of a 'diffuse' sound field and consequently to only be significant at low frequencies. Darlington *et al.* argued that the transmission path is a composite of several independently excited paths. Subsequently Rodriguez and Lim [8] confirmed these results even for a single loudspeaker. They found that with gradient microphones, for which low frequency noise attenuation occurs anyway, virtually no improvement took place. This led Rodriguez and Lim to conclude that for the cockpit problem ANC may be essentially useless.

It would not be appropriate to end this section on such a downbeat note. This application is a salutary lesson, and a clear example of the defeat of signal processing by basic physical laws. It would not be appropriate, however, to claim that ANC itself is not helpful, there remain plenty of applications where ANC has proved its value. Even for the application described above, Darlington *et al.* point out that the more recent simulations may be unduly pessimistic, and research into the problem continues. The point generally is simply that great care must be taken when designing such a system to ensure the model takes account of the physical situation.

5.2 Adaptive Line Enhancement

5.2.1 Enhancing Narrowband Signals Embedded in Broadband Noise

Adaptive Line Enhancement (ALE) is a development of the adaptive noise cancellation method described in the previous section. In this application, the adaptive algorithm is directed towards the problem of enhancing one or more narrowband signals ('spectral lines') of unknown and possibly drifting amplitudes and frequencies which are embedded in broadband noise. One of the primary applications for ALE arises in passive sonar, in the detection of low-level 'target' signatures. The targets are ships and submarines, and the signatures are acoustic emissions from propulsion systems and auxiliary machinery, as well as hydrodynamic noise resulting from the passage of the vessel through the ocean [9]. The signature typically consists of a number of narrowband components (spectral lines) superimposed on a broadband spectrum. The overall level of the spectrum increases as the speed of the vessel increases and, in the case of submarines, as the depth decreases. The frequencies and intensity of the propulsion system generated lines may also vary with speed and depth, although those due to auxiliary systems are usually quite stable [9]. Apart from source fluctuations, the received signature typically fluctuates as propagation conditions vary, and the line frequencies are influenced by varying Doppler shifts.

Measurement of the target signature is complicated by the presence of ambient as well as local hydrodynamic and instrumentation noise. The amplitude of the signature at the receiving platform is typically very low relative to these noise sources and thus the measured Signal-to-Noise Ratio (SNR) in such problems is very low, often less than 0 dB. A useful idealization of this problem, therefore, involves a model for the sampled received signal $s(n)$, as

$$s(n) = \sum_{i=0}^{M-1} A_i \cos(\omega_i n T + \theta_i) + w(n) \,, \qquad (5.2.1)$$

where A_i, ω_i, θ_i, are the amplitudes, frequencies and phase angles, respectively, associated with the sinusoidal components, and $w(n)$ is the noise measurement. As an analytical simplification it is usual to assume that $w(n)$ is white with variance σ_w^2, say.[4]

The adaptive line enhancement system is depicted in Figure 5.2.1. The system, which was introduced by Widrow *et al.* in [1], uses the measured signal as desired response and a delayed version of itself as input. The principle is that the delay should decorrelate the noise while leaving the narrowband components correlated. When functioning in an ideal way, the adaptive filter output is an enhanced version of the sinusoidal components. In line enhancement it is thus the filter output which is required, rather than the system error as was the case in the noise cancelling problem.

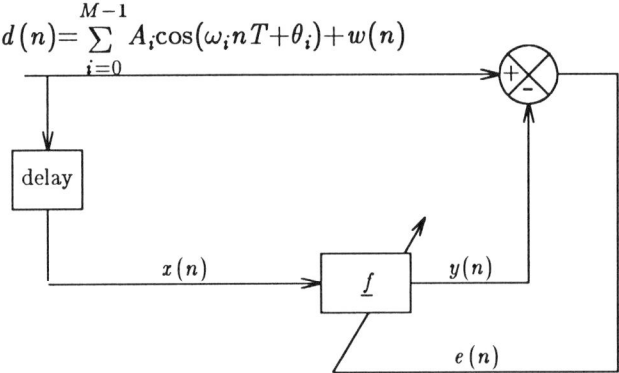

$$d(n)=\sum_{i=0}^{M-1} A_i\cos(\omega_i nT+\theta_i)+w(n)$$

Figure 5.2.1 Adaptive Line Enhancement (ALE) system. The ALE structure is similar to ANC but the roles of signal and noise are reversed — the delay, Δ is intended to decorrelate broadband noise but will not decorrelate narrowband signals.

Before considering the adaptive line enhancer in detail, let us examine the fixed optimal solution to this problem. For the sake of simplicity we restrict attention initially to the case of a single sinusoid at frequency ω_0. With this condition equation (5.2.1) becomes

---

4. As a practical matter the less restrictive condition

$$E\{w(n)w(n-\Delta)\}\approx 0 \,,$$

where Δ is a finite delay, is enough to allow the system to function.

$$s(n) = A_0\cos(\omega_0 nT + \theta_0) + w(n) , \qquad (5.2.2)$$

where, as above we assume that $w(n)$ is white with variance σ_w^2. As usual, we restrict attention to linear processing, and hence we generate an output $y(n)$ which is obtained by applying a linear filter to the measured signal $s(n)$

$$y(n) = f(n) * s(n) = y_s(n) + y_w(n) , \qquad (5.2.3)$$

where $y_s(n)$, $y_w(n)$ are the signal and noise components of the output, respectively. If the aim of processing is purely to detect the presence of signal, we may maximize the output SNR at each sample. We may define this output SNR as

$$SNR_{out} = \frac{output\ signal\ power}{output\ noise\ power}$$

$$= \frac{E\{y_s^2(n)\}}{E\{y_w^2(n)\}} .$$

The optimal linear solution for this problem is provided by the so-called **matched filter** [10]. For a single sinusoid at frequency ω_0 embedded in white noise, the matched filter can be shown to have the form (see Appendix 5A)

$$f(i) = C\cos(\omega_0 iT + \gamma) , \qquad (5.2.4)$$

where C is a constant,[5] and γ is a phase shift which, when added to the phase of the input sinusoid, produces a coherent signal output.[6] That is γ is chosen to ensure peak signal output at each instant n. Thus, the matched filter produces the peak SNR at each sample, but does *not* preserve the signal waveform in the output. The matched filter solution provides the best SNR gain obtainable by linear processing. On the other hand, we see that this solution can only be constructed given prior knowledge of the sinusoidal frequency ω_0. Moreover, the matched filter does not

---

5. Note that the value of C is arbitrary in the sense that its magnitude has no impact on the output SNR.

6. γ is dependent on n (see Appendix 5A) and therefore so is f. We suppress this dependence here to avoid confusion with the adaptive filter which follows.

preserve the signal waveform. If, in addition to detection, we are interested in **classification**, then preservation of the waveform shape is important. The ALE method, introduced in 1975, is a technique designed to approximate the SNR gain obtained by the matched filter solution for this problem, while preserving the waveform. A further advantage of ALE is that no prior knowledge of sinusoidal frequencies, amplitudes or phases, or even of the number of narrowband components present is required. As usual, the adaptive system attempts to iterate towards the least-squares optimum, and to track variations as they occur. We therefore begin our analysis of the ALE with the fixed least-squares solution.

5.2.2 The Least-Squares Solution for the Line Enhancer

Consider the ALE system depicted in Figure 5.2.1. The input signals may be written

$$d(n)=s(n), \quad x(n)=s(n-\Delta) , \tag{5.2.5}$$

where $s(n)$ is defined by equation (5.2.1) and Δ is a delay. We consider the fixed least-squares solution to this problem. In particular, we will demonstrate that the fixed least-squares finite causal filter for the single sinusoid model of equation (5.2.2) is given by:

$$f^*(i)=B\cos(\omega_0 iT+\phi) \quad ; i=0,1,...,L-1 , \tag{5.2.6}$$

where $\phi=\Delta\omega_0 T$ is a phase angle which ensures coherent cancellation and where

$$B=\frac{(2/L)\sigma_s^2}{(2/L)\sigma_w^2+\sigma_s^2} . \tag{5.2.7}$$

We can verify this result, and the values of B and ϕ as follows: Recall the normal equations of Section 3.2, equations (3.2.7a)

$$\sum_{i=0}^{L-1} r(j-i)f^*(i)=g(j) \quad ; j=0,1,...,L-1 , \tag{3.2.7a}$$

where for the signal definitions of equation (5.2.5) we have

$$g(j)=E\{d(n)x(n-j)\}$$
$$=E\{s(n)s(n-j-\Delta)\} \ ,$$

or

$$g(j)=r(j+\Delta) \ . \tag{5.2.8}$$

Hence, equations (3.2.7a) become

$$\sum_{i=0}^{L-1} r(j-i)f^{*}(i)=r(j+\Delta) \ , \tag{5.2.9}$$

and

$$r(m)=\sigma_{w}^{2}\delta(m)+\sigma_{s}^{2}\cos(\omega_{0}mT) \ , \tag{5.2.10}$$

where $\sigma_{s}^{2}=A_{0}^{2}/2$ is the power of the sinusoid, and σ_{w}^{2} is the power of the noise.[7] Now, substituting $r(m)$ and the assumed form for $f^{*}(n)$ as given by (5.2.6) into the normal equations (5.2.9) gives

$$\sum_{i=0}^{L-1}\left[\sigma_{w}^{2}\delta(j-i)+\sigma_{s}^{2}\cos(\omega_{0}(j-i)T)\right]B\cos(\omega_{0}iT+\phi)=$$

$$=\sigma_{w}^{2}\delta(j+\Delta)+\sigma_{s}^{2}\cos(\omega_{0}(j+\Delta)T) \quad ; j=0,1,...,L-1 \ .$$

Recognizing that with $\Delta>0$, $\delta(j+\Delta)=0$ for $j\geq0$, we have

$$\sigma_{w}^{2}B\cos(\omega_{0}jT+\phi)+\sigma_{s}^{2}B\sum_{i=0}^{L-1}\cos(\omega_{0}(j-i)T)\cos(\omega_{0}iT+\phi)=$$

$$=\sigma_{s}^{2}\cos(\omega_{0}(j+\Delta)T) \ . \tag{5.2.11}$$

Now, expanding the second term of the left-hand side of this equation into sums and differences, and assuming that

$$\sum_{i=0}^{L-1}\cos(2\omega_{0}iT+\phi)\to0 \ ,$$

that is, assuming the filter length 'spans' an integer number of periods at frequency $2\omega_{0}T$ (see below), we have

---

7. The form of the correlation of $A_{0}\cos(\omega_{0}nT+\theta_{0})$, can be obtained from the signal model by the assumption that θ_{0} is an initially random phase (see also Problem 2.3).

$$\left[\sigma_w^2 B + \frac{\sigma_s^2 BL}{2}\right]\cos(\omega_0 jT + \phi) = \sigma_s^2 \cos(\omega_0(j+\Delta)T) \quad ; \, j=0,1,...,L-1 \ .$$

A solution to this set of equations is provided by

$$\Delta \omega_0 T = \phi \ , \qquad\qquad\qquad (5.2.12)$$

and

$$B = \frac{\sigma_s^2}{\sigma_w^2 + \dfrac{\sigma_s^2 L}{2}} = \frac{(2/L)\sigma_s^2}{(2/L)\sigma_w^2 + \sigma_s^2} \ ,$$

which is the result of (5.2.7). The solution of (5.2.6), with B given by equation (5.2.7) is unique because the matrix R with elements given by equation (5.2.10) is positive definite (see Problem 5.10). When $f^*(n)$ is convolved with $x(n) = \cos(\omega_0(n-\Delta)T)$, the phase angle ϕ acts to produce a coherent output as expected. From equation (5.2.7), the amplitude B, can also be written [11]

$$B = \frac{\dfrac{\sigma_s^2}{\sigma_v^2}}{(2/L) + \dfrac{\sigma_s^2}{\sigma_v^2}}\left[\frac{2}{L}\right] \ .$$

The input SNR, defined as the ratio of the power of the sinusoid to that of the noise, is given by

$$SNR_{in} = \frac{A_0^2}{2\sigma_w^2} = \frac{\sigma_s^2}{\sigma_w^2} \ . \qquad\qquad (5.2.13)$$

In terms of the input SNR we have

$$B = \frac{SNR_{in}}{(2/L) + SNR_{in}}\left[\frac{2}{L}\right] \ . \qquad\qquad (5.2.14)$$

Comparing the result of (5.2.6) with the matched filter of equation (5.2.4), we see that the general form of the two filters is similar, though for the least-squares filter, the amplitude B is determined by equation (5.2.14). Also, crucially for waveform preservation, in the least-squares case the phase angle is not dependent on the sample index. Note that the size of the constant B does not affect the output SNR, but does produce imperfect signal cancellation in the error. The great advantage of the least-squares method, apart from preserving the waveform, is the lack of prior knowledge required. All that is needed for the least-squares solution are the auto and cross-correlation coefficients of (5.2.10) and (5.2.8), which can themselves be estimated from the data.

Let us examine the gain in SNR which may be obtained with the least-squares filter. In particular, we will concentrate on the simpler case of $\theta_0=0$, since the phase has no impact on the SNR of the solution. Accordingly, equation (5.2.6) becomes

$$f(i)=B\cos(\omega_0 iT) . \tag{5.2.15}$$

The output can be computed via the convolution of the input with the least-squares filter f given by (5.2.15)

$$y(n)=f(n)*s(n) ,$$

$$=B\cos(\omega_0 nT)*\left[A_0\cos(\omega_0 nT)+w(n)\right] ,$$

$$=\frac{BA_0L}{2}\cos(\omega_0 nT)+B\sum_{i=0}^{L-1} w(n-i)\cos(\omega_0 iT) . \tag{5.2.16}$$

So,

$$y(n)=y_s(n)+y_w(n) , \tag{5.2.3}$$

where

$$y_s(n)=\frac{BA_0L}{2}\cos(\omega_0 nT) ,$$

$$y_w(n)=B\sum_{i=0}^{L-1} w(n-i)\cos(\omega_0 iT) .$$

Hence, the signal component of the output remains proportional to $\cos(\omega_0 nT)$.[8]

---

8. The signal component being proportional to $\cos(\omega_0 nT)$ in (5.2.3), and the final equality in (5.2.17), depend on the approximation $\sum_{i=0}^{L-1}\cos(\omega_0 2iT)\approx 0$. As we

The output noise power is obtained by squaring the second term of equation (5.2.3) and taking the expectation (the cross-terms arising in the squaring operation are zero in expectation). This gives

$$output\ noise\ power = E\{y_w^2(n)\} = \left[B^2 \sum_{i=0}^{L-1} \cos^2(\omega_0 i T) \right] \sigma_w^2 ,$$

$$= B^2 \sigma_w^2 \frac{L}{2} . \qquad (5.2.17)$$

Hence the output SNR is

$$SNR_{out} = \frac{E\{y_s^2(n)\}}{E\{y_w^2(n)\}} = \frac{A_0^2 L}{2\sigma_w^2} = \frac{L\sigma_s^2}{\sigma_w^2} . \qquad (5.2.18)$$

Comparing (5.2.13) and (5.2.18) we see that

$$Output\ SNR = L \times Input\ SNR . \qquad (5.2.19)$$

Note that, the least-squares result can be generalized to the case of a set of M sinusoids in white noise, with solution for $f^*(n)$ given by

$$f^*(n) = \sum_{i=0}^{M-1} B_i \cos(\omega_i n T + \phi_i) . \qquad (5.2.20)$$

For L sufficiently large the B_i's can be approximated by

$$B_i = \left(\frac{\sigma_i^2}{\sigma_w^2} \right) \frac{(2/L)}{(2/L) + \left(\frac{\sigma_i^2}{\sigma_w^2} \right)} , \qquad (5.2.21)$$

where $\sigma_i^2 = A_i^2/2$, and where the ϕ_i's act to cancel the delay at each frequency, and thus ensure coherent cancellation in the error (see Problem 5.9).

---

have observed this holds with strict equality if L spans a multiple of π radians at frequency $\omega_0 T$. In other cases, we anticipate that the approximation becomes increasingly accurate as L increases.

5.2.3 The Performance of the Adaptive Line Enhancer

The least-squares solution requires less prior knowledge compared to the matched filter. We remove even the need for auto and cross-correlation estimates by replacing the fixed least-squares filter with the ALE system. Of course, the performance of this system is bound to be degraded by the adaptive nature of the filter. We will divide our consideration of the behavior of the system into convergence and steady-state aspects, and will then illustrate ALE performance with some computer simulations.

Convergence of the Filter and Nonuniform Convergence Effects

The performance of the ALE system can be evaluated using the independence theory described in Section 4.2.[9] Note that the use of this theory can certainly not be justified in this case because of the highly correlated nature of the sinusoidal inputs. Even so, predictions derived from such analysis have been reported to agree closely with simulations [12]. We may treat the analysis as a further example of the statistical approach of Section 4.2.2. We begin with equation (4.2.7)

$$E\{\underline{f}_{n+1}\}=(I-\alpha R)E\{\underline{f}_n\}+\alpha \underline{g} , \qquad (4.2.7)$$

which, it will be recalled, describes the mean behavior of the filter coefficients assuming independence of successive input vectors. We begin by considering the single sinusoid problem with inputs in the form of equation (5.2.5):

$$d(n)=s(n), \qquad x(n)=s(n-\Delta) . \qquad (5.2.5)$$

With $s(n)$ in the form of (5.2.2), we have[10]

$$r(m)=\sigma_w^2\delta(m)+\frac{A_0^2}{2}\cos(\omega_0 m T) . \qquad (5.2.10)$$

---

9. References [11]-[14] contain further detail on the properties of the LMS-ALE system described here.

10. As usual, we view the signal $x(n)=A_0\cos(\omega_0 n T+\theta_0)$ has having a uniformly distributed starting phase in order to obtain the correlation function (5.2.10), (see also Problem 2.3).

Then the correlation matrix R has the form

$$R=(\sigma_w^2 I + R_c) , \tag{5.2.22}$$

where R_c is Toeplitz with elements $r_c(i)=(A_0^2/2)\cos(\omega_0 iT)$ and I is the identity matrix. Following the analysis of Section 4.2 we define the mean error vector:

$$\underline{u}_n = E\{\underline{f}_n\} - \underline{f}^* . \tag{4.2.9}$$

We can decouple the equations (4.2.7) by employing the decomposition $R=Q\Lambda Q^t$, where Q is the orthonormal modal matrix. Writing

$$\underline{u}_n' = Q^t \underline{u}_n , \tag{4.2.15}$$

and then applying (4.2.9) and (4.2.15) to (4.2.7) we have

$$\underline{u}_{n+1}' = (I - \alpha\Lambda)\underline{u}_n' . \tag{4.2.16}$$

The decoupled error equation for the i-th coefficient is therefore

$$u_{n+1}'(i)=(1-\alpha\lambda_i)u_n'(i) \qquad ; i=0,1,...,L-1 , \tag{4.2.17}$$

where the development has exactly paralleled that of Section 4.2, but where λ_i is the i-th eigenvalue of $R=(\sigma_w^2 I + R_c)$. The eigenvalues of R are related to those of R_c by:

$$\lambda_i = \sigma_w^2 + \lambda_i^{(c)} \qquad ; i=0,1,...,L-1 , \tag{5.2.23}$$

where $\lambda_i^{(c)}$ is the i-th eigenvalue of R_c. We know from Section 4.2 that the iteration (4.2.17) is stable and that $E\{\underline{f}_n\}$ converges to $\underline{f}^*$ provided

$$0<\alpha<\frac{2}{\lambda_{\max}} ,$$

where $\lambda_{\max}$ is the largest eigenvalue of R. Thus, for eigenvalues in the form (5.2.23) we have

$$0<\alpha<\frac{2}{\sigma_w^2 + \lambda_{\max}^{(c)}} , \tag{5.2.24}$$

where $\lambda_{\max}^{(c)}$ is the largest eigenvalue of R_c. We also know that convergence is characterized by time-constants:

$$t_i = \frac{1}{\alpha \lambda_i} \; . \tag{4.2.27}$$

Further progress depends on explicit knowledge of the eigenvalues of R_c. We may write the correlation matrix R_c as

$$R_c = \frac{A_0^2 P}{2} \; , \tag{5.2.25}$$

where P is an $(L \times L)$ matrix with coefficients $p(m) = \cos(\omega_0 m T)$. It can be shown [12] that P has eigenvalues given by

$$\lambda_1 = \frac{1}{2} \left[L + \frac{\sin(\omega_0 L T)}{\sin(\omega_0 T)} \right] \; ,$$

$$\lambda_2 = \frac{1}{2} \left[L - \frac{\sin(\omega_0 L T)}{\sin(\omega_0 T)} \right] \; , \tag{5.2.26}$$

and $\lambda_3, \lambda_4, \ldots, \lambda_L = 0$.[11] Note that if $\omega_0 = \pi i / L T$ for any integer i,

then $$\frac{\sin(\omega_0 L T)}{\sin(\omega_0 T)} = 0,$$

and from (5.2.26), $\lambda_1 = \lambda_2 = L/2$. In fact this is a reasonable approximation for the eigenvalues for most frequencies ω_0 [12], and using this result the eigenvalues of R_c are given by

$$\lambda_1^{(c)} = \lambda_2^{(c)} = \frac{A_0^2 L}{4} \; , \quad \lambda_3^{(c)} = \lambda_4^{(c)} = \ldots = \lambda_L^{(c)} = 0. \tag{5.2.27}$$

Let us now return to the adaptive system which is driven by input $x(n)$ with correlation matrix R and eigenvalues $\sigma_w^2 + \lambda_i^{(c)}$. From (5.2.24), stability in the mean requires that

---

11. For $L > 2$, the matrix R_c does not have full rank. By contrast the matrix R has eigenvalues $\sigma_w^2 + \lambda_i^{(c)}$, which are strictly positive and hence R does have full rank.

$$0 < \alpha < \cfrac{2}{\sigma_w^2 + \cfrac{A_0^2 L}{4}} \ .$$

We see that the uncoupled modes now adapt with time-constants t_i

$$t_i = \frac{1}{\alpha(\sigma_w^2 + \lambda_i^{(c)})} \ . \tag{5.2.28}$$

For $\lambda_1^{(c)}, \lambda_2^{(c)}$ of (5.2.27) this gives,

$$t_{1,2} = \cfrac{1}{\alpha \left(\sigma_w^2 + \cfrac{A_0^2 L}{4} \right)} \ , \tag{5.2.29}$$

and for the remaining modes

$$t_i = \frac{1}{\alpha \sigma_w^2} \quad ; \ i = 3, 4, ..., L \ . \tag{5.2.30}$$

The convergence is certainly non-uniform. Another interesting point is that the growth (learning) and decay (forgetting) time-constants generally differ [12]. In particular, if a sinusoid present in signal $s(n)$ is suddenly removed, the presence of this component in the filter response decays at a rate determined by

$$t = \frac{1}{\alpha \sigma_w^2} \ , \tag{5.2.31}$$

since this is now the input power driving this mode of the filter. This time-constant is certainly higher than the time-constant (5.2.29), and thus the decay of this component is *slower* than the growth.

The generalization of these results in the case of M sinusoids with independent starting phases is straightforward. The eigenvalues of the correlation matrix become [12]:

$$\lambda_{2i}^{(c)} = \lambda_{2i+1}^{(c)} = \frac{A_i^2 L}{4} \quad ; \ i = 0, 1, ..., M-1 \ , \tag{5.2.32}$$

and $\qquad \lambda_i^{(c)}=0 \qquad ; i=2M, 2M+1, ..., L,$

where A_i is the amplitude of sinusoid i, and where we assume $M<L/2$. For input $x(n)$, with correlation matrix $R=\sigma_w^2 I + R_c$ the result is

$$t_{2i}=t_{2i+1}=\cfrac{1}{\alpha \left(\sigma_w^2 + \cfrac{A_i^2 L}{4} \right)} \qquad ; i=0,1,...,M-1 , \qquad (5.2.33)$$

$$t_i = \frac{1}{\alpha \sigma_w^2} \qquad ; i=2M, 2M+1,...,L . \qquad (5.2.34)$$

Again, convergence is non-uniform. Examination of (5.2.33) shows that the modes associated with the sinusoids will converge at different rates with the *time-constant reducing as the power of the component increases*. Thus we observe that the speed of convergence of a particular mode increases as the spectral power of the associated component increases. The same phenomenon can be seen in the noise-free case.

As we saw in Section 4.2, a better limit for overall stability of the adaptive algorithm is given by

$$0 < \alpha < \frac{2}{tr\{R\}} , \qquad (4.2.23)$$

or, with eigenvalues of the form (5.2.32)

$$0 < \alpha < \cfrac{2}{L \left(\sigma_w^2 + \cfrac{1}{4} \sum_{i=0}^{M-1} A_i^2 \right)} .$$

Note, however, in line enhancement where the SNR is often very low, it is necessary to use much lower values for α than suggested by these limits in order to control misadjustment effects.

Misadjustment

The general picture of the performance of the ALE system is disturbed by the fact that in practice the filter coefficients are perturbed by the effects of misadjustment noise. From Section

4.2.3, for small α the steady-state mean-squared error J_∞ may be approximated by

$$J_\infty \approx J_{\min}\left(1 + \frac{\alpha L}{2} \text{ power of the input}\right) . \qquad (4.2.52)$$

In general, as we have stated, input SNR is usually very low in ALE problems, often less than 0dB. Consequently, the minimum mean-squared error $J_{\min}$ is typically a large proportion of the total input power. Hence, misadjustment is generally a significant factor in ALE problems. The misadjustment can only be reduced by reducing one or more of the factors in equation (4.2.52). For a fixed filter length L, $J_{\min}$ is fixed. Similarly, L controls the resolution of the filter, and is therefore constrained by the operational requirements of the system. In this application, these requirements are determined by the desired resolution of narrowband components. For a fixed input power, therefore, we can only reduce the misadjustment by reducing α. Of course, reducing α also reduces the convergence rate and thus diminishes the tracking capacity of the ALE system.

Simulations

The principles outlined above will be illustrated using some simple simulations. The data employed in these trials consisted of fixed and linearly swept sinusoids embedded in Gaussian white noise. The sample rate was set arbitrarily at $500Hz$ and frequencies quoted below should be referenced to this figure. In each of the simulations the figures are **time-frequency** plots (also known as **A-scans**). The frequency displays are calculated from raw, non-overlapping, 512 point Discrete Fourier Transforms. These transforms are taken sequentially from the data at intervals of 1024 points, and the numbers on the left in the plots refer to the ending sample index (in time). The top figure in each display is the decomposition of the input, and the bottom is the corresponding display for the output.

1. Single sinusoid plus white noise (overall SNR = - 23.1dB)

Figure 5.2.2 shows the result for a single sinusoid with frequency $50Hz$, with an overall SNR of -23.1dB. A filter length of 200 and an adaptation step-size of $\alpha = 1 \times 10^{-12}$ were employed. The plot demonstrates that enhancement is attained for a sufficiently small value of the adaptation constant α, though the improvement may not become apparent for a significant number of iterations. Figure 5.2.3 shows a similar trial but with the filter length reduced to $L=40$ (and with a commensurate increase in the step-size). Again enhancement of the sinusoid is attained, but it it is clear that reducing the length of the filter reduces the resolution of the narrowband signal in the output. Note that the enhancement achieved is essentially independent of f_0.

2. Multiple sinusoids plus white noise (overall SNR= -15.4dB)

Figure 5.2.4 shows a trial of the ALE system when confronted with four separate sinusoids at frequencies 50, 60, 75 and 150Hz. The filter length in this trial was $L=200$ and the step-size was chosen as $\alpha=1\times10^{-9}$. The result shows that the ALE provides enhancement for each sinusoid. The only limitation in such a trial would be the ability or inability of the filter to resolve closely spaced sinusoids, which is determined by the filter length.

3. Swept sinusoid plus noise, (overall SNR = -23.1dB)

This simulation indicates one of the major limitations of the ALE method − tracking ability at these extremely low SNR's is very limited. In the simulation shown in Figure 5.2.5, even the very low sweep range (50 Hz per 20,000 samples) is sufficient to badly impair enhancement.

Referring back to Figure 5.2.1, we may observe one of the principle weaknesses of ALE − at low SNR's the error is almost totally noise. Hence, misadjustment is a major problem. As usual, this problem can be reduced by reducing either α or L. However, as we have already indicated we cannot reduce L because this parameter controls the filter resolution. (The filter can be thought of as infinite in length but viewed through an L point rectangular data window − the window length controls the resolution). α can be reduced, and this is reflected in the very low adaptation constants employed in the trials described above. Unfortunately, this results in very slow adaptation, (as we saw in the simulations). In practical problems such as sonar, it is sometimes necessary to use adaptation constants which only produce satisfactory enhancement with adaptation periods on the order of a minute or more.

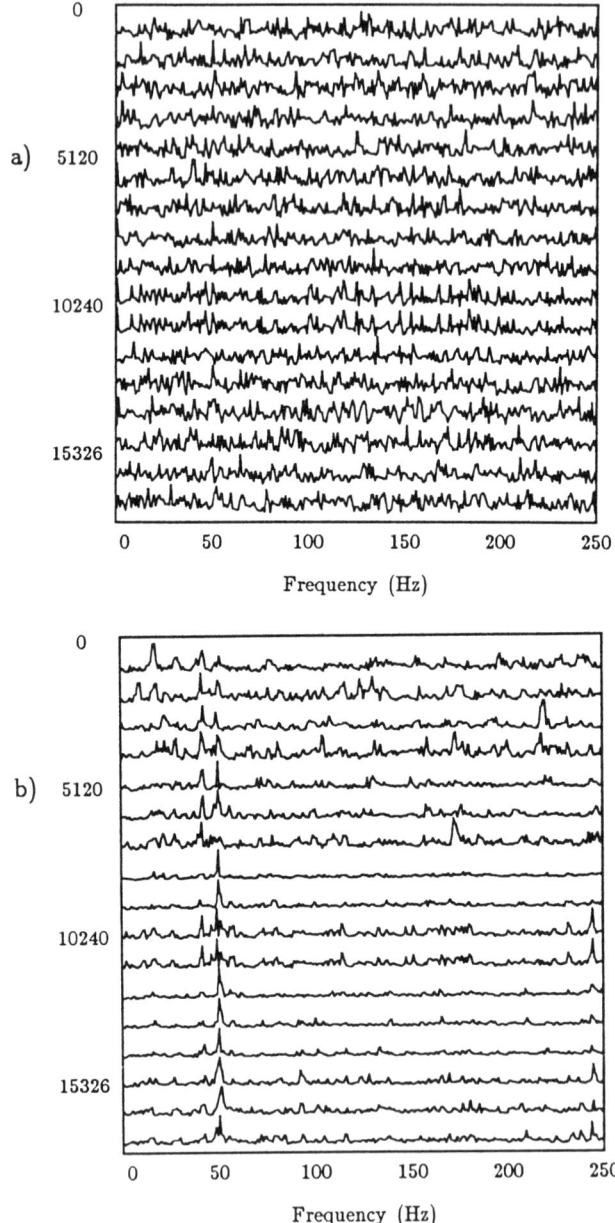

Figure 5.2.2 Adaptive line enhancement trial: single sinusoid plus white noise. Filter length L=200, f_0=50Hz, SNR=-23.1 dB, and α=1×10^{-12}. Each line represents a raw 512 point amplitude spectrum.

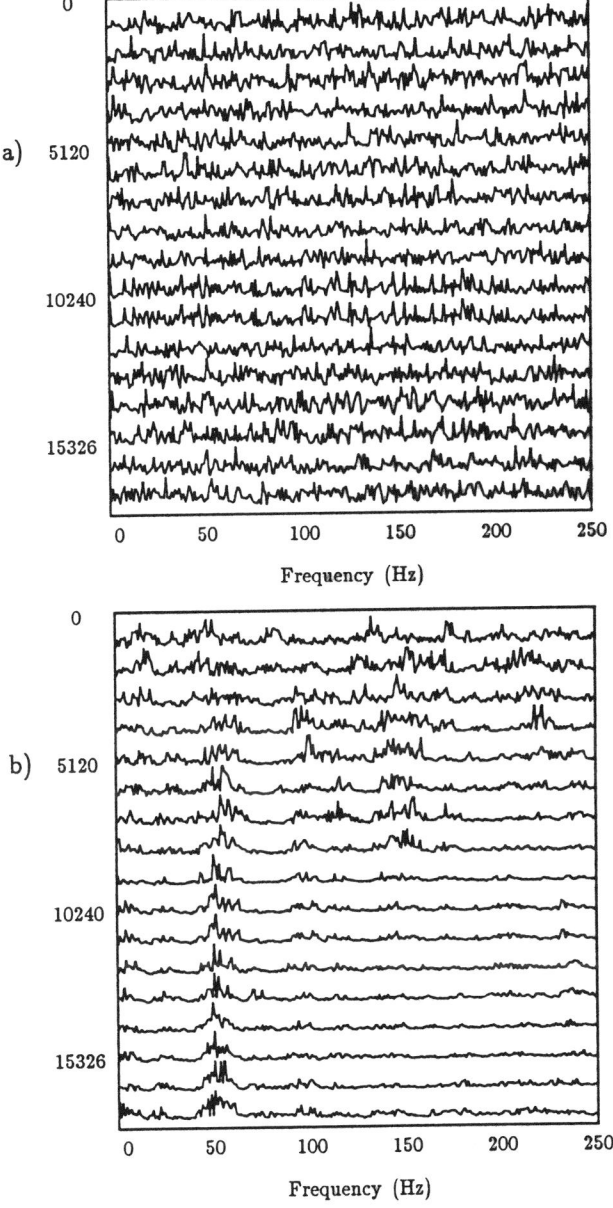

Figure 5.2.3 Adaptive line enhancement trial: single sinusoid plus white noise. Filter length $L=40$, $f_0=50Hz$, SNR=-23.1 dB, and $\alpha=2.5\times10^{-11}$. Each line represents a raw 512 point amplitude spectrum.

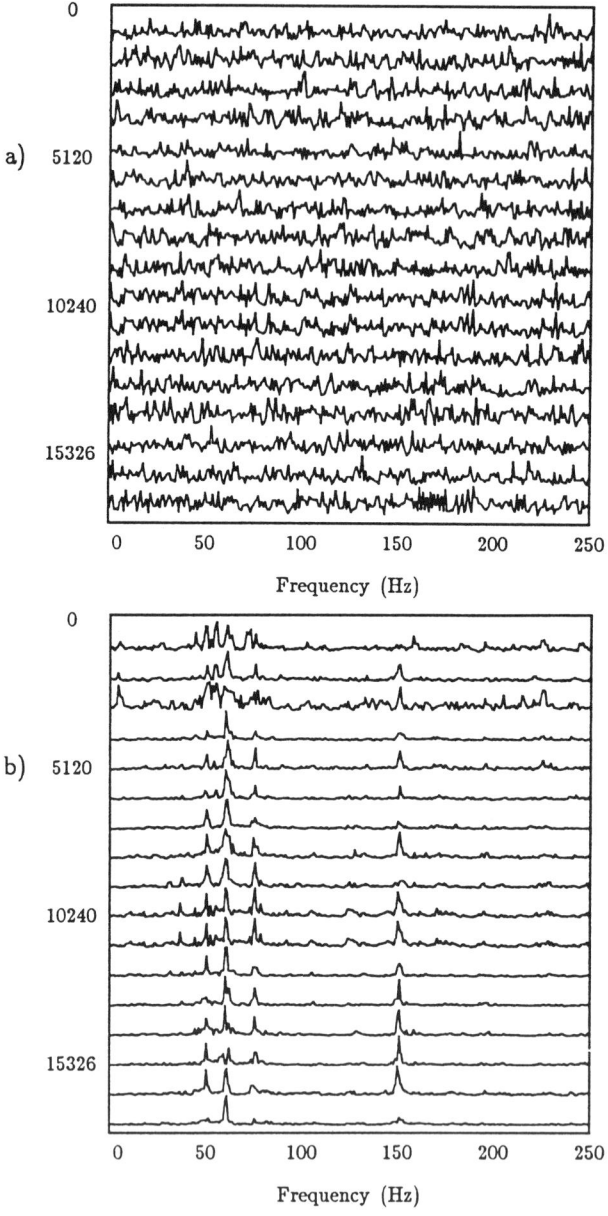

Figure 5.2.4 Adaptive line enhancement trial: multiple sinusoids plus white noise. Filter length $L=40$, sinusoid frequencies 50, 60, 75, and $150Hz$, SNR=-15.4 dB, and $\alpha=1\times10^{-9}$. Each line represents a raw 512 point amplitude spectrum.

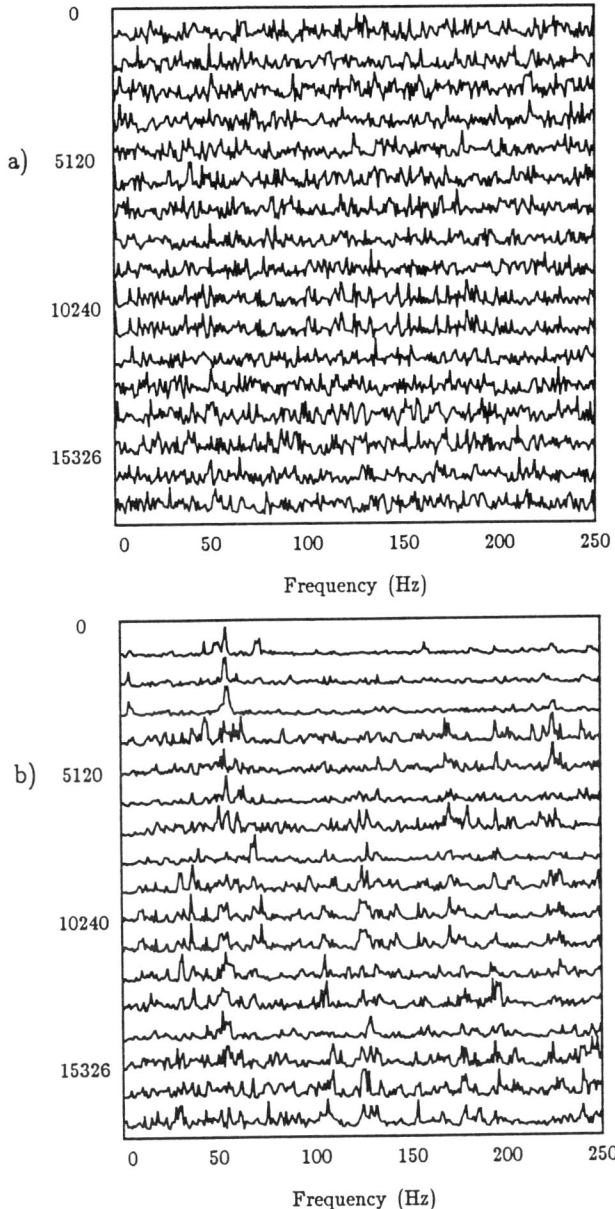

Figure 5.2.5 Adaptive line enhancement trial: swept sinusoid plus white noise. Filter length $L=200$, sweep from 50 to $100Hz$ over 20,000 samples. $\alpha=2.5\times10^{-11}$. Each line represents a raw 512 point amplitude spectrum.

5.3 Adaptive Filters for Time-Delay Estimation

5.3.1 Introduction to Time-Delay Estimation

In this section we consider the use of the LMS adaptive filter for the estimation of the time-delay between two measured signals. This is a problem which occurs in a diverse range of applications including radar, sonar, geophysics and biomedical signal analysis, among others. In the simplest arrangement, measurements $x_1(t)$, $x_2(t)$ are made at sensors P_1, P_2, say, separated by a distance d (see Figure 5.3.1).

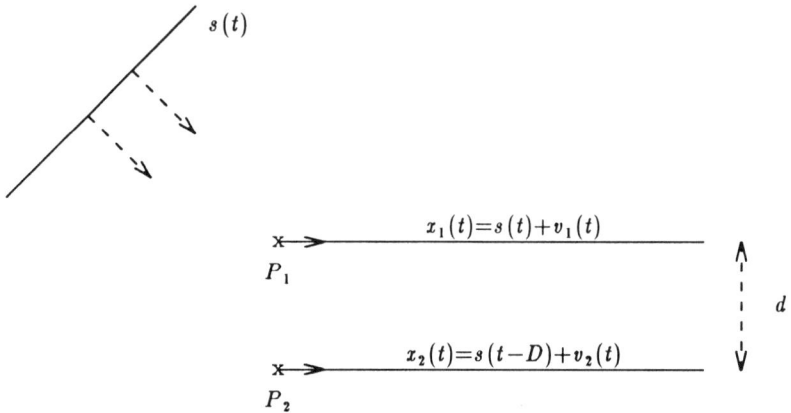

Figure 5.3.1 Plane wave signal $s(t)$ incident at sensors P_1, P_2 separated by (vertical) distance d. Measurement noises $v_1(t)$, $v_2(t)$ are assumed to be present.

The form of measurement sensor varies according to the application. In sonar, for example, the measurements are generated at the output of two hydrophones (or at the output of two steered half-beams of hydrophones with phase centers d apart). A simple model for the received signals is

$$x_1(t)=s(t)+v_1(t) \ ,$$

and

$$x_2(t)=s(t-D)+v_2(t) \ , \tag{5.3.1}$$

where $s(t)$ is the signal,[12] which is derived from some distant source, and where we assume the signal can be represented as travelling through the medium at constant speed along **plane wavefronts** as indicated in Figure 5.3.1. (We consider plane propagating waves more carefully in Chapter 8.) Here, $v_1(t)$ and $v_2(t)$ are additive noise terms measured at the receiver, and D is the delay. For convenience, $v_1(t)$ and $v_2(t)$ are assumed to be zero-mean, stationary, and mutually uncorrelated. As in the noise cancellation problem, we also assume that these measurement noises are uncorrelated with the signal $s(t)$. Note that the delay is constant as a consequence of the assumptions about the propagation of the signal.

The delay D, can be related to the angle of arrival of the signal through simple geometry. Referring to Figure 5.3.1, we have

$$D = \left(\frac{d}{c}\right)\sin(\theta) , \qquad (5.3.2)$$

where θ is the arrival or 'bearing' angle to the source, and c is the propagation velocity of the signal through the medium. Thus, the estimation of bearing angle is reduced to the estimation of delay D given the noisy measurements $x_1(t)$ and $x_2(t)$.

The delay estimation problem has been treated extensively in the literature (see, for example, [15]-[18]). One approach to the problem consists of simply computing the cross-correlation between $x_1(t)$ and $x_2(t)$ and taking the peak value as the estimate of the time-delay. The correlation for the stationary continuous signals $x_1(t)$, $x_2(t)$, may be defined in a similar manner to that of stationary discrete signals.[13] By analogy with Section 2.2, equation (2.2.8), we have

$$r_{x_1 x_2}(\tau) = E\{x_1(t)x_2(t+\tau)\} , \qquad (5.3.3)$$

---

12. Note that here we show the dependence on t to indicate that, at this stage, the signals are continuous rather than sampled. This avoids problems arising from values of D which do not correspond to integer multiples of the sample interval.

13. The assumptions of stationary signal and noise, and constant delay are included to facilitate the discussion and analysis. In practice, one of the primary reasons to introduce an adaptive structure is to allow processing when such assumptions do not hold, or are only approximately correct. In fact, the only assumption which should be considered inviolate is that of mutually uncorrelated signal and noise. That said, we retain the assumptions so as to simplify the discussion of the filter.

where τ is the correlation lag. As in the discrete case, the correlation is often estimated using a time average. The peak value for the correlation lag $\tau=\hat{D}$, say, provides the required estimate of the delay (see Figure 5.3.2).

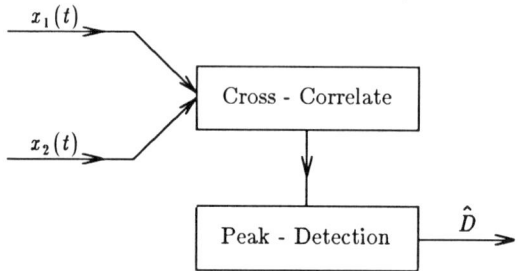

Figure 5.3.2 Schematic of delay estimation using cross-correlation. Delay estimate $\hat{D}$ is taken as the peak value of the short-time correlation between $x_1(t)$ and $x_2(t)$.

Referring to the model of equations (5.3.1), intuition tells us to expect a peak in the correlation at the lag corresponding to $\tau=D$, that is, when the signals are aligned. It is obvious, however, that the nature of this peak will be strongly dependent on the properties of $s(t)$, and the measurement noises $v_1(t)$ and $v_2(t)$.

While simple correlation may have an intuitive appeal, a more organized framework for time-delay estimation is provided by the **generalized correlation method** of Knapp and Carter [16]. The weighted or generalized cross-correlation is defined as

$$r_{y_1 y_2}(\tau) = w(\tau) \, {}^* r_{x_1 x_2}(\tau) \, , \tag{5.3.4}$$

where $w(\tau)$ is a (continuous time) weighting function. Of course, simple correlation may be obtained as a special case of (5.3.4) by choosing $w(\tau)=\delta(\tau)$. More generally, a number of estimators can be accommodated within this framework corresponding to different weighting functions $w(\tau)$. These include an 'optimal' Maximum Likelihood (ML) weighting $w_{ML}(\tau)$. As we might anticipate in the light of the discussion in Section 3.1, the variance associated with this estimate asymptotically achieves the Cramer-Rao Lower Bound (CRLB) as the estimation time increases.

The ML weighting function is expressed in the frequency domain in terms of the **magnitude squared coherence (msc)** function between $x_1(t)$ and $x_2(t)$. This function is defined by

$$msc = |\gamma_{12}(\omega)|^2 = \frac{|R_{z_1 z_2}(\omega)|^2}{R_{z_1 z_1}(\omega) R_{z_2 z_2}(\omega)} , \qquad (5.3.5)$$

where $R_{z_i z_i}(\omega)$ and $R_{z_i z_j}(\omega)$ are the power and cross spectra, respectively, for signals $x_i(t)$ and $x_j(t)$.[14] These spectra are defined as the Fourier transforms of $r_{ij}(\tau)$ by direct analogy with the definitions of discrete spectra given in Section 2.2. With these definitions for magnitude squared coherence and spectra, the maximum likelihood weighting can be shown to have the form [16]

$$W_{ML}(\omega) = \frac{|\gamma_{12}(\omega)|^2}{|R_{z_1 z_2}(\omega)|\,[1 - |\gamma_{12}(\omega)|^2]} , \qquad (5.3.6)$$

where $|\gamma_{12}(\omega)|^2 < 1$. This weighting may be expanded using the signal model of equations (5.3.1) as

$$W_{ML}(\omega) = \frac{\dfrac{R_{ss}(\omega)}{R_{v_1 v_1}(\omega) R_{v_2 v_2}(\omega)}}{1 + \dfrac{R_{ss}(\omega)}{R_{v_2 v_2}(\omega)} + \dfrac{R_{ss}(\omega)}{R_{v_1 v_1}(\omega)}} , \qquad (5.3.7)$$

where the derivation of this equation depends on the assumed properties for the correlations of $s(t)$, $v_1(t)$, $v_2(t)$ given above. This result may be simplified somewhat if we assume that the noises at each sensor have identical spectra $R_{vv}(\omega)$, say. Equation (5.3.7) then becomes

$$W_{ML}(\omega) = \frac{\dfrac{R_{ss}(\omega)}{R_{vv}^2(\omega)}}{1 + \dfrac{2 R_{ss}(\omega)}{R_{vv}(\omega)}} . \qquad (5.3.8)$$

The derivation of this and the previous equation is explored in Problem 5.15. Two points are apparent from (5.3.8): *i)* Even in this simplified case, to effect the optimal weighting we need prior knowledge of both signal and noise spectra. *ii)* The solution is

---

14. We differentiate between spectra of continuous and discrete signals by writing $R(\omega)$ in the former case and $R(e^{j\omega})$ in the latter.

fixed, having no flexibility in the face of time-varying delays caused by source and/or receiver motion.

Knapp and Carter showed that this and other generalized cross-correlation methods can be viewed as a prefiltering operation followed by cross-correlation in which individual inputs are filtered or weighted with transforms $H_1(\omega)$ and $H_2(\omega)$, prior to the correlation operation (see Figure 5.3.3). The weighting function in (5.3.4) is then equivalent to

$$W(\omega)=H_1(\omega)H_2^*(\omega) . \tag{5.3.9}$$

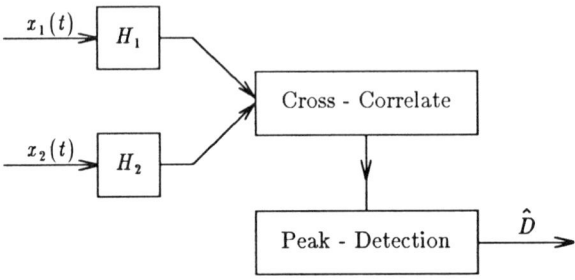

Figure 5.3.3 Schematic of generalized cross-correlation for delay estimation. Raw correlation scheme has been augmented by pre-filters $H_1(\omega)$, $H_2(\omega)$ applied to $x_1(t)$, $x_2(t)$, respectively.

5.3.2 The LMS Time-Delay Estimator (LMSTDE)

One approach to the time-delay estimation problem that is particularly attractive when limited *a priori* information about signal and noise spectra is available, or when the delay estimate is subject to variation, is to utilize an adaptive filter. The LMS Time-Delay Estimator (LMSTDE) [19]-[22] is a scheme which attempts to do just that. The setup for the adaptive filter is depicted in Figure 5.3.4. Referring to the figure, the two inputs $x_1(t)$ and $x_2(t)$ are sampled at intervals T, say, giving $x_1(n)$, $x_2(n)$. These sampled signals are then supplied as reference and primary, respectively, to the adaptive filter. The filter employed is usually the LMS algorithm, and for this application the LMS update formula becomes

$$\underline{f}_{n+1}=\underline{f}_n+\alpha e(n)\underline{x}_n^{(1)} , \tag{5.3.10}$$

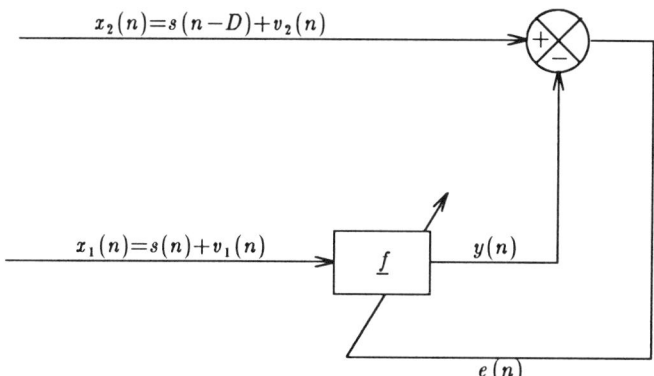

Figure 5.3.4 Adaptive filtering for time-delay estimation. $\underline{f}$ acts to cancel the delay between $x_1(n)$ and $x_2(n)$. The peak in $\underline{f}$ gives the delay estimate. Here D is taken as an integer multiple of the sample interval.

where $\underline{x}_n^{(1)} = [x_1(n), x_1(n-1), ..., x_1(n-L+1)]^t$. The desired input is formed from $x_2(n)$, and thus the error is

$$e(n) = x_2(n) - y(n) = x_2(n) - \underline{f}_n^t \underline{x}_n^{(1)} , \qquad (5.3.11)$$

and where, as usual $\underline{f}_n = [f_n(0), f_n(1), ..., f_n(L-1)]^t$ and α is the adaptation constant. In this application it is the filter impulse response function which provides the required information rather than the output or error. The delay estimate is taken as the peak value from the filter response. As we have indicated, the adaptive filter employed is the usual discrete form, operating on sampled signals $x_1(n)$, $x_2(n)$. This is in contrast to the continuous domain cross-correlation and generalized cross-correlation methods described above.[15] The major impact of this change is that if the time-delay does not correspond exactly to an integer multiple of the sample interval, then the peak in the filter response cannot correspond to the exact delay. To mitigate this problem an interpolation between adjacent samples in the filter response is used to improve the delay estimate. For simplicity in the discussion which follows, we avoid any problems arising from this

---

15. A continuous form of the LMS algorithm does exist, and is actually used fairly extensively in array processing (see, for example, Compton [23]). Though this continuity refers to the updating of the coefficients and so is not useful here.

interpolation by assuming that D *is* an integer multiple of T.

The LMSTDE system described by equations (5.3.10), (5.3.11), acts as an adaptive cross-correlator. From Figure 5.3.4 we have

$$e(n)=s(n-D)+v_2(n)-f(n)*[s(n)+v_1(n)] . \qquad (5.3.12)$$

Given $v_1(n)$, $v_2(n)$ mutually uncorrelated, it seems apparent that a good solution would be $f(n)=\delta(n-D)$ producing

$$e(n)=v_2(n)-v_1(n-D) , \qquad (5.3.13)$$

with

$$E\{e^2(n)\}=\sigma_2^2+\sigma_1^2 , \qquad (5.3.14)$$

where σ_1^2, σ_2^2 are the variances of the noise terms $v_1(n)$, $v_2(n)$, respectively. In fact, such a solution is by no means guaranteed to correspond to the least-squares optimum both because of the noise terms $v_1(n)$ and $v_2(n)$, and because of the spectral character of $s(n)$.

As we have seen, when the input signals are statistically stationary the steady-state behavior of the LMS filter may be approximated by the infinite two-sided Wiener filter (see Section 4.2). For the input signals $x_1(n)$ and $x_2(n)$ we have

$$F(e^{j\omega})=\frac{R_{x_1 x_2}(e^{j\omega})}{R_{x_1 x_1}(e^{j\omega})} . \qquad (5.3.15)$$

From (5.3.1), the sampled signals are

$$x_1(n)=s(n)+v_1(n) ,$$

$$x_2(n)=s(n-D)+v_2(n) . \qquad (5.3.16)$$

In the noise-free case $(v_1(n)=v_2(n)=0)$, equation (5.3.15) reduces to

$$F(e^{j\omega})=\frac{R_{ss}(e^{j\omega})e^{-j\omega D}}{R_{ss}(e^{j\omega})}=e^{-j\omega D} , \qquad (5.3.17)$$

and hence

$$f(n) = \delta(n - D) \ . \tag{5.3.18}$$

In other cases, the noise modifies the response as

$$F(e^{j\omega}) = \frac{R_{ss}(e^{j\omega})}{R_{ss}(e^{j\omega}) + R_{v_1 v_1}(e^{j\omega})} e^{-j\omega D} \ , \tag{5.3.19}$$

so that the energy in the filter is spread and the time-domain resolution is reduced. This fixed least-squares solution has itself been proposed for time-delay estimation. The method uses data based estimates of the power and cross spectra and is known as the Roth processor [16]. The LMS filter attempts to mimic this least-squares solution.

Example – We may illustrate the performance of the LMSTDE using some simple computer examples. We consider the system with input signals in the form (5.3.16), with measurement noises $v_1(n)$, $v_2(n)$ zero, and with four forms for the input signals:

i) $s(n) = w(n)$,
ii) $s(n) = h_1(n) * w(n)$,
iii) $s(n) = h_2(n) * w(n)$,
iv) $s(n) = h_3(n) * w(n)$,

where $w(n)$ is a zero-mean Gaussian iid sequence, and where $h_1(n)$, $h_2(n)$ and $h_3(n)$ are unit gain linear phase bandpass filters with nominal passbands:

$$
\begin{aligned}
H_1(e^{j2\pi fT}) &= 1 &&; \ 150 \leq f \leq 350 \, Hz, \\
H_2(e^{j2\pi fT}) &= 1 &&; \ 200 \leq f \leq 300 \, Hz, \\
H_3(e^{j2\pi fT}) &= 1 &&; \ 225 \leq f \leq 275 \, Hz \ ,
\end{aligned}
$$

where the bandwidths are given relative to a sampling rate of $1000 \, Hz$.[16] For these trials, in each case, the delay D was set to 10

---

16. The filters were 32 point FIR linear phase, designed using a Remez-Exchange algorithm, using a design program available in *Selected Programs for Digital Signal Processing* [24]. The transition bands adjoining the passband were chosen arbitrarily and resulted in approximately 30 dB attenuation in the stopband of the filter.

samples. Thus we are concerned with the performance of the LMSTDE with broadband and with bandpass signals.

Let us consider first case *i)* − the sequence with no bandpass filtering, that is $s(n)$ white. In this case the optimal least-squares solution as given by the normal equations (3.2.12) is particularly simple. The correlation matrix is diagonal with $r(0)=\sigma_w^2 I$ and the cross-correlation vector has a single nonzero element $g(D)=\sigma_w^2$. Accordingly, the least-squares solution is

$$\underline{f}^* = R^{-1}\underline{g} = \frac{1}{\sigma_w^2}[0,0,...,0,\sigma_w^2,0,...,0]^t .$$

Hence,

$$f^*(i)=0 \quad ; i\neq D$$
$$f^*(i)=1 \quad ; i=D . \tag{5.3.20}$$

Moreover, since this is a homogeneous example (see also Problem 4.10), the filter should approach $\underline{f}^*$ without misadjustment. The performance of the filter is illustrated in Figure 5.3.5a) which shows the final set of filter coefficients after a run of 1000 iterations. It is clear that the adaptive filter approximates the least-squares solution with a very high degree of accuracy. These simulations used a filter length $L=21$, and adaptation constant $\alpha=0.01$. The input white noise had unit variance. The rate of adaptation is illustrated in Figure 5.3.6a) which shows the track of $f_n(10)$ for $n=0,1,..,999$, confirming the expected convergence.

The final sets of filter coefficients obtained for the bandpass inputs from cases *ii)*, *iii) and iv)* are shown in Figure 5.3.5b), c), and d) respectively. From the diagram, we see that as the bandwidth decreases, the energy within the filter becomes less concentrated at one point and more distributed. As a result, the peak becomes less and less distinct, and hence more ambiguous. The limit occurs when the input is a sinusoid in which case the filter response is also sinusoidal and the peak value is not unique.

We may use bandpass signals such as these to illustrate the eigenvalue effect in the LMS filter. One way to approach this is to take 'snapshots' of the filter response at different iterations and display the Fourier transforms of these responses. We recall from Section 5.2 that the eigenvalue disparity effect is such that we expect the filter to respond more quickly to stronger frequency components in the input. This is illustrated in Figure 5.3.7 which

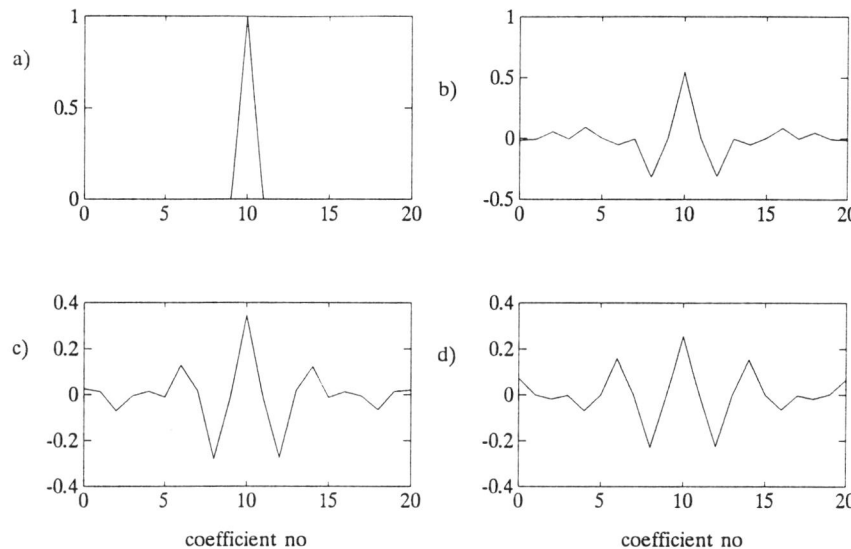

Figure 5.3.5 Filter vectors for four signal conditions: a) white input, b) bandpass white input, bandwidth $200Hz$, c) bandpass white noise, bandwidth $100Hz$, d) bandpass white noise, bandwidth $50Hz$. Delay=10, f_s=1000 for all cases.

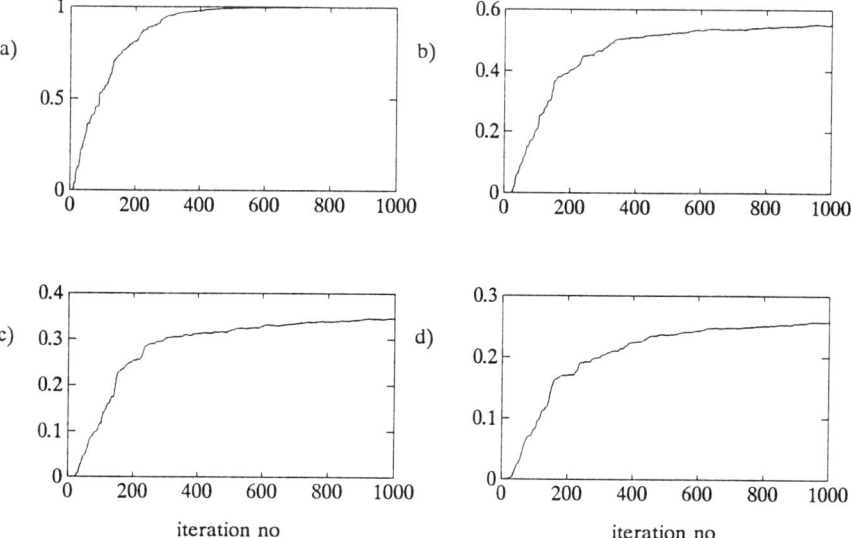

Figure 5.3.6 $f_n(10)$ for iterations $0,1,...,1000$, for four signal conditions: a) white input, b) bandpass white input, bandwidth $200Hz$, c) bandpass white noise, bandwidth $100Hz$, d) bandpass white noise, bandwidth $50Hz$. Delay=10, f_s=1000 for all cases.

shows a time-frequency decomposition of the adaptive filter impulse response. In the diagram each line (in time) is a Fourier transform of an impulse response, obtained by zero-padding the response to 256 points.

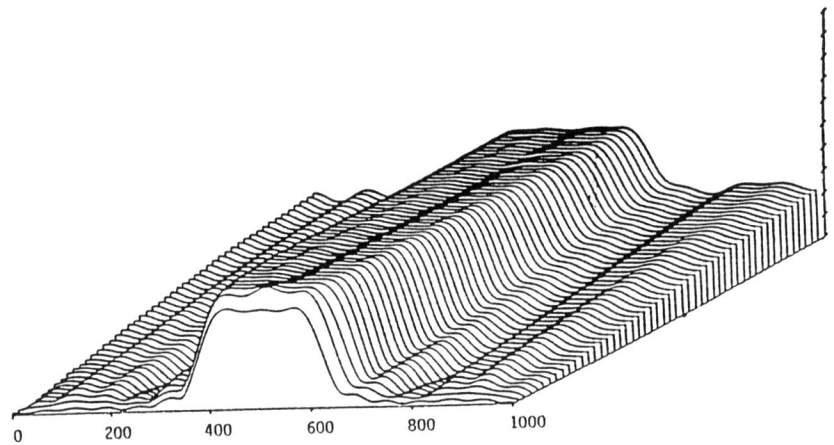

Figure 5.3.7 Time-frequency decomposition of the LMS adaptive filter response. Each line represents the amplitude spectrum obtained by zero-padding the response to 256 points.

The responses are taken at intervals of $N{=}50$ iterations, so that the 20 individual plots span a total of 1000 iterations. As expected, we see a rapid growth in the filter within the passband of the data. The out-of-band response, however, develops much more slowly, clearly illustrating slower modes of convergence associated with those frequency components.[17] Even so, we should note that in this *noise-free* environment the delay estimate obtained by taking the peak value is correct within the first few iterations.

As we have indicated, in a noisy environment $\left(v_1(n), v_2(n){\neq}0\right)$ the least-squares filter is not equivalent to the Maximum Likelihood (ML) estimator and consequently in a noisy environment cannot be expected to achieve the minimum variance for the delay estimator as expressed by the CRLB. Clearly, if the fixed processor does not achieve the CRLB, the adaptive scheme

---

17. We look at this problem again in the next Chapter, after first describing algorithms capable of operating with reduced eigenvalue disparity

cannot be expected to do so. Studies have shown, however, that for low signal-to-noise ratios the performance of the LMSTDE is within 0.5 dB of this limit [19].

Krolik *et al.* [25],[26] performed an evaluation of the LMSTDE and computed its mean-squared error performance for a fixed observation time (4 seconds). They compared unsmoothed estimates (the peak in the LMS filter response), interpolated delay estimates, and estimates interpolated assuming the signal bandwidth is known. The variance of the delay estimate was compared with that predicted by the CRLB. It was found that the unsmoothed estimate was the least effective with some improvement in the estimate resulting from simple interpolation and further substantive improvement resulting when the interpolation is related to the signal bandwidth. In each case, however, the results were significantly worse than those predicted by the CRLB. On the other hand, the advantages of the adaptive implementation, as in other applications, are the very limited *a priori* knowledge required, the simplicity of the implementation, and the ability to track time-varying input conditions.

5.4 Some Applications of Adaptive Filtering in Communications

5.4.1 Echo Cancellation in Voiced Channels

A significant problem in long-distance telephone links is the generation of echos. Echos can arise in a number of ways but are primarily caused by impedance mismatch at the point where the two-wire local subscriber loop becomes a four-wire trunk. In two-wire transmission, speech moving in both directions is superimposed. That is, all the speech is carried by the same wire pair. In the four-wire section, speech moving in each direction is carried separately. This four-wire arrangement is necessary if amplifiers, switches and other devices are to be used.

The conversion from two-wire to four is performed using a device called a **hybrid**. The overall setup is shown in Figure 5.4.1. The hybrid is intended to direct arriving energy from the four-wire section into the local two-wire loop. Unfortunately, because of impedance mismatches some of this energy is returned to the outgoing leg of the four-wire system and is transmitted along with the outgoing speech. Consequently, after a transmission delay, this speech is returned to the original speaker and is perceived as an echo.

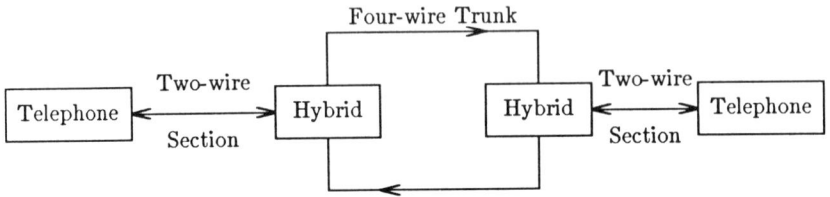

Figure 5.4.1 Schematic of a telephone communication system. Echos are created by energy reflected due to impedance mismatch at the hybrids.

The impact of echos depends on the length of the loop, and hence on the two-way propagation delay. For short delays, the echo will have no subjective impact (in fact, for delays less than 50ms, the subjective quality of the signal may actually be improved by the echo). Transmission over long paths, for example with satellite transmission, may involve delays in excess of 0.5s, however. Such echos are very disquieting for the speaker and can make conversation difficult to conduct.

Echo cancellation can be achieved by fitting voice-operated switches which open the transmission path only when the local caller is speaking. Such solutions are unsatisfactory, however, both because the voice-operated switches are imperfect, and because they do not cancel echos during periods when both speakers are talking (doubletalk). An alternative approach is to cancel the echos using an adaptive filter.[18] Such a system is depicted in in Figure 5.4.2.[19] Referring to the diagram, speech is received at the hybrid from a remote speaker (speaker 1, say). This signal forms the input to the adaptive filter. The output from the hybrid is used as the desired signal (the terms $x(n)$, $d(n)$ are used in the diagram to help clarify the relation between this configuration and the usual adaptive filtering arrangement). The error $e(n)$ is the signal transmitted to the remote speaker.

---

18. Echo cancellation can also be used in an acoustic framework (see, for example, [27],[28]). The problem arises in remote teleconferencing, for example, and is closely related to the acoustic ANC problem described in Section 5.1.

19. To be effective for both speakers, echo cancellation is required at both ends of the loop. In the diagram for clarity only one end of the loop is shown.

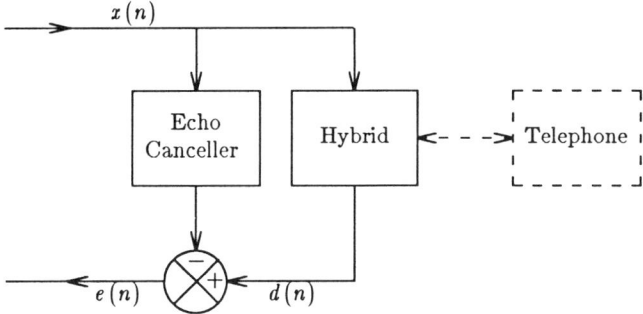

Figure 5.4.2 Echo cancellation in telephone communications. The canceller models the reflection at the hybrid and cancels echos. This leaves the desired transmission in $e(n)$.

If we imagine temporarily that speaker 2 is silent, then the role of the adaptive filter is to identify the transfer function of the echo generation process within the hybrid. If this is successful the filter output will equal the echo, which will therefore be cancelled in the error.

In fact, the use of the adaptive filter in a system identification mode is not unique to the echo cancellation problem. Adaptive filters have also been used for system identification in geophysical source signature modelling and in FIR filter design [29]. Thus, we may generalize this discussion if we consider the use of an adaptive filter to model an unknown system, $h(i)$, say. The basic configuration is shown in Figure 5.4.3. It is assumed that measurements of the input and output for the unknown system are available. The input is supplied as reference to the adaptive filter and the output as desired signal. It is intuitively clear from the diagram that by minimizing the mean-squared error between the output of the system and that of the adaptive filter, the coefficients of the adaptive filter represent an estimate of the unknown system. It is interesting to note that this is an example of a potentially homogeneous application of the adaptive filter — that is, where the desired signal is matched exactly by the adaptive filter output. For the system identification problem, when this exact match occurs the situation is referred to as **input-output equivalence**. Even if the unknown system and the adaptive filter are input-output equivalent, however, it is not certain that the filter coefficients will match the unknown system. There are several reasons why this can occur:

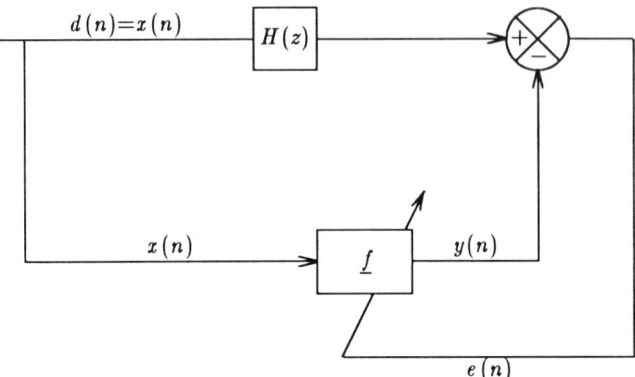

Figure 5.4.3 Adaptive filtering for system identification.

i) The nature of the unknown system may be different from that of the adaptive filter. For example, $h(i)$ might be a recursive implementation while $f_n(i)$ is a tapped delay line. More significantly, $h(i)$ may or may not be linear. In terms of the filtering operation at least, $f_n(i)$ is linear.

ii) Even if the structure of $h(i)$ and $f_n(i)$ are the same, the order (that is length) of the filters may differ.

iii) The excitation may not be broadband. If this occurs, the adaptive filter can only model the unknown system at frequencies which are excited at the input. At other frequencies, there is no input, no output, and therefore no information available for the identification.

iv) The measurement of input, output, or both may be corrupted by noise.

The point is that even in the face of any or all of these problems, the adaptive filter may still be able to match the system output. That is, it may be input-output equivalent, because of the extra flexibility provided by the time-variation in the adaptive filter.

Returning to the echo cancellation problem, a simple model for the transfer function of a hybrid is a delay and amplitude scale [30]. In reality, there is no guarantee that the hybrid transfer function will be simple, or even linear, and in many cases it is markedly nonlinear [31]. This does not necessarily stop the LMS from effective cancellation, however, because as we have noted the time-variation within the adaptive filter provides an extra degree of freedom which can enable the filter to mimic nonlinear effects.

The echo cancellation system is complicated by the presence in $d(n)$ of speech from speaker 2, $s_2(n)$, say. While this is uncorrelated with $x(n)$ and therefore should have limited impact on the adaptive solution, $s_2(n)$ will contribute to $e(n)$ and therefore will produce a non-zero minimum mean-squared error. Hence, through adaptation effects the system will exhibit misadjustment. To minimize this, the echo canceller should converge as rapidly as possible after the connection is made, before speaker 2 starts to talk [30]. More sophisticated cancellers usually have a built-in switch whose objective is to detect the onset of (local) speech. Once such speech is detected, adaptation can be frozen until a further period of silence occurs.

5.4.2 Adaptive Differential Pulse Code Modulation (ADPCM)

Adaptive filters are often employed in digital communications systems. Consider a communication channel in which digital data is transmitted using B bits to represent each sample. For a uniform sample rate f_s, this implies a total **bit rate** of $R = f_s B$ bits per second (bps). Obviously, the cost of transmission is proportional to R, and lowering R is one of the principal objectives of processing. The sample rate is fixed by the bandwidth of the signal, hence only through reduction of B can reductions in data rate be obtained. On the other hand, B, must be large enough to keep **quantization noise** at reasonable levels. The problem, therefore, is to reduce B while maintaining acceptable quality for the received data. The study of such problems is many faceted and involves consideration of both aspects related to the nature of the quantization operation, and of the coding of the information to be transmitted (see, for example, [32],[33]).

One of the more popular approaches is is that of Adaptive Differential Pulse Code Modulation (ADPCM). (Useful reviews of ADPCM are to be found in [34],[35] as well as in [32],[33] cited above). Such systems are based upon the following premise: The number of bits required to transmit a signal with a given fidelity is proportional to the power of the signal transmitted. This is a rather approximate, but intuitively appealing statement. This principle is exploited in Differential Pulse Code Modulation

(DPCM) in which the difference between adjacent samples is transmitted, rather than the data itself. Speech has a fairly high short-term correlation, consequently the power of the difference is substantially lower than that of the original signal. At the next level of sophistication, a linear prediction of the signal is used (sometimes with just a single predictor coefficient) and the error is transmitted. In ADPCM an adaptive prediction of the signal is employed and it is the error or residual which is transmitted.[20] Since the power of the prediction error is less than that of the speech, less bits are required for transmission.

An ADPCM system has adaptive predictors at the transmitter *and* at the receiver. At the transmitter, the predictor uses the speech $s(n)$ as a desired signal. A prediction, $\hat{s}(n/n-1)$ is formed using previous predictions and the prediction errors $e(n)$

$$e(n)=s(n)-\hat{s}(n/n-1) \; . \tag{5.4.1}$$

Referring to Figure 5.4.4a), the block labelled Q denotes the quantization of the error to B bits

$$e_q(n)=Q\{e(n)\} \; , \tag{5.4.2}$$

where $e_q(n)$ is the quantized error signal. It is this quantized error which is transmitted. The prediction $\hat{s}(n/n-1)$ has the form

$$\hat{s}(n/n-1)=\underline{f}_{n-1}^{t}\underline{e}_q(n-1) \; , \tag{5.4.3}$$

where $\underline{e}_{q(n)}=[e_q(n),e_q(n-1),...,e_q(n-L+1)]^t$. There are many ways in which the quantization may be achieved, but a discussion of the properties of the various quantizers and quantization laws would take us outside the scope of this text. The interested reader is referred to the texts by Jayant and Noll [32] and by Rabiner and Schafer [33] for discussion of these matters.

At the receiver, an identical prediction scheme is used to reconstruct $s(n)$ from the transmitted $e_q(n)$ (see Figure 5.4.4b)). The reconstruction is completed by forming the sum

---

20. In its original form it was the quantizer rather than the filter which was adaptive in ADPCM. Now the name is usually taken to imply an adaptive quantizer *and* an adaptive prediction.

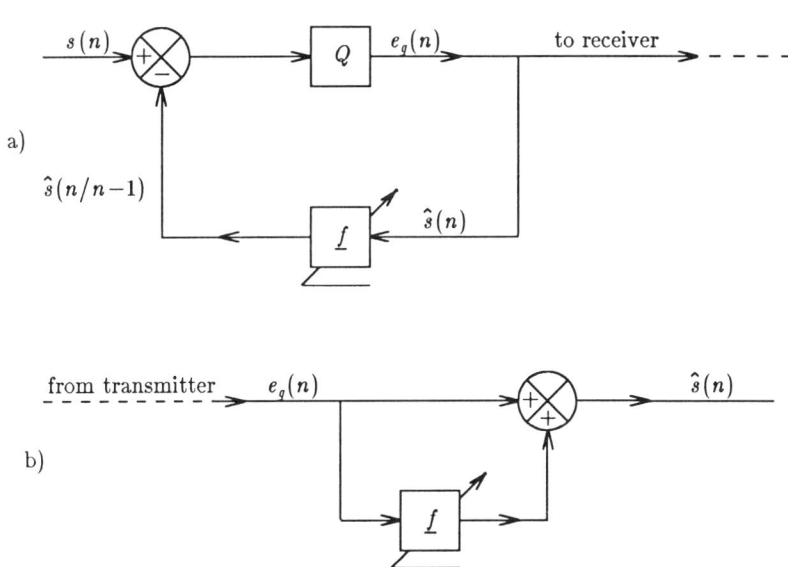

Figure 5.4.4 ADPCM System. a) Transmitter, b) Receiver. Coefficients $\underline{f}$ are driven by quantized error $e_q(n)$. Note that a single sample delay is implicit in the filtering system.

$$\hat{s}(n) = \hat{s}(n/n-1) + e_q(n) \ . \tag{5.4.4}$$

It is easy to see that if no quantization is present (that is if $e_q(n) = e(n)$), then $\hat{s}(n) = s(n)$ and the transmission is perfect. Moreover, this result holds irrespective of the nature of the prediction. This illustrates one of the important special features of this particular application – the prediction is essentially secondary to the quantization. Prediction is only helpful in so far as it assists the performance of the quantizer. In the absence of quantization the predictor becomes irrelevant. In general, the quantization scheme is the primary determinant of the performance of the system. Usually, this is measured as a Signal-to-Noise (SNR) for a given bit rate, and the adaptive predictor increases the total SNR by only a few decibels. Even so, the effect of the predictor makes a significant difference to the perceptual quality of the result, and considerable energy has been expended on optimizing the adaptive predictor, leading to an international (CCITT) standard for 32,000 bps system [36].

An LMS type predictor for an L point FIR implementation is provided by the equations:

$$\hat{s}(n/n-1)=\underline{f}_{n-1}^{t}\underline{e}_{q}(n-1) \ , \qquad (5.4.3)$$

$$e(n)=s(n)-\hat{s}(n/n-1) \ , \qquad (5.4.1)$$

$$e_{q}(n)=Q\{e(n)\} \ , \qquad (5.4.2)$$

$$\underline{f}_{n}=\underline{f}_{n-1}+\alpha\underline{e}_{q}(n-1)\,e_{q}(n) \ . \qquad (5.4.5)$$

This algorithm has a similar structure to the usual LMS, but the data vector has been replaced by the quantized error vector $\underline{e}_{q}(n)$ and the error has been replaced by the quantized prediction error $e_{q}(n)$. The same update is used at both transmitter and receiver. $e_{q}(n)$ need not be calculated at the receiver, this is the information received from the transmitter. Instead the receiver computes the reconstructed signal $\hat{s}(n)$ via (5.4.4). It is apparent that given no channel errors and the same initial conditions, the adaptive filter coefficients at both transmitter and receiver are identical for all iterations (they operate on identical input data). In this way the filter coefficients do not need to be transmitted, they can be reproduced purely from the error $e_{q}(n)$. Note that, if one or more bit errors *do* occur in transmission, the transmitted and received $e_{q}(n)$ sequences will no longer be identical. Hence, the transmitted and received filter coefficients will also diverge. More disturbingly, the adaptive filter at the receiver has no feedback mechanism by which such errors can be corrected. This is because the sequence $e_{q}(n)$ is provided at the end of the transmission process and is not generated by the receiver. For this and other reasons, ADPCM systems almost never employ the simple LMS type update of equation (5.4.5). In practice the algorithm is normalized (Section 6.2), and a leakage factor is incorporated (Section 6.2). Additionally, in most cases (including the CCITT system) the filter employed is an IIR structure (Section 6.4) rather than the simple FIR form discussed in this Chapter. We shall return to this application in the next Chapter, once these algorithmic issues have been reviewed.

Appendix 5A — Matched Filter for Sinusoid in White Noise

Consider the signal

$$x(n)=A_0\cos(\omega_0 nT+\theta_0)+w(n)=s(n)+w(n) , \qquad \text{(A1)}$$

where $s(n)=A_0\cos(\omega_0 nT+\theta_0)$, and where $w(n)$ is a zero-mean iid noise sequence with unit variance. When such a signal is filtered using a linear filter $\underline{f}$, the output Signal-to Noise Ratio (SNR_{out}) may be defined by

$$SNR_{out}=\frac{output\ signal\ power}{output\ noise\ power} ,$$

$$SNR_{out}=\frac{(\underline{f}^t\underline{s}_n)^2}{E\{(\underline{f}^t\underline{w}_n)^2\}}=\frac{\underline{f}^t(\underline{s}_n\underline{s}_n^t)\underline{f}}{\underline{f}^t\underline{f}} , \qquad \text{(A2)}$$

where $\underline{s}_n=[A_0\cos(\omega_0 nT+\theta_0),...,A_0\cos(\omega_0(n-L+1)T+\theta_0)]^t$ is the signal vector, and where $\underline{w}_n=[w(n),w(n-1),...,w(n-L+1)]^t$ is a vector of noise samples. The maximization of SNR_{out} is therefore equivalent to solving the constrained problem:

$$Maximize\ \{\underline{f}^t(\underline{s}_n\underline{s}_n^t)\underline{f}\}\ subject\ to\ \underline{f}^t\underline{f}=k , \qquad \text{(A3)}$$

where k is a constant. Using Lagrange multipliers, we may write a modified functional J as

$$J=\underline{f}^t R_s\underline{f}-\lambda(\underline{f}^t\underline{f}-k) , \qquad \text{(A4)}$$

where $R_s=\underline{s}_n\underline{s}_n^t$ and where λ is the Lagrange multiplier. Differentiating (A4) gives

$$2R_s\underline{f}-2\lambda\underline{f}=\underline{0} ,$$

and hence

$$(R_s-\lambda I)\underline{f}=0 . \qquad \text{(A5)}$$

Note that λ is an eigenvalue of $R_s=\underline{s}_n\underline{s}_n^t$ and $\underline{f}$ is the corresponding eigenvector. Also, from (A5)

$$\underline{f}^t R_s \underline{f} - \lambda \underline{f}^t \underline{f} = 0 \ ,$$

or

$$\lambda = \frac{\underline{f}^t R_s \underline{f}}{\underline{f}^t \underline{f}} = SNR_{out} \ , \qquad\qquad (A6)$$

where the second equality follows from (A2). Clearly from (A6) SNR_{max} corresponds to λ_{max}, the largest eigenvalue, and the optimum filter is given by the corresponding eigenvector. *viz*

$$SNR_{max} = \lambda_{max} = \frac{\underline{f}^{*t} R_s \underline{f}^{*}}{\underline{f}^{*t} \underline{f}^{*}} \ . \qquad\qquad (A7)$$

Now, from (A5)

$$(SNR_{max})\underline{f}^{*} = R_s \underline{f}^{*} \ ,$$

so that combining with (A7) we have

$$\frac{\underline{f}^{*t} \underline{s}_n \underline{s}_n^t \underline{f}^{*}}{\underline{f}^{*t} \underline{f}^{*}} \underline{f}^{*} = \underline{s}_n \underline{s}_n^t \underline{f}^{*} \ .$$

Cancelling the scalar terms $\underline{s}_n^t \underline{f}^{*}$ we may write

$$\left(\frac{\underline{f}^{*t} \underline{s}_n}{\underline{f}^{*t} \underline{f}^{*}} \right) \underline{f}^{*} = \underline{s}_n \ , \qquad\qquad (A8)$$

but

$$\left(\frac{\underline{f}^{*t} \underline{s}_n}{\underline{f}^{*t} \underline{f}^{*}} \right) = \frac{1}{c_1} \ ,$$

a scalar value. Hence, from (A8)

$$\underline{f}^{*} = c_1 \underline{s}_n \ , \qquad\qquad (A9)$$

with components

$$f^{*}(i) = c_1 A_0 \cos[\omega_0 (n-i) T + \theta_0] \quad ; \ i = 0,1,...,L-1 \ ,$$

or

$$f^{*}(i) = C \cos(\omega_0 i T + \gamma) \quad ; \ i = 0,1,...,L-1 \ , \qquad\qquad (A10)$$

where $C=c_1 A_0$, $\gamma=-\theta_0-\omega_0 nT$.[21] This gives the form indicated by equation (5.2.4). Also, the magnitude of C has no impact on SNR_{out}, but to satisfy a particular constraint $\underline{f}^{*t}\underline{f}^{*}=k$, we have from (A9)

$$\underline{s}_n^t c_1^2 \underline{s}_n = k \; ,$$

or

$$c_1^2 = \frac{k}{\underline{s}_n^t \underline{s}_n} = 2\frac{k}{A_0^2 L} \; . \tag{A11}$$

Problems

5.1 A signal $s(n)$ is corrupted by an additive, zero-mean stationary measurement noise $v(n)$. A second measurement is made which has the form

$$x(n)=av(n)+bv(n-1) \; ,$$

where a and b are constants. Set up an adaptive noise cancellation system to recover the signal $s(n)$. Find the infinite two-sided Wiener approximation to the adaptive filter and show that for this idealized filter, the signal is exactly recovered.

5.2 Consider signals

$$d(n)=s(n)+v_0(n) \; ,$$

$$x(n)=g(n)*s(n)+h(n)*v_0(n) \; ,$$

where $g(n)$, $h(n)$ are linear shift-invariant systems, and $s(n)$, $v_0(n)$ are zero-mean stationary signal and noise, respectively. Use the infinite two-sided Wiener approximation

---

21. Note that strictly the matched filter $\underline{f}^{*}$, depends on n through the phase angle γ. We have suppressed this dependence in the notation used here in the interests of avoiding confusion with the adaptive filter.

$$F(e^{j\omega}) = \frac{R_{xd}(e^{j\omega})}{R_{xx}(e^{j\omega})} \ ,$$

to demonstrate that

$$\rho_{out}(e^{j\omega}) = \frac{1}{\rho_{ref}(e^{j\omega})} \ , \qquad (5.1.16)$$

where ρ_{out} and ρ_{ref} are the signal-to-noise density ratios for the output and reference, respectively (as defined by equations (5.1.14) and (5.1.15)).

5.3 Consider measurements $d(n)$, $x(n)$ in the following (simplified) noise cancellation problem:

$$d(n) = s(n) + v(n) * h_1(n) \ ,$$
$$x(n) = v(n) \ ,$$

where $v(n)$ is a zero-mean iid sequence and $h_1(n)$ is a causal FIR N point transmission system.

a) Find the L point FIR optimal least squares noise canceller $\underline{f}^*$. (Differentiate between the cases $N > L$ and $N \leq L$).

b) Find the minimum mean-squared error for the LMS filter and comment on the uniformity, or otherwise, of convergence.

5.4 In noise cancellation, one performance measure is **Noise Reduction Number (NRN)** [37], which may be defined by:

$$NRN = 10\log_{10}\left[\frac{E\{e_r^2(n)\}}{E\{v_1^2(n)\}}\right] \ ,$$

where $e_r(n)$ is the residual noise (after cancellation), and $v_1(n)$ is the original (primary) noise. Given measurements

$$d(n) = v(n) * h_1(n) \ ,$$
$$x(n) = v(n) \ ,$$

where $v(n)$ is a zero-mean iid unit variance signal, and $h_1(n)$

is a causal linear shift-invariant system, show that

$$NRN=10\log_{10}\left[\frac{\displaystyle\sum_{j=L}^{\infty}h_1^2(j)}{\displaystyle\sum_{j=0}^{\infty}h_1^2(j)}\right].$$

5.5 A zero-mean stationary random signal $s(n)$ is corrupted by an additive zero-mean stationary noise $u(n)$ which is uncorrelated with $s(n)$. A second measurement $x(n)$ of the form

$$x(n)=s(n)+0.5s(n-1)+v(n),$$

is made, where $v(n)$ is a zero-mean stationary measurement noise (uncorrelated with $u(n)$ and $s(n)$):

a) Draw a block diagram to illustrate how an LMS adaptive filter can be used to enhance the signal $s(n)$.

b) Find the infinite two-sided, Wiener approximation for the filter $F(z)$.

c) Derive an expression for the filter output spectrum $R_{yy}(e^{j\omega})$ and the error spectrum $R_{ee}(e^{j\omega})$.

d) If $v(n)=0$, find an explicit form for the filter.

5.6 A ship tows a 750m cable which has a hydrophone attached to the lower end. The array is to be used for ambient ocean noise measurements. The measured signal is corrupted by noise from the towing ship. A second hydrophone is mounted close to the tow-ship to obtain a reference of the tow-ship noise. The resulting signals are modelled as:

$$x_1(n)=s(n)+h(n)*v(n),$$
$$x_2(n)=v(n),$$

where $h(n)$ is a causal FIR linear structure which is designed to allow for the effects of delay, transmission losses, and extra reflections. The noise $v(n)$ is assumed to be zero-mean stationary and white with variance σ_w^2, and the signal and noise are assumed mutually uncorrelated.

a) Design a least-squares filter to remove the noise from the signal measurement. Find the analytic solution for the filter (using the autocorrelation form normal equation solution), and discuss the conditions under which the signal will be exactly recovered.

b) Assuming that in addition to the direct transmission path, only a single bottom reflection is significant, that transmission losses are negligible, and that reflection can be characterized by a single reflection coefficient, what is the minimum length of the filter?

(Assume the sampling frequency $f_s=2kHz$, the distance between the sensors is 750m, the depth of the sea is 900m and that the tow-cable leaves the ship at an angle 60 degrees from the horizontal. Assume further that the velocity of sound in water is 1500m/s.)

5.7 Consider the LMS ANC system shown in the diagram.

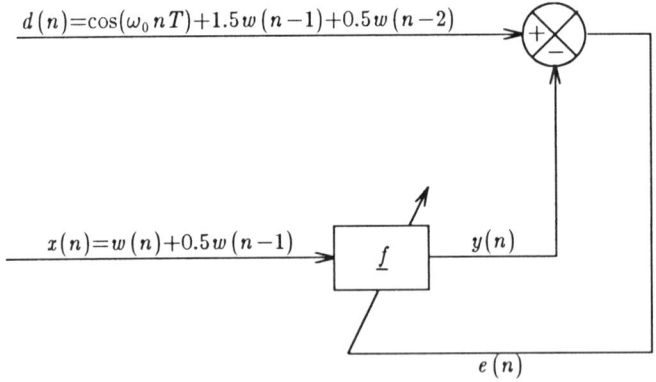

$$d(n)=\cos(\omega_0 nT)+1.5w(n-1)+0.5w(n-2)$$

$$x(n)=w(n)+0.5w(n-1)$$

$y(n)$

$e(n)$

Assume that $w(n)$ is a stationary, zero-mean white noise sequence with variance $\sigma_w^2=1$:

a) Find the $L=2$ point causal least-squares optimal filter $f^*(n)$ and comment on the performance of the ANC system with this optimal filter.

b) Find approximate time-constants for the adaptive system assuming an adaptation constant α.

c) Suppose the ANC primary input is changed to

$$d(n)=\cos(\omega_0 n T)+w(n)+1.5w(n-1) \ ,$$

find the optimal filter $f^*(n)$ in this case. Comment on the change in cancellation performance, and on the adaptive system performance.

5.8 A transducer mounted on the casing of a rotating gear wheel records a measurement of the form

$$x(n)=s(n)+w(n) \ ,$$

sketch a block diagram illustrating how an adaptive digital filter could be used to enhance the signal $s(n)$ (without making a second measurement). What properties must $s(n)$ and $w(n)$ have in order for the enhancement system to be effective.

5.9 Consider the line enhancement problem with input signals

$$d(n)=A_1\cos(\omega_0 n T)+A_2\cos(\omega_1 n T)+w(n) \ ,$$
$$x(n)=d(n-\Delta) \ ,$$

where $w(n)$ is a zero-mean white noise sequence with variance σ_w^2. Show that the causal, finite L point least-squares filter $\underline{f}^*$ which maps $x(n)$ to $d(n)$, can be approximated by:

$$f^*(n)=B_1\cos(\omega_1 n T)+B_2\cos(\omega_2 n T) \ ,$$

where

$$B_i=\frac{\sigma_i^2(2/L)}{(2/L)\sigma_w^2+\sigma_i^2} \quad ; i=1,2 \ ,$$

where $\sigma_i=(A_i^2/2)$. State carefully any assumptions you make.

5.10 Show that the correlation matrix with elements

$$r(i)=A\cos(\omega i T)+\sigma_w^2\delta(i) ,$$

is positive definite.

5.11 Consider an input $d(n)$ consisting of a single sinusoid with frequency ω_0, and amplitude A, embedded in white noise with variance σ_w^2, with $x(n)=d(n-\Delta)$. Find the mean-squared error associated with the least-squares solution

$$f^*(i)=B\cos(\omega_0 i T) ,$$

where

$$B=\frac{\sigma_s^2(2/L)}{(2/L)\sigma_w^2+\sigma_s^2} \quad ; i=1,2 ,$$

where $\sigma_s^2=(A^2/2)$.

5.12 Consider the ALE system with inputs

$$d(n)=\cos(\omega_0 n T)+w(n) ,$$
$$x(n)=d(n-5) .$$

If $\omega_0=2\pi/3T$, and $L=3$:

a) Find $\underline{f}^*$ and J_{min}.

b) Identify time-constants t_1, t_2 and t_3 for the LMS system.

5.13 Consider the adaptive line enhancer with inputs

$$d(n)=\sum_{i=0}^{M-1} A_i\cos(\omega_i n T)+w(n) ,$$
$$x(n)=d(n-\Delta) ,$$

where $w(n)$ is zero-mean iid with variance σ_w^2. Use the final value theorem for z transforms to show that

$$\lim_{n\to\infty}\{f_n'(i)\}=\frac{g'(i)}{\sigma_w^2+\lambda_i} \quad ; i=0,1,...,L-1 ,$$

where λ_i is the i-th eigenvalue of the correlation matrix derived from the sinusoidal components of $d(n)$, where $g(i)$ and $f(i)$ are the i-th components of the cross-correlation and filter respectively, and where $'$ denotes rotation using the orthogonal matrix of eigenvectors Q [12].

5.14 The diagram shows an LMSTDE system used to estimate the delay, Δ, between two signal measurements $s(n)$, $s(n-\Delta)$.

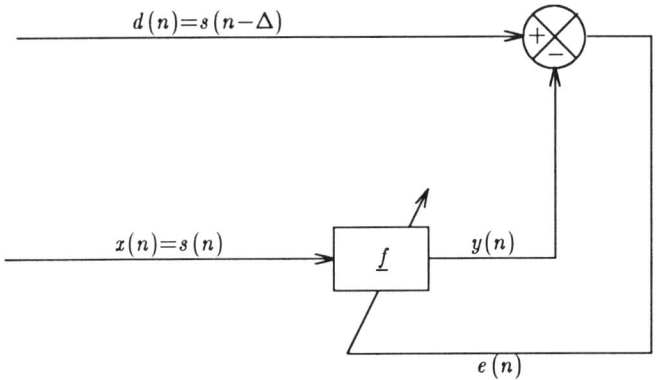

If the input signal $s(n)$ is a stationary, zero-mean white noise sequence with variance σ_s^2:

a) Find the two-sided infinite (fixed) Wiener filter $F(z)$.

b) Find the fixed, finite, causal least-squares filter $\underline{f}^*$. What is the minimum error energy in this case?

c) Assuming that $E\{\underline{x}_k \underline{x}_k^t \underline{f}_k\} = E\{\underline{x}_k \underline{x}_k^t\} E\{\underline{f}_k\}$, show that the LMS adaptive filter with update equation given by (4.1.14) converges in the mean to $\underline{f}^*$ provided

$$0 < \alpha < \frac{2}{\sigma_s^2} .$$

5.15 Consider the time-delay estimation problem with the signal model of equations (5.3.1)

$$x_1(t) = s(t) + v_1(t) ,$$

and

$$x_2(t) = s(t-D) + v_2(t) . \tag{5.3.1}$$

Use the definition of magnitude squared coherence (msc) given by equation (5.3.5)

$$msc = |\gamma_{12}(\omega)|^2 = \frac{|R_{z_1 z_2}(\omega)|^2}{R_{z_1 z_1}(\omega) R_{z_2 z_2}(\omega)} , \qquad (5.3.5)$$

where $R_{z_i z_i}(\omega)$ and $R_{z_i z_j}(\omega)$ are the power and cross-spectra, respectively, for signals $x_i(t)$ and $x_j(t)$, to show that the maximum likelihood delay estimate

$$W_{ML}(\omega) = \frac{|\gamma_{12}(\omega)|^2}{|R_{z_1 z_2}(\omega)| [1 - |\gamma_{12}(\omega)|^2]} , \qquad (5.3.6)$$

can be written as

$$W_{ML}(\omega) = \frac{\dfrac{R_{ss}(\omega)}{R_{v_1 v_1}(\omega) R_{v_2 v_2}(\omega)}}{1 + \dfrac{R_{ss}(\omega)}{R_{v_2 v_2}(\omega)} + \dfrac{R_{ss}(\omega)}{R_{v_1 v_1}(\omega)}} . \qquad (5.3.7)$$

Show further that if the noises v_1, v_2 have identical spectra $R_{vv}(\omega)$, then this result can be simplified to

$$W_{ML}(\omega) = \frac{\dfrac{R_{ss}(\omega)}{R_{vv}^2(\omega)}}{1 + \dfrac{2 R_{ss}(\omega)}{R_{vv}(\omega)}} . \qquad (5.3.8)$$

5.16 Consider the LMS TDE system with inputs

$$x(n) = w(n) + 1.5 w(n-1) ,$$
$$d(n) = x(n-1) ,$$

where $w(n)$ is zero-mean, iid with variance $\sigma_w^2 = 1$. Assuming that the filter length $L = 3$:

a) Find the optimal least-squares solution $\underline{f}^* = R^{-1} \underline{g}$.

b) Find, the LMS stability limits corresponding to (4.2.20) and

(4.2.23).

c) Find the time-constants for each mode of convergence.

d) Find expressions for the orthogonalized error coefficients $u_n'(0)$, $u_n'(1)$ and $u_n'(2)$.

e) Find the steady-state mean-squared error J_∞.

5.17 Repeat **5.16**, changing the desired input to

$$d(n)=x(n-1)+v(n) \ ,$$

where $v(n)$ is a zero-mean stationary input (uncorrelated with $w(n)$), with variance σ_v^2. Indicate which parts of **5.16** have a different solution as a result of the new form for $d(n)$.

5.18 An LMS system modelling structure is depicted in the diagram

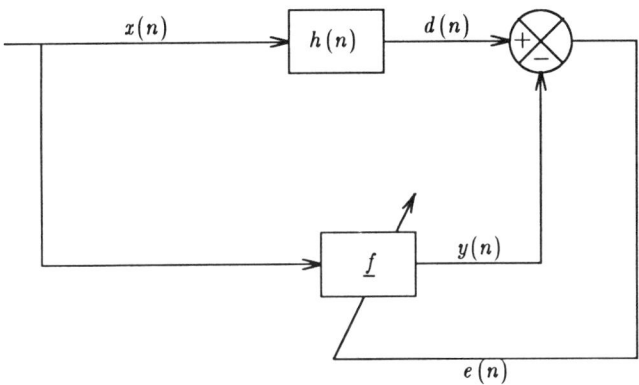

a) If $h(0)=1$, $h(1)=a$, $h(2)=b$ and $x(n)=w(n)$ is a zero-mean white noise, find $\underline{f}^*$, the L point fixed causal least-squares solution.

b) Suppose now that $h(0)=0$, $h(1)=0$, $h(2)=0$, $h(3)=1$, $h(4)=a$, $h(5)=b$. If the number of initial zero values in the impulse response is known, draw a block diagram illustrating how the solution can still be obtained using $L=3$.

c) If $x(n)=\cos(\omega_0 nT)$, will the system work? Explain your answer.

d) Repeat part **c)** if $x(n)=\delta(n)$.

5.19 Consider an LMS system used to cancel echos created by a simple echo path with impulse response

$$h(0)=1 , \qquad h(10)=0.5 .$$

Assume an input of the form

$$x(n)=\sum_{k=0}^{\infty} \delta(n-kN) ,$$

where $N\gg L$, the filter length, and a filter length $L=20$:

a) Find closed form expressions for $f_n(i)$ for $n=0,1,...,L-1$.

b) What is the stability limit for the algorithm to converge?

c) Find the steady-state error associated with the algorithm.

5.20 Repeat question **5.19** using $L=5$.

Computer Problem 1 – Narrowband ANC

Implement a single input ANC system with

$$\underline{f}_{n+1}=\underline{f}_{n}+\alpha e(n)\underline{x}_n ,$$
$$e(n)=d(n)-y(n) ,$$

where

$$x(n)=d(n-\Delta) ,$$

where $\underline{x}_n=[x(n),x(n-1),...,x(n-L+1)]^t$,
$\underline{f}=[f_n(0),f_n(1),...,f_n(L-1)]^t$.

Use $f_s=10kHz$, $L=20$, $\Delta=5$, $\alpha=0.001$ in all trials:

a) Create an input $x(n)=A\cos(\omega_0 nT)$ for $n=0,1,...,4999$, where $\omega_0=2\pi500$. Confirm that the LMS algorithm performance is exactly predicted by the transfer function of equation (4.2.70). Verify the stability limit (4.2.72) for this data. Examine the filter coefficients $f_n(i)$ for $n=4999$. The coefficients have the form of a damped sinusoid. Why? Are the transfer function, stability limit, or filter coefficient solutions influenced by the choice of *i)* delay, *ii)* sinusoidal phase?

b) Repeat the experiment using identical parameters, but with $\omega_0=2\pi510$. What differences in the filter behavior result?

c) Repeat the trial using $x(n)$ formed as the sum of two sinusoids with frequencies $\omega_0=2\pi500$, $\omega_1=2\pi1000$. Confirm that the LMS behavior is equivalent to a parallel combination of notches at the reference frequencies.

d) Examine the frequency response of the filter error (obtained using a 1024 point transform from $n=0,1,...,1023$) as the frequency ω_1 is reduced successively to $2\pi900$, $2\pi600$, $2\pi520$ and $2\pi505$. How do you account for your results?

Computer Problem 2 – Adaptive Line Enhancement

Implement an LMS filter with

$$\underline{f}_{n+1}=\underline{f}_n+\alpha e(n)\underline{x}_n ,$$

$$e(n)=d(n)-y(n) ,$$

where

$$x(n)=d(n-\Delta) .$$

Assume that the input $x(n)$ consists of 10000 points of data consisting of

$$x(n)=\cos(\omega_0 nT)+v(n) \quad ; n=0,1,...,9999 .$$

Here, $\omega_0=2\pi500Hz$, $f_s=10kHz$, and $v(n)$ is a zero-mean, stationary noise. In each of the following trials perform an ad-hoc optimization by varying α to achieve the highest output SNR.

a) Take $v(n)=w(n)$, where $w(n)$ is a zero-mean iid unit variance Gaussian sequence.

b) Form $v(n)$ as an MA(2) process with

$$v(n)=(w(n)+0.5w(n-1)+0.25w(n-2))^*C ,$$

where $w(n)$ is the iid sequence of a), and where C is a constant which should be chosen so as to ensure $v(n)$ has unit variance. Examine the SNR gain achievable using *i)* $\Delta=1$, and *ii)* $\Delta=3$.

c) Form $v(n)$ as an AR(1) process with

$$x(n)=ax(n-1)+w(n) ,$$

where a is constant. Optimize the output SNR for *i)* $a=0.01$, *ii)* $a=0.1$, *iii)* $a=0.8$. In each case use $\Delta=100$. What conclusions can you draw?

Note: SNR should be calculated by forming the residual noise $e_{res}(n)$

$$e_{res}(n)=e(n)-\cos(\omega n T) ,$$

and then forming

$$SNR=\frac{\{Sinusoidal\ Power\}}{\{Residual\ Noise\ Power\}} .$$

The sinusoidal noise power is easily obtained. The residual noise power is best obtained by forming an "ensemble" average over a large number of repetitions of the experiment, with $n=9999$. Construct a suitable procedure for this estimation. Do you anticipate any problems, or possible bias in your results?

Computer Problem 3 – Time-Delay Estimation

Part 1 – For a system with inputs

$$x(n)=s(n)+v_1(n) ,$$
$$d(n)=s(n-\Delta)+v_2(n) ,$$

implement the LMSTDE adaptive filter using the update equation:

$$\underline{f}_{n+1}=\underline{f}_n+\alpha e(n)\underline{x}_n^{(1)},$$

where $$e(n)=x_2(n)-y(n),$$

and $$y(n)=\underline{f}_n^t\underline{x}_n^{(1)},$$

where $\underline{f}_n=[f_n(0),f_n(1),...,f_n(L-1)]^t,$
$\underline{x}_n^{(1)}=[x_1(n),x_1(n-1),...,x_1(n-L+1)]^t.$

You may assume $\underline{f}_0=\underline{0}$ throughout. The inputs to this program should be:

   i) Signal $s(n)$.
   ii) Noise files $v_1(n)$ and $v_2(n)$.
   iii) Filter length L.
   iv) Adaptation Constant α.
   v) Delay Δ.
   vi) Gain g applied to both additive noises.

$x_2(n)$ should be generated using $s(n)$ and Δ. Use zero-mean iid data for both signal and noises, and work with a total of 8,000 samples for each trial. You will need to generate (independently) three sets of 8,000 samples of each of the following:

i) Uniformly distributed random numbers with range $[-1,1]$.

ii) Gaussianly distributed random numbers with distribution $N(0,1)$.

a) Test the filter using the Gaussian data, choose $\Delta=5$, $L=20$, $g=0$ (noise-free) and $\alpha=0.01$. What is the form of the final set of filter coefficients? Is this in line with your expectations?

b) For the same data, plot the trajectory of filter coefficient $f_n(5)$ versus n. By examining this trajectory against a logarithmic scale confirm that the convergence of the coefficient is *exponential* and try to relate the rate of convergence to that predicted by equation (4.2.17) in Section 4.2.

c) Repeat the last trial but with $g\neq0$ (noisy-data). Examine the trajectory of the filter coefficient as g increases. Try to estimate the minimum SNR (in dB) for which the final delay estimate (at iteration 8,000) is unambiguous.

Part 2 — In the noisy condition ($g{\neq}0$) the delay estimate derived from the peak value of the final set of filter coefficients will not always equal the true delay D. In this case, the quality of the delay estimate is determined by the variance of that estimate about the true value as given by

$$v = \frac{1}{N} \sum_{i=1}^{N} (D_i - D)^2 \; ,$$

where D_i is the delay estimate obtained from the i-th independent trial from a total of N such trials.

a) Program an LMSTDE system which generates the random signal and noise data internally. This program should be capable of performing N independent repetitions of the trial. Each of these trials should proceed sequentially with independently generated data, each trial should consist of 8,000 iterations and the filter coefficients should be reset after each trial. The parameters g, α, L and D should be common to all of the trials. The output from each trial is the delay estimate D_i, obtained as the largest value from the final set of filter coefficients. The program should then calculate the delay estimate variance v defined above.

b) Evaluate the delay estimate variance using $N{=}200$ trials, for SNR's and with varying values for α. Use the same data type, length and delay values as in **Part 1**. How does the delay variance compare with your intuition as to the minimum SNR for unambiguous estimates?

References

1. B.Widrow *et al.*, "Adaptive noise cancellation: principles and applications," *Proc. IEEE*, vol. 63, pp. 1691-1717, 1975.

2. S.F.Boll and D.C.Pulispher, "Suppression of acoustic noise in speech by two microphone adaptive noise cancellation," *IEEE Trans. Acoust., Speech, Signal Processing*, vol. ASSP-28, pp. 752-753, 1980.

3. D.M.Chabries, R.W.Christiansen, R.H.Brey and M.S. Robinette, "Application of the LMS adaptive filter to improve speech communication in the presence of noise," *Proc. IEEE Int. Conf. Acoust., Speech, Signal Processing*, pp. 148-151, 1982.

4. J.B.Allen and D.A.Berkley, "Image method for efficiently simulating small-room acoustics," *J. Acoust. Soc. Amer.*, vol. 65, pp. 943-950, 1979.

5. W.Harrison, J.S.Lim and E.Singer, "A new application of adaptive noise cancellation," *IEEE Trans. Acoust., Speech, Signal Processing*, vol. ASSP-34, pp. 173-176, 1986.

6. P.Darlington, P.D.Wheeler and G.A.Powell, "Adaptive noise reduction in aircraft communication systems," *Proc. IEEE Int. Conf. Acoust., Speech, Signal Processing*, pp. 19.2.1-19.2.4, 1985.

7. G.A.Powell, P.Darlington and P.D.Wheeler, "Practical adaptive noise reduction in the aircraft cockpit environment," *Proc. IEEE Int. Conf. Acoust., Speech, Signal Processing*, pp. 173-176, 1987.

8. J.J.Rodriguez and J.S.Lim, "Adaptive noise reduction in aircraft communication systems," *Proc. IEEE Int. Conf. Acoust., Speech, Signal Processing*, pp. 169-172, 1987.

9. W.S.Burdic, *Underwater Acoustic System Analysis*, Prentice-Hall, 1984.

10. A.Papoulis, *Signal Analysis*, McGraw-Hill, 1977.

11. J.R.Zeidler, E.H.Satorius, D.M.Chabries and H.T.Wexler, "Adaptive enhancement of multiple sinusoids in uncorrelated noise," *IEEE Trans., Acoust., Speech, Signal Processing*, vol. ASSP-26, pp. 240-254, 1978.

12. J.R.Treichler, "Transient and convergent behavior of the adaptive line enhancer," *IEEE Trans., Acoust., Speech, Signal Processing*, vol. ASSP-27, pp. 53-62, 1979.

13. J.T.Rickard and J.R.Zeidler, "Second-order output statistics of the adaptive line enhancer," *IEEE Trans. Acoust., Speech, Signal Processing*, vol. ASSP-27, pp. 31-39, 1979.

14. N.J.Bershad, P.L.Feintuch, F.A.Reed and B.Fisher, "Tracking characteristics of the LMS adaptive line enhancer − response to a linear chirp signal in noise," *IEEE Trans. Acoust., Speech, Signal Processing*, vol. ASSP-28, pp. 504-516, 1980.

15. *IEEE Trans. Acoust., Speech, Signal Processing*, − Special Issue on Time-Delay Estimation, vol. ASSP-29, June 1981.

16. C.H.Knapp and G.C.Carter, "The generalized correlation method for the estimation of time-delay," *IEEE Trans. Acoust., Speech, Signal Processing*, vol. ASSP-24, pp. 320-327, 1976.

17. G.C.Carter, A.H.Nuttall and P.G.Cable, "The smoothed coherence transform," *Proc. IEEE*, vol. 61, pp. 1497-1498, 1973.

18. E.K.Al-Hussaini and S.A.Kassam, "Robust Eckhart filters for time-delay estimation," *IEEE Trans. Acoust., Speech, Signal Processing*, vol. ASSP-32, pp. 1052-1063, 1984.

19. F.A.Reed, P.Feintuch and N.J.Bershad, "Time-delay estimation using the LMS adaptive filter − static behavior," *IEEE Trans. Acoust., Speech, Signal Processing*, vol. ASSP-29, pp. 561-571, 1981.

20. F.A.Reed, P.Feintuch and N.J.Bershad, "Time-delay estimation using the LMS adaptive filter − dynamic behavior," *IEEE Trans. Acoust., Speech, Signal Processing*, vol. ASSP-29, pp. 571-576, 1981.

21. D.H.Youn, N.Ahmed and G.C.Carter, "On using the LMS algorithm for time-delay estimation," *IEEE Trans. Acoust., Speech, Signal Processing*, vol. ASSP-30, pp. 798-801, 1982.

22. D.H.Youn, N.Ahmed and G.C.Carter, "An adaptive approach for time-delay estimation of band-limited signals," *IEEE Trans. Acoust., Speech, Signal Processing*, vol. ASSP-32, pp. 780-784, 1983.

23. R.T.Compton, Jr. *Adaptive Antennas: Concepts and Performance*, Prentice-Hall, 1988.

24. *Selected Programs for Digital Signal Processing*, IEEE Press, 1979.

25. J.Krolik, M.Joy, S.Pasupathy and M.Eizenman, "A comparative study of the LMS adaptive filter versus generalized correlation methods for time-delay estimation," *Proc. IEEE Int. Conf. Acoust., Speech, Signal Processing*, pp. 15.11.1-15.11.4, 1984.

26. J.Krolik, M.Eizenman and S.Pasupathy, "Time-delay estimation of signals with uncertain spectra," *IEEE Trans. Acoust., Speech, Signal Processing*, vol. ASSP-36, pp. 1801-1811, 1988.

27. A.Gilloire, "Experiments with sub-band acoustic echo cancellers for teleconferencing," *Proc. IEEE Int. Conf. Acoust., Speech, Signal Processing*, pp. 2141-2144, 1987.

28. F.Amano and H.Perez, "A new subband echo canceler structure," *Proc. IEEE Int. Conf. Acoust., Speech, Signal Processing*, pp. 3585-3588, 1991.

29. B.Widrow and S.D.Stearns, *Adaptive Signal Processing*, Prentice-Hall, 1985.

30. S.T.Alexander, *Adaptive Signal Processing: Theory and Applications*, Springer-Verlag, 1986.

31. O.Agazzi, D.G.Messerscmitt and D.A.Hodges, "Nonlinear echo cancellation of data signals," *IEEE Trans. Commun.*, vol. COM-30, pp. 2421-2433, 1982.

32. N.S.Jayant and P.Noll, *Digital Coding of Waveforms: Principles and Applications to Speech and Video*, Prentice-Hall, 1984.

33. L.R.Rabiner and R.W.Schafer, *Digital Processing of Speech Signals*, Prentice-Hall, 1978.

34. J.D.Gibson, "Adaptive prediction in speech differential encoding systems," *Proc. IEEE*, vol. 68, pp. 488-525, 1980.

35. J.D.Gibson, "Adaptive prediction for speech encoding," *IEEE ASSP Magazine*, pp. 12-26, July 1984.

36. "32 kbit/s Adaptive differential pulse code modulation (ADPCM)," CCITT standard recommendation G.721, Melbourne 1988.

37.   W.B.Mikhael and P.D.Hill, "Performance evaluation of a real-time TMS320-10 based adaptive noise canceller," *IEEE Trans., Acoust., Speech, Signal Processing*, vol. ASSP-36, pp. 411-412, 1988.

chapter six

Adaptive Signal Processing: Algorithms and Structures

6.1 General Remarks

So far we have focused on just one adaptive algorithm (the LMS), derived from one particular criterion (least-squares), using one iterative philosophy (steepest descent) and implemented in one particular way (as a tapped delay line). In this Chapter, we consider some alternative possibilities for adaptive filtering. In the first part of the Chapter we shall concentrate on alternative algorithms. The aim here is not to provide an exhaustive review since there are far too many possibilities. Instead, we shall try to indicate the principle directions in which adaptive algorithms can be developed, and point out some of the main themes of recent work.

Figure 6.1.1 shows a tree structure giving an overview of adaptive algorithm development. Recall that in Chapter 4, in choosing the LMS algorithm we opted for a least-squares optimization, a gradient search algorithm using steepest descent, and estimation of the gradient using the instantaneous value of the error. From the diagram it is clear that while this set of choices leads uniquely to the LMS algorithm, this was in fact only one of many options. While the LMS has certainly been the most widely used adaptive algorithm, it is not the only possibility. In this Chapter we seek to redress the balance by considering some of the other possibilities for algorithm development.

Referring to Figure 6.1.1, the first choice which must be made in deciding which adaptive algorithm to use is whether or not to employ a least-squares optimization policy. The disparity in the amount of detail under each of the main directions in the figure is related to the amount of research which has been directed at each, rather than to their respective merits. Relatively little research in adaptive filtering has been directed towards optimization of criteria other than minimum mean-squared error. One exception is the

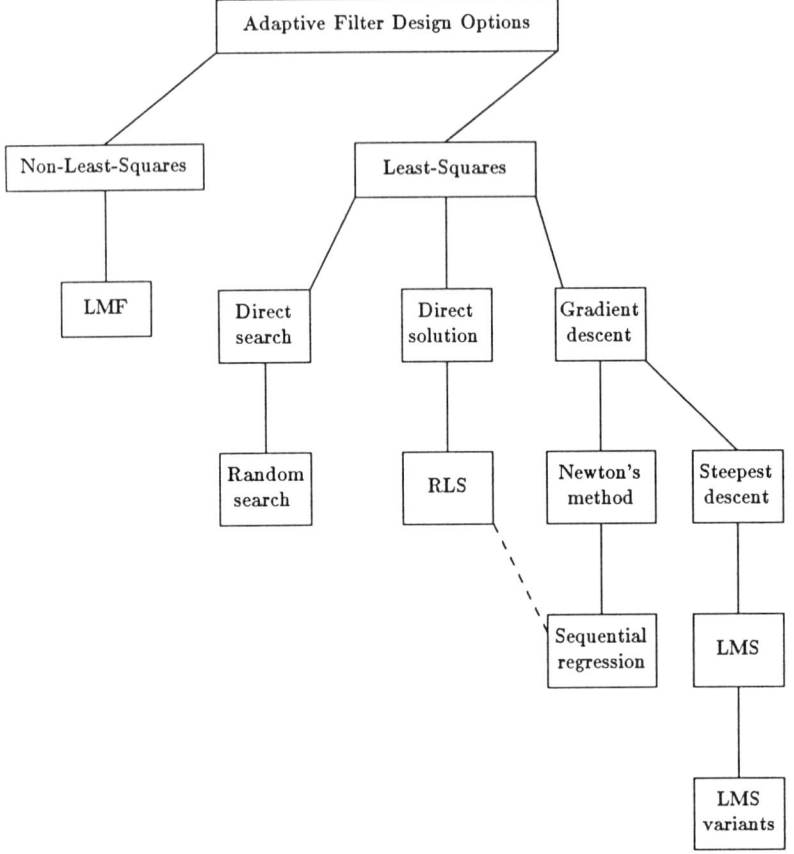

Figure 6.1.1 Options for adaptive algorithm design.

Least Mean Fourth (LMF) adaptive algorithm [1], in which adaptive structures are derived from minimization of a class of functions of the form:

$$J^{(N)} = E\{e^{2N}(n)\} \quad ; N \geq 1 , \qquad (6.1.1)$$

where N is an integer constant, and where as usual

$$e(n) = d(n) - \underline{f}_n^t \underline{x}_n .$$

It can be shown that the functional $J^{(N)}$ has no local minima (it is a **convex function**) and we may therefore use a gradient based adaptation scheme without fear of convergence to a local minimum. Replacing the expectation of $e^{2N}(n)$ with the instantaneous value as in the LMS, and taking the derivative, we have

$$\nabla(e^{2N}(n))=-2Ne^{2N-1}(n)\underline{x}_n \ . \tag{6.1.2}$$

Hence using the steepest descent update

$$\underline{f}_{n+1}=\underline{f}_n-\frac{\alpha}{2}\nabla(e^{2N}(n)) \ , \tag{6.1.3}$$

and we obtain

$$\underline{f}_{n+1}=\underline{f}_n+\alpha Ne^{2N-1}(n)\underline{x}_n \ . \tag{6.1.4}$$

Different values of N give rise to a family of adaptive filters. In fact, the name Least Mean Fourth is only appropriate when $N=2$. Another special case arises when $N=1$ for which the algorithm reduces to the conventional LMS.

Widrow and Walach [1] examined the properties of the LMF algorithm in the context of the noisy system identification problem:

$$d(n)=\underline{f}^{*t}\underline{x}_n+w(n) \ ,$$

where $w(n)$ is a zero-mean iid noise (uncorrelated with $x(n)$) representing both observation and modelling errors. They gave conditions for convergence in the mean and in mean-square for the general class of algorithms. The same authors showed that an algorithm obtained with $L>1$ may outperform LMS for some data distributions. On the other hand they found that when the unmodelled noise $w(n)$ is Gaussian, then LMS is the best choice.[1]

With the exception of algorithms like LMF, most adaptive filter designs have been derived from least-squares minimizations.[2] Referring again to Figure 6.1.1 we see that within the least-squares

---

1. These tests were comparisons of mean-squared errors obtained using adaptation constants chosen to give equal convergence rates for the different algorithms.

2. In fact, some of the LMS modifications discussed in Section 6.2 can be derived from non least-squares, or at least modified least-squares criteria. One example is the so-called 'leaky' algorithm, described below, which is designed to protect adaptive filter coefficients from the impact of noise, but which can be derived from a constrained least-squares functional.

framework there are a number of distinct paths — all leading to different algorithms. Perhaps the simplest are those referred to as LMS variants which we discuss in the next Section.

6.2 LMS Variants

A number of algorithms have been developed which can be classified as **LMS variants**. These include a wide range of possibilities but have the common feature that they start from the LMS algorithm, and attempt to improve on this by incorporating some modification, usually designed on an ad-hoc basis, to achieve some specified objective. This objective may be improved convergence, reduced steady-state mean-squared error, reduced computational requirement, or other application specific objective.

6.2.1 Normalized LMS

We have seen that the stability, convergence and steady-state properties of the LMS algorithm are all directly influenced by the length of the filter and the power of the input signal $x(n)$ (see Section 4.2). Perhaps the most obvious modification to the LMS, in view of this dependence, is to normalize the adaptation constant with respect to these factors. That is, we seek to replace α in the LMS update (4.1.14) by

$$\alpha' = \frac{\alpha}{L \times E\{x^2(n)\}} = \frac{\alpha}{T_p} \ . \tag{6.2.1}$$

In practice, the power T_p is estimated using a time average of the measured data. A convenient estimate is provided by

$$\hat{T}_p = \sum_{j=0}^{L-1} x^2(n-j) = \underline{x}_n^t \underline{x}_n \ . \tag{6.2.2}$$

For a stationary and ergodic input $x(n)$, $\hat{T}_p$ is an unbiased and consistent estimator for T_p (see Section 2.6). Incorporation of this simple normalization into the LMS update gives the modified form

$$\underline{f}_{n+1} = \underline{f}_n + \alpha \frac{e(n)\underline{x}_n}{\underline{x}_n^t \underline{x}_n} \ . \tag{6.2.3}$$

This modified algorithm, which is generally referred to as the **Normalized LMS (NLMS)** algorithm, is used in many situations. Indeed, in applications where the input signals are subject to widely fluctuating power levels, normalization is essential. The normalization does carry a computational overhead, though this can be limited to a single multiplication, division and addition, at the expense of an increased storage requirement. It is also important to avoid division by zero in (6.2.3). The probability of such zero division can be reduced by simply extending the data length over which the power estimate is made. This has the disadvantage of increasing the response time of the algorithm when power changes do occur. Another approach is to add a small constant to the divisor in equation (6.2.3). Thus the update becomes

$$\underline{f}_{n+1} = \underline{f}_n + \alpha \{ \frac{e(n)\underline{x}_n}{c + \underline{x}_n^t \underline{x}_n} \} \ , \tag{6.2.4}$$

where c is a small positive constant. For the homogeneous problem $d(n) = \underline{f}^{*t}\underline{x}_n$, with this normalized algorithm it can be shown [2] that provided $0 < \alpha < 2$:

i) $\lim_{n \to \infty} \{e(n)\} \to 0.$

ii) The error vector $\underline{v}_n = \underline{f}_n - \underline{f}^*$, converges to a (finite) constant as $n \to \infty$.

We note that these results are essentially independent of the input data $x(n)$. In the more general problem where homogeneity is not assumed, the analysis of both the convergence and steady-state behavior of the algorithm are greatly complicated by the normalizing factor.[3] However, for zero-mean Gaussian inputs and using the independence assumption, the basic normalized algorithm (6.2.3) can be shown to converge in the mean to the least-squares solution $\underline{f}^*$ [3].

---

3. By contrast, for deterministic inputs the transfer function approach of Section 4.2.4 may easily be extended to give a complete performance description of the normalized algorithm, at least for some inputs (see Problem 4.13 for the explicit transfer function derivation in this case).

6.2.2 *LMS Algorithms with a Variable Adaptation Rate*

Another straightforward extension of the LMS algorithm is obtained by replacing the constant α by a time or data-dependent step-size $\alpha(n)$. The objective of this modification is to improve the initial convergence of the algorithm by using a high adaptation rate. Once the algorithm has converged, a smaller value for α is employed to produce a lower steady-state mean-squared error. For stationary data, $\alpha(n)$ is typically chosen to reduce in magnitude with time. For example Ristow and Kosbahn [4] used

$$\alpha(n) = \frac{1}{n+c} \ , \tag{6.2.5}$$

where c is a positive constant. With this policy, $\alpha(n)$ reduces monotonically as (hopefully) the optimum solution is approached. Such solutions have *apparently* obvious advantages for stationary data, though as we shall see, there may not be any advantage even when the data *is* stationary. For nonstationary data such a formula would clearly be inappropriate since the reduction in steady-state error would be accompanied by inability of the algorithm to respond to changes in the optimum solution $\underline{f}^*$.

A natural extension of this approach is to develop some method of control whereby the adaptation can be decreased in steady-state, and increased to respond to possible nonstationarity. Such an algorithm, dubbed the **Variable Step (VS)** algorithm, has been proposed by Harris *et al.* [5]. The algorithm is characterized by update equation:

$$\underline{f}_{n+1} = \underline{f}_n + e(n) M_n \underline{x}_n \ , \tag{6.2.6}$$

where

$$M_n = \begin{bmatrix} \alpha_0(n) & 0 & . & 0 \\ 0 & \alpha_1(n) & 0 & . \\ . & 0 & . & 0 \\ 0 & . & . & \alpha_{L-1}(n) \end{bmatrix} . \tag{6.2.7}$$

This represents a generalization of LMS in two ways: Firstly, each coefficient of the filter is updated using an independent adaptation coefficient, and secondly constant adaptation is replaced by a time-variable step-size. Hence, the i-th coefficient is updated using $\alpha_i(n)$. The step-sizes are determined according to an ad-hoc strategy designed to force $\alpha_i(n)$ to decrease as the error approaches steady-state, and to increase as the solution moves away from the region of the optimum solution. This is achieved by monitoring sign changes in the instantaneous gradient estimate $-2e(n)x(n-i)$. $\alpha_i(n)$ is decreased by some factor c_1 if m_1 successive sign changes are observed in $e(n)x(n-i)$, and conversely $\alpha_i(n)$ is increased by a factor c_2 if $e(n)x(n-i)$ has the same sign on m_2 successive updates. Harris *et al.* argued that successive changes in the sign of the gradient estimate indicate that the algorithm is close to the optimum solution (for which $\nabla J = 0$), and conversely when no sign changes occur, the filter coefficient vector is distant from $\underline{f}^*$. Imposing hard limits on $\alpha_{\max}$ and $\alpha_{\min}$ in the procedure is enough to guarantee stability (see Problem 6.4). m_1 and m_2 are parameters which can be adjusted to optimize performance, as can the factors c_1, c_2. Harris *et al.* demonstrated that this variable-step LMS algorithm can achieve considerable improvements in convergence rate for only a modest increase in computation. (The behavior of this algorithm is examined more fully in Computer Problem 6.1.)

As it turns out, even in the stationary case, there are certainly some applications where time or data modification of α offers no advantages. In particular, when the steady-state error represents a high proportion of the total energy, as for example in the line enhancement application, then no advantage accrues. The problem arises because the adaptation mechanism must allow the filter to 'forget' old data as well as to acquire new data. The point is that if the filter is operated at an initially high adaptation rate, then the error in the filter coefficients due to the noise in the input will also be high. If α subsequently falls too quickly, the filter may contain erroneous components which it is unable to forget. The filter has acquired a set of **bad habits** which it is subsequently stuck with!

6.2.3 The Leaky LMS Algorithm

The **leaky LMS algorithm** employs the LMS update modified by the presence of a constant 'leakage' factor γ

$$\underline{f}_{n+1} = \gamma \underline{f}_n + \alpha e(n) \underline{x}_n , \tag{6.2.8}$$

where $0<\gamma<1$, but where typically, γ is close to one.[4] Leakage allows the impact on the filter coefficient vector of any single input sample to decay with time. Also, if the gradient estimates are consistently zero then f_n decays to zero. These properties are useful in several situations. For example, when the adaptive filter has modes associated with zero eigenvalues which do not converge (see also Section 4.2). In this case, these marginally stable modes can be driven unstable in a practical implementation (with finite word length representation) [6]. Another important problem arises in the ADPCM system outlined in Section 5.4. In practical ADPCM [7], the transmission is often corrupted by bit errors. Typical bit error rates (BER's) may range from lows of 10^{-6} to up to 10^{-2} and beyond. In DPCM coding systems, such errors create problems for both the quantizer and predictive segments of the system. In relation to the predictor coefficients, the leaky algorithm allows the impact of these channel errors to gradually decay (see also Section 6.7). Other applications where the leaky algorithm has been employed include adaptive equalization [8] and side-lobe cancellation [9]. The use of this algorithm is not without cost, however. In the noise-free system, the performance of the leaky LMS will be inferior to conventional LMS. This is clear from consideration of the steady-state solution (6.2.10) for the leaky algorithm. In particular, rearranging the update (6.2.8) we may write

$$\underline{f}_{n+1}=(\gamma I-\alpha\underline{x}_n\underline{x}_n^t)\underline{f}_n+\alpha d(n)\underline{x}_n \ . \tag{6.2.9}$$

Assuming independence of filter and data vectors we have

$$E\{\underline{f}_{n+1}\}=(\gamma I-\alpha R)E\{\underline{f}_n\}+\alpha\underline{g} \ , \tag{6.2.10}$$

or

$$E\{\underline{f}_{n+1}\}=\left[I-\alpha[R+\frac{(1-\gamma)}{\alpha}I]\right]E\{\underline{f}_n\}+\alpha\underline{g} \ .$$

From this, it can be shown (see Problem 6.5) that if the algorithm is stable then

---

4.   In fact, the leakage update can be written in several slightly different forms
     depending on the definition of the leakage factor γ and adaptation constant α.

$$\lim_{n\to\infty}\{E\{\underline{f}_n\}\}\to\left[R+\frac{(1-\gamma)}{\alpha}I\right]^{-1}\underline{g}\,,\tag{6.2.11}$$

which is clearly biased from $R^{-1}g$. Thus, the use of leakage can be thought of as a compromise between bias and coefficient protection. The closer γ is to 1, the less the leakage. Note also that the mean filter vector is driven towards the optimum via the **transition matrix**

$$\left[I-\alpha[R+\frac{(1-\gamma)}{\alpha}I]\right]\,,$$

instead of the usual $(I-\alpha R)$. We may observe that only the diagonal values of the matrix are different compared to LMS. Moreover, it is easy to show that if R has eigenvalues λ_i for $i=1,2,...,L$, then $(R+[(1-\gamma)/\alpha]I)$ has eigenvalues $\lambda_i+(1-\gamma)/\alpha$ for $i=1,2,...,L$. Thus, all of the eigenvalues of the transition matrix for the leaky algorithm are increased by a constant amount compared to LMS. This has three consequences:

i) The stability requirement (in the mean) is

$$0<\alpha<\frac{2}{\lambda_{\mathrm{max}}+\frac{(1-\gamma)}{\alpha}}\,,\tag{6.2.12}$$

for $0<\gamma<1$ where λ_{max} is the largest eigenvalue of R. Clearly, the implicit nature of this formula complicates the determination of stability limits.

ii) The matrix $R+[(1-\gamma)/\alpha]I$ is strictly positive definite (for $\alpha>0$, $0<\gamma<1$) and thus has no zero eigenvalues.

iii) The time-constants for the algorithm are:

$$t_i^{(L)}=\frac{1}{\alpha\lambda_i+(1-\gamma)}\leq t_i\,,\tag{6.2.13}$$

where $t_i^{(L)}$ is the time-constant for the i-th mode for the leaky algorithm and t_i is the corresponding constant for the conventional LMS (see also Section 4.2.2, equation (4.2.27)).

The bias of the leaky algorithm is illustrated numerically in Figure 6.2.1, which shows the filter coefficient $f_n(0)$, for $n=0,1,...,999$, for the leaky and regular LMS algorithms. In this experiment, the signals were

$$d(n)=x(n)+0.5x(n-1) , \qquad\qquad (6.2.14)$$

where $x(n)$ is zero-mean, unit variance iid Gaussian. We note immediately, that $f^*(0)=1$, $f^*(1)=0.5$, and $f^*(i)=0$, $i=2,3,...$ In these experiments the parameters used were $L=10$, and $\alpha=0.01$.

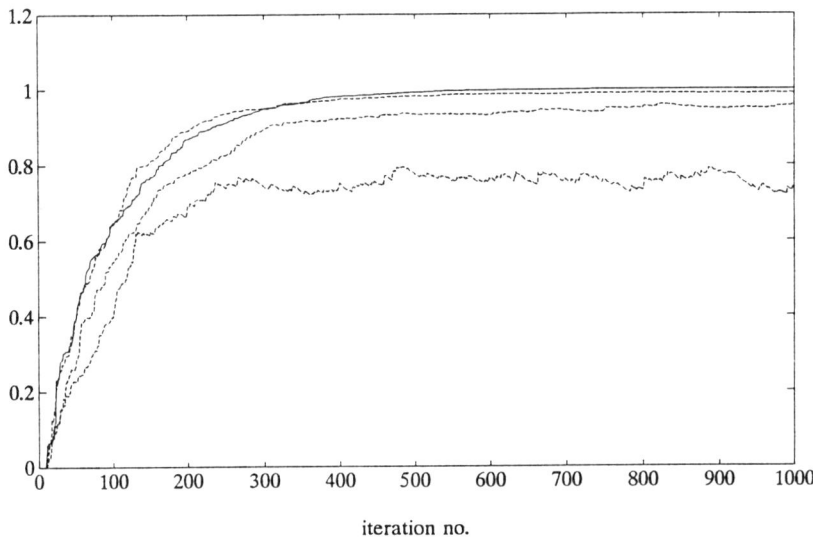

iteration no.

Figure 6.2.1 Leaky LMS trial. Plot shows $f_n(0)$ versus n, for leakage factors γ of 1 (solid line), 0.9999, 0.9995 and 0.997 (dashed lines).

Figure 6.2.1 illustrates leakage factors of 1 (LMS), 0.9999, 0.9995 and 0.997. Note that, given $R=\sigma^2 I$, with $\sigma=1$, then from equation (6.2.11) for $f_n(0)$ the leaky algorithm should converge in the mean to 1, 1/1.01, 1/1.05, and 1/1.3, respectively. Inspection of Figure 6.2.1 shows that experimental results are consistent with theory. Note, however, that because of the dependence on α of the bias, apparently small leakage factors can cause significant bias. Note also that apart from the bias, we also observe increasing fluctuations about the steady-state value as γ is reduced. This is due to the loss of the homogeneity in the problem when $\underline{f}_\infty$ is biased from $\underline{f}^*$.

Finally, we note that the leaky LMS algorithm can be derived directly [10] by consideration of the modified functional

$$J' = e^2(n) + a\underline{f}_n^t\underline{f}_n \ , \tag{6.2.15}$$

where a is a constant. In Problem 6.7 we show that minimization of (6.2.15) leads to the leaky update

$$\underline{f}_{n+1} = (1-\alpha a)\underline{f}_n + \alpha e(n)\underline{x}_n \ . \tag{6.2.16}$$

Re-defining $\gamma = (1-\alpha a)$ confirms the equivalence of this scheme and that of equation (6.2.8).

6.2.4 Sign Algorithms

In spite of the relative computational simplicity of the LMS algorithm, there are some applications, particularly in high speed communications, where even this computational load is too high. Efforts have been made to develop simplified algorithms with a reduced computational requirement. One such method is the Block LMS described in Section 6.2.9 below, in which the coefficients are updated at a reduced rate compared to the filtering operation. Another approach designed to reduce computation and simplify hardware requirements is provided by the **sign algorithms**. Three sign based LMS algorithms have been proposed. These quantize data, error or both in the LMS update to a single level representing the sign of the quantity only. The algorithms are:

i) The Pilot LMS, or Signed Error (SE), or simply Sign Algorithm (SA)

$$\underline{f}_{n+1} = \underline{f}_n + \alpha sgn[e(n)]\underline{x}_n \ . \tag{6.2.17}$$

ii) The Clipped LMS, or Signed Regressor (SR)

$$\underline{f}_{n+1} = \underline{f}_n + \alpha e(n)sgn[\underline{x}_n] \ . \tag{6.2.18}$$

iii) The Zero-Forcing LMS, or Sign-Sign (SS)

$$\underline{f}_{n+1} = \underline{f}_n + \alpha sgn[e(n)]sgn[\underline{x}_n] \ . \tag{6.2.19}$$

In each case, the 'sign' operation is defined for a scalar variable r by

$$sgn\,[r] = 1 \qquad ; r > 0 \,, \qquad\qquad (6.2.20)$$
$$= 0 \qquad ; r = 0 \,,$$
$$= -1 \quad ; r < 0 \,,$$

and the operation is defined component by component for vector quantities. The variety of names for each algorithm is simply a reflection of more-or-less independent development work on these algorithms reported by various authors. The advantage of the use of sign representations is that multiplications in the update are reduced to single bit operations. The filtering requirement is unchanged in the sign algorithms, but the update is reduced to simple shifting and addition.[5] On the other hand, compared to LMS the update mechanism is degraded by the crude quantization of the gradient estimates, and this is manifested as either decreased convergence rate or increased steady-state error. The impact of the sign operation in the Signed Regressor is potentially more profound than in the Signed Error algorithm. As pointed out by Treichler *et al.* [6], the SE only modifies the *size* of each iterative step, the SR also changes the *direction* of the update (in the space of filter coefficients). This is reflected in less predictable behavior for the SR algorithm, although using the independence assumption and zero-mean Gaussian inputs, mean convergence can be demonstrated. However, Sethares *et al.* [11] have shown that sequences exist which produce stable results for LMS but are *unstable* for the SR algorithm. Moreover, this instability has nothing to do with the size of the adaptation constant α. It is caused by a change of direction produced by replacing $\underline{x}_n$ with $sgn\{\underline{x}_n\}$. For some input sequences, this change drives the filter coefficients *away* from $\underline{f}^*$ (we examine examples of such behavior in the sign algorithms in Problems 6.9-6.11). In spite of these occasional, rather pathological examples, the algorithm remains popular in applications. (Actually before such divergent examples were first uncovered at least one hardware implementation of the SR algorithm had been developed, and enjoyed some success [12]).

---

5. For signed error, the update requires the further constraint that the adaptation constant α be an integer power of two if multiplications are to be avoided in the update.

Although the three algorithms were conceived as LMS variants, in fact the SE can be derived directly (see Problem 6.8) as an LMS type update using the modified functional

$$J = E\{|e(n)|\} . \qquad (6.2.21)$$

The Sign-Sign (SS) algorithm is the simplest of all, and for that reason the most widely used (including for example, in the CCITT ADPCM standard [13] referred to in Section 5.4). The update is very crude in this case, and the algorithm is consequently the least robust of all. For the SS algorithm, simple periodic sequences of length three can be found which will force the system to be unstable [14], (see Problem 6.9).

6.2.5 Linear Smoothing of the LMS Gradient Estimates

An obvious drawback of the LMS algorithm is the use of a noisy instantaneous gradient estimate in the update equation. A possible improvement consists of smoothing this estimate by averaging over several consecutive instantaneous values. Such an average may be constructed using a moving window of the N most recent gradient estimates [15]. Thus the update equation for this **Averaged LMS (ALMS)** algorithm has the form

$$\underline{f}_{n+1} = \underline{f}_n + \frac{\alpha}{N} \sum_{j=n-N+1}^{n} e(j)\underline{x}_j . \qquad (6.2.22)$$

More generally Proakis [16] and later Glover [17], proposed that the noisy gradient estimate be subjected to a general lowpass filtering operation. The update equation for this modified algorithm is then

$$\underline{f}_{n+1} = \underline{f}_n + \alpha \underline{b}_n , \qquad (6.2.23)$$

where $\underline{b}_n = [b_n(0), b_n(1), ..., b_n(L-1)]^t$, where $b_n(i)$ is the output of a lowpass filtering operation applied to the i-th component of the gradient estimate. This lowpass filtering is interpreted element-by-element. That is,

$$b_n(i) = LPF\{e(n)x(n-i), e(n-1)x(n-i+1), ...\} , \qquad (6.2.24)$$

where *LPF* represents a lowpass filter which is restricted to causal form. The ALMS algorithm of (6.2.22) is one example of this class in which the lowpass filter has impulse response $\{1/N, 1/N, ..., 1/N\}$. There is, however, no particular reason to restrict attention to filters with uniform weighting, or even to Finite Impulse Response (FIR) structures. For example, consider the first-order recursive equation

$$\underline{b}_n = \underline{b}_{n-1} + \gamma(e(n)\underline{x}_n - \underline{b}_{n-1}) \ , \tag{6.2.25}$$

where γ is a constant with $0 < \gamma < 1$. Here, each element of $e(n)\underline{x}_n$ is filtered using a transfer function of the form

$$G(z) = \frac{\gamma}{1 - (1-\gamma)z^{-1}} \ . \tag{6.2.26}$$

This transfer function is a first order lowpass filter (see Figure 6.2.2) with a single pole at $z_p = (1-\gamma)$.

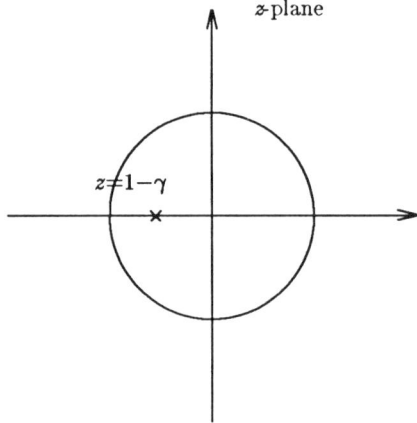

Figure 6.2.2 Pole-zero plot for the lowpass operation of equation (6.2.26).

The LMS update has an effective feedback loop of

$$\frac{\alpha}{1 - z^{-1}} \ , \tag{6.2.27}$$

and combining this with (6.2.26) we obtain an overall update equation

$$\underline{f}_{n+1}=\underline{f}_n+(1-\gamma)(\underline{f}_n-\underline{f}_{n-1})+\alpha\gamma e(n)\underline{x}_n \ . \tag{6.2.28}$$

This algorithm is also known as the **Momentum LMS**. Glover [17] compared the behavior of this second-order scheme with the LMS and with a similar third-order scheme. He found, however, that reducing misadjustment with such algorithms also reduces the convergence rate. Similar conclusions have been reached more recently by Roy and Shynk [18].

6.2.6 Nonlinear Smoothing of the LMS Gradient Estimates

There are certainly some situations where some form of smoothing can greatly improve the performance of the LMS algorithm. For example, the presence of impulsive interference in either $d(n)$ or $x(n)$ generally has a degrading effect on the filter, and in extreme cases can lead to instability. One solution to this problem, described in [19], subjects the gradient estimate to a *nonlinear* smoothing using the sample **median** of the N most recent instantaneous gradient estimates. The median of N samples is obtained by algebraically ranking the samples from smallest to largest and then selecting the middle value. Thus, given a sequence of N signal samples $s_1, s_2, ..., s_N$, we rank the samples as $s_{(1)}, s_{(2)}, ..., s_{(N)}$, with $s_{(1)} \le s_{(2)} ... \le s_{(N)}$. The median is then

$$s_{med}=Med\{s_i\}_N=s_{(N+1)/2} \ , \tag{6.2.29}$$

where, for convenience we have assumed N is odd so that the middle point is an integer index. **Median filters** [20] are widely used in signal and image processing. A median filter is defined[6] as a running median applied to a window of N data points:

$$y(n)=Med\{x(n)\}_N=median\{x(n), x(n-1), ..., x(n-N+1)\}. \tag{6.2.30}$$

Such filters are not linear and hence cannot be characterized in terms of a simple transfer function. They are characterized by two key properties (see Problems 6.14 and 6.15):

i) Median filters completely remove short duration impulses from a signal.

---

6. Actually, there are both recursive and non-recursive median filters. That given here is a non-recursive form.

ii) Median filters pass 'edges' without smearing.

Both of these properties contrast sharply with linear filters which generally smear edges and can only partially attenuate impulses.

A **Median LMS (MLMS) Algorithm** [21] is defined by applying the median operation to the elements of the most recent N gradient estimates in the LMS algorithm. That is, the MLMS algorithm is characterized by an update of the form

$$f_{n+1}(i) = f_n(i) + \alpha Med\{e(n)x(n-i)\}_N \quad ; \, i=0,1,...,L-1 \, , \quad (6.2.31)$$

where the operation $Med\{e(n)x(n)\}_N$ denotes the application of the median to the N most recent gradient estimates:

$$Med\{e(n)x(n-i)\}_N = median\{e(n)x(n-i),...$$
$$...,e(n-N+1)x(n-i-N+1)\} \, .$$

In the MLMS algorithm, it is the impulse removal property of the median filter which is of interest. Consider a simple example with input signals (see Figure 6.2.3)

$$d(n) = \underline{h}^t \underline{w}_n + \eta(n) \, ,$$
$$x(n) = w(n) + \varsigma(n) \, , \quad\quad\quad\quad\quad\quad (6.2.32)$$

where $w(n)$ is a zero-mean, Gaussian iid sequence, $\varsigma(n)$ and $\eta(n)$ are sparse impulsive interferences, $\underline{h} = [h(0),...,h(L-1)]^t$ is a linear time-invariant system, and where $\underline{w}_n = [w(n),...,w(n-L+1)]^t$. In this experiment we are interested in identifying $\underline{h}$, given only input and output measurements that are corrupted by sparse impulsive interference. Let us assume for simplicity that $\underline{h} = [-1,0,...,0]^t$ and that $\underline{f}_0 = \underline{0}$. We will examine the behavior of the adaptive filter for two cases: a) $\varsigma(n) \equiv 0$ and b) $\eta(n) \equiv 0$, in order to separately examine the influence of input and output impulsive disturbance. Referring to Figure 6.2.3, the input $w(n)$ is a zero-mean iid sequence with samples uniformly distributed on $[-1,1]$. The disturbances $\eta(n)$, $\varsigma(n)$, are 'sparse' sequences with non-zero values (impulses) occurring with a mean interval of 100 samples. The impulse amplitudes are Gaussianly distributed with mean zero and standard deviation equal to 8. Figure 6.2.4 shows the coefficient error $v_n(0) = f_n(0) - f_n^*(0)$ for $n=0,1,...,3999$ for case a) using the MLMS algorithm. For reference, the coefficient track obtained

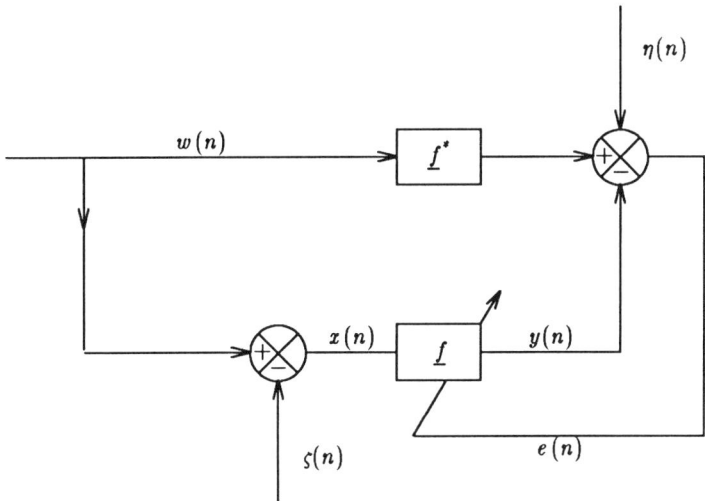

Figure 6.2.3 Adaptive system with additive impulsive noises applied to input and desired signals.

using conventional LMS is also included. The plots were obtained using $L=5$, and $\alpha=0.01$ in both cases. For MLMS, the median window length N in this trial was set as 3. The influence of the sparse impulsive noise is to perturb the coefficient solution for LMS. For MLMS, the coefficient track remains smooth and largely unaffected by impulses. Figure 6.2.5 shows the corresponding results for case b) using a similar sparse impulsive disturbance. Here, the performance of LMS is even more seriously degraded, fluctuating wildly, and in fact it can be shown (see Problem 6.20) that the mean convergence in this case is to a solution which is biased away from $\underline{h}$. MLMS, by contrast, again exhibits smooth convergence towards $\overline{\underline{h}}$.

Even though MLMS has obvious advantages in such environments, we should note that this gain is offset by slightly slower convergence when the inputs are non-impulsive [21]. Also, while the performance of the algorithm has been convincingly demonstrated in many practical situations, analytic performance predictions have only proved possible for very restrictive inputs, and subject to a number of simplifying assumptions (see Problem 6.19). There are also known to be a few pathological cases (as with the sign algorithms) where instability can occur in the MLMS algorithm [22] (see Problem 6.18).[7]

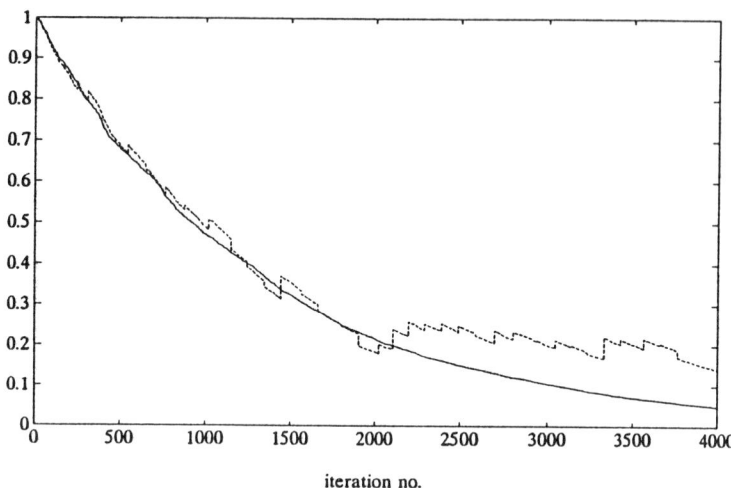

Figure 6.2.4 Adaptive system modelling with impulsive noise in the system output. Plot shows filter coefficient error $v_n(0)=f_n(0)-f_n^*(0)$ for MLMS (solid line) and LMS (dashed line).

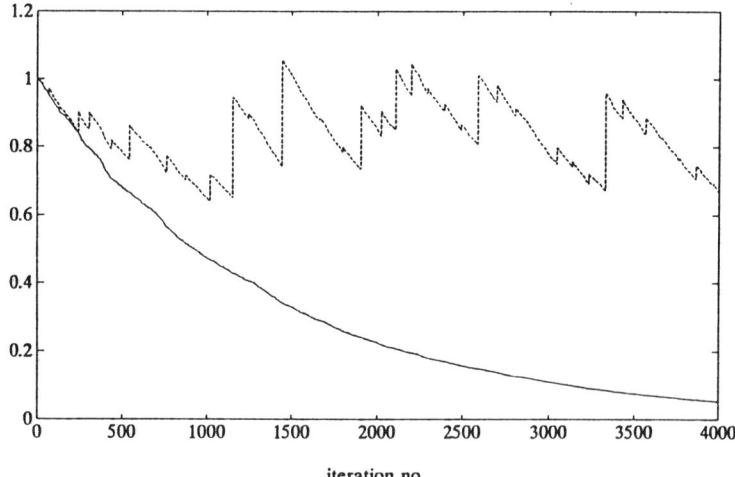

Figure 6.2.5 Adaptive system modelling with impulsive noise in the system input. Plot shows filter coefficient error $v_n(0)=f_n(0)-f_n^*(0)$ for MLMS (solid line) and LMS (dashed line).

7. We note that the median LMS is in fact just one member of a class of **Order Statistic LMS** algorithms [23] where the median operation is replaced by various **Order Statistic** filters (see also Problems 6.16, 6.17).

6.2.7 Double Updating

Yet another LMS variant is provided by the 'double updating' procedure of Ristow and Kosbahn [4]. The basic update is provided by the simple normalized LMS equation:

$$\underline{f}_{n+1}=\underline{f}_n+\alpha\frac{e(n)\underline{x}_n}{\underline{x}_n^t\underline{x}_n}\ ,\qquad\qquad (6.2.3)$$

where, as usual

$$e(n)=d(n)-y(n)\ ,\qquad y(n)=\underline{f}_n^t\underline{x}_n\ .$$

The difference between this procedure and the LMS is that in this algorithm, the updated coefficients $\underline{f}_{n+1}$ are applied to the same data, that is to the data at time n. This produces

$$\hat{y}(n)=\underline{f}_{n+1}^t\underline{x}_n\ ,\qquad\qquad (6.2.33)$$

and $\hat{y}(n)$ represents the final output at time n. An interesting feature of this algorithm emerges if we consider the application of the algorithm with $\alpha=1$. The final output $\hat{y}(n)$ is then given by

$$\hat{y}(n)=\underline{f}_{n+1}^t\underline{x}_n=\underline{f}_n^t\underline{x}_n+e(n)\frac{\underline{x}_n^t\underline{x}_n}{\underline{x}_n^t\underline{x}_n}\ ,$$

where the second equality was obtained by substituting (6.2.3) into (6.2.33). Hence, we have

$$y^{'}(n)=y(n)+(d(n)-y(n))\ ,$$
$$y^{'}(n)=d(n)\ .\qquad\qquad (6.2.34)$$

In words, the output of the filter is equal to the desired input, and the result is a perfect algorithm! Paradoxically this is generally *not* a desirable result. Consider, for example, the noise cancellation problem of Section 5.1. The underlying principal was that the filter should adapt to cancel the noise but be *unable* to cancel the signal component. With $\alpha=1$, this algorithm applied to the ANC problem would not only cancel the noise but would also remove the signal! In fact, further consideration suggests that this algorithm does not in general solve the least-squares problem since it can always produce J equal to 0. Clearly, in many applications this is not desirable.

6.2.8 Complex LMS Algorithm

Thus far we have dealt strictly with real signals. A version of the LMS algorithm designed to operate with complex inputs has been developed [24]. This algorithm is useful for adaptive arrays which are discussed in Chapter 8, and for frequency domain implementations which are introduced in Section 6.6.

As a starting point for our consideration of the complex algorithm we define, analogously to the real case, a complex data vector

$$\underline{x}_n = \underline{x}_n^R + j\underline{x}_n^I , \qquad (6.2.35)$$

where $\underline{x}_n^R$ and $\underline{x}_n^I$ are the real and imaginary parts of the data vector and where

$$\underline{x}_n^R = [x^R(n), x^R(n-1), ..., x^R(n-L+1)]^t ,$$

with $\underline{x}_n^I$ defined similarly. The complex weight vector $\underline{f}_n$ can be similarly decomposed into real and imaginary components

$$\underline{f}_n = \underline{f}_n^R + j\underline{f}_n^I . \qquad (6.2.36)$$

Likewise, definitions of the complex output, desired signal and error may be given as

$$y(n) = \underline{f}_n^t \underline{x}_n = y^R(n) + jy^I(n) , \qquad (6.2.37)$$

$$= (\underline{f}_n^{R^t} \underline{x}_n^R - \underline{f}_n^{I^t} \underline{x}_n^I) + j(\underline{f}_n^{I^t} \underline{x}_n^R + \underline{f}_n^{R^t} \underline{x}_n^I) ,$$

$$d(n) = d^R(n) + jd^I(n) , \qquad (6.2.38)$$

and

$$e(n) = e^R(n) + je^I(n) = d(n) - y(n) . \qquad (6.2.39)$$

For real signals we minimize

$$J = E\{e^2(n)\} .$$

For complex signals this is replaced by

$$J = E\{|e(n)|^2\} = E\{e(n)e^*(n)\} = E\{[e^R(n)]^2 + [e^I(n)]^2\} , \quad (6.2.40)$$

where * denotes complex conjugate.[8] An LMS type algorithm, derived from a steepest descent iteration applied to the instantaneous version of the functional (6.2.40), can be found. We can obtain the form for the update equation without invoking any results from the calculus of complex vectors, by using the decomposition into real and imaginary parts given above. From (6.2.37) and (6.2.38), $e(n)$ has the form

$$e(n) = d^R(n) + j d^I(n) - \left[(\underline{f}_n^{R^t} \underline{x}_n^R - \underline{f}_n^{I^t} \underline{x}_n^I) + j(\underline{f}_n^{I^t} \underline{x}_n^R + \underline{f}_n^{R^t} \underline{x}_n^I) \right] , \quad (6.2.41)$$

so that $|e(n)|^2$ is given by

$$|e(n)|^2 = (d^R(n) - \underline{f}_n^{R^t} \underline{x}_n^R + \underline{f}_n^{I^t} \underline{x}_n^I)^2 + j(d^I(n) - \underline{f}_n^{I^t} \underline{x}_n^R - \underline{f}_n^{R^t} \underline{x}_n^I)^2$$

$$= [e^R(n)]^2 + [e^I(n)]^2 . \quad (6.2.42)$$

The LMS algorithm uses the derivative with respect to the instantaneous version of the functional. As we saw in Chapter 4, the LMS algorithm is derived from the steepest descent equation

$$\underline{f}_{n+1} = \underline{f}_n - \frac{\alpha}{2} \frac{\partial \hat{J}}{\partial \underline{f}_n} .$$

Hence, for the complex LMS the algorithm is obtained from

$$\underline{f}_{n+1} = \underline{f}_n - \frac{\alpha}{2} \frac{\partial |e(n)|^2}{\partial \underline{f}_n} , \quad (6.2.43)$$

or, splitting into real and imaginary components

$$\underline{f}_{n+1}^R = \underline{f}_n^R - \frac{\alpha}{2} \frac{\partial |e(n)|^2}{\partial \underline{f}_n^R} ,$$

and

---

8. Care should be taken to avoid confusion with the use of * to denote the optimal filter coefficients.

$$\underline{f}_{n+1}^I = \underline{f}_n^I - \frac{\alpha}{2} \frac{\partial |e(n)|^2}{\partial \underline{f}_n^I} \ .$$

In total

$$\underline{f}_{n+1} = (\underline{f}_n^R + j\underline{f}_n^I) - \frac{\alpha}{2} \left[\frac{\partial |e(n)|^2}{\partial \underline{f}_n^R} + j \frac{\partial |e(n)|^2}{\partial \underline{f}_n^I} \right] \ . \tag{6.2.44}$$

Now from (6.2.42)

$$\frac{\partial |e(n)|^2}{\partial \underline{f}_n^R} = -2e^R(n)\underline{x}_n^R - 2e^I(n)\underline{x}_n^I \ ,$$

$$\frac{\partial |e(n)|^2}{\partial \underline{f}_n^I} = 2e^R(n)\underline{x}_n^I - 2e^I(n)\underline{x}_n^R \ .$$

Substituting into (6.2.44) gives

$$\underline{f}_{n+1} = \underline{f}_n + \alpha \left[(e^R(n)\underline{x}_n^R + e^I(n)\underline{x}_n^I) - j(e^R(n)\underline{x}_n^I - e^I(n)\underline{x}_n^R) \right] \ ,$$

$$= \underline{f}_n + \alpha \left[(e^R(n) + je^I(n))(\underline{x}_n^R - j\underline{x}_n^I) \right] \ . \tag{6.2.45}$$

Hence

$$\underline{f}_{n+1} = \underline{f}_n + \alpha e(n)\underline{x}_n^* \ . \tag{6.2.46}$$

This is the update for the complex LMS, which as we can see is identical to that for LMS with real inputs, except for the conjugation of the data vector.

6.2.9 The Block LMS Algorithm

Another algorithm motivated by a desire to reduce the computational burden of LMS is the Block LMS (BLMS) algorithm [25]. The BLMS is identical to the LMS algorithm except that in the BLMS the filter weights are updated only once every N data samples. That is, the filter coefficients are held constant over each block of N data points. The system output and error at each point within a block are calculated using the filter vector for that block. The filter is updated using the average of the gradient estimates for that block. Thus, the BLMS update has the form:

$$\underline{f}_{(j+1)N}=\underline{f}_{jN}+\frac{\alpha_B}{N}\sum_{i=0}^{N-1}e(jN+i)\underline{x}_{jN+i} \quad ; i=0,1,...,N-1 , (6.2.47)$$

$$; j=0,1,... ,$$

where

$$y(jN+i)=\underline{f}_{jN}^{t}\underline{x}_{jN+i} \tag{6.2.48}$$

$$e(jN+i)=d(jN+i)-y(jN+i)=d(jN+i)-\underline{f}_{jN}^{t}\underline{x}_{jN+i} , \tag{6.2.49}$$

and where $\underline{x}_{jN+i}=[x(jN+i),x(jN+i-1),...,x(jN+i-L+1)]^{t}$ as in the usual data vector. Here, $\underline{f}_{jN}$ is the filter vector corresponding to the j-th block of N data points. α_B is the adaptation constant for the algorithm.

The BLMS algorithm generally has a computational advantage compared to LMS. The BLMS algorithm requires $NL+L$ multiplications per block for the updates. This compares to $NL+N$ per block of N points for the LMS. (Both algorithms have an equal number of additions.) For a direct implementation, both algorithms have an equal computational requirement for the convolution (NL multiplications per block). Thus, the total computation is less for BLMS whenever $L<N$. However, the major advantage for BLMS arises when efficient techniques for the implementation of block digital filters using parallel processing (or serial processing with the FFT algorithm) are employed.

Analytically, the BLMS algorithm is also of interest because of its relation to frequency domain implementations of the LMS which are discussed in Section 6.6. The algorithm is not, of course, identical to the LMS. To see this we need only apply repeated substitution to the usual LMS and thus write

$$\underline{f}_{(j+1)N}=\underline{f}_{jN}+\alpha\sum_{i=0}^{N-1}e(jN+i)\underline{x}_{jN+i} , \tag{6.2.50}$$

which is identical to (6.2.47), but where in the LMS

$$e(jN+i)=d(jN+i)-\underline{f}_{jN+i}^{t}\underline{x}_{jN+i} . \tag{6.2.51}$$

Substituting into (6.2.50) gives

$$\underline{f}_{(j+1)N} = \underline{f}_{jN} + \alpha \sum_{i=0}^{N-1} \left[d(jN+i)\underline{x}_{jN+i} - \underline{x}_{jN+i}\underline{x}_{jN+i}^t \underline{f}_{jN+i} \right] . \quad (6.2.52)$$

The equivalent result for the Block LMS can be obtained by substituting (6.2.49) into (6.2.47) to give

$$\underline{f}_{(j+1)N} = \underline{f}_{jN} + \frac{\alpha_B}{N} \sum_{i=0}^{N-1} \left[d(jN+i)\underline{x}_{jN+i} - \underline{x}_{jN+i}\underline{x}_{jN+i}^t \underline{f}_{jN} \right] , \quad (6.2.53)$$

which is clearly not the same as (6.2.52). Apart from the factor of N which, as we have indicated, could easily be normalized out by setting $\alpha = \alpha_B/N$, the difference is that the BLMS updates are produced by the average of the gradient values over a block, obtained with a single (fixed) filter coefficient vector. In the LMS algorithm, the filter is updated at every data point, and the updated information is fed back into the gradient estimation process. In fact, as shown in Problem 6.23, the BLMS algorithm can be derived as a steepest descent algorithm with functional

$$J = \frac{1}{N} E\{ \sum_{i=0}^{N-1} e^2(jN+i) \} . \quad (6.2.54)$$

While the adaptive algorithm derived from this functional may produce BLMS, the direct solution for the fixed least-squares operator (assuming, of course, that $x(n)$ and $d(n)$ are jointly stationary) is the usual form

$$\underline{f}^* = R^{-1}\underline{g} , \quad (6.2.55)$$

where the definitions of R and g are unchanged from those of Section 3.2 (see also Problem 6.23).

The algorithms may not be precisely equivalent, nevertheless for stationary data, and using the usual independence assumption for the data and filter vector, it is easy to demonstrate similar convergence properties for the two algorithms. Starting from the BLMS update (6.2.47) and error (6.2.49) we have

$$\underline{f}_{(j+1)N} = \underline{f}_{jN} + \frac{\alpha_B}{N} \sum_{i=0}^{N-1} [d(jN+i) - \underline{f}_{jN}^t \underline{x}_{jN+i}] \underline{x}_{jN+i}$$

$$= [I - \frac{\alpha_B}{N} \sum_{i=0}^{N-1} \underline{x}_{jN+i}\underline{x}_{jN+i}^t] \underline{f}_{jN} + \frac{\alpha_B}{N} \sum_{i=0}^{N-1} d(jN+i)\underline{x}_{jN+i} . \quad (6.2.56)$$

Now

$$E\{\sum_{i=0}^{N-1} \underline{x}_{jN+i}\underline{x}_{jN+i}^t\}=NR$$

and

$$E\{\sum_{i=0}^{N-1} d(jN+i)\underline{x}_{jN+i}\}=N\underline{g} \ .$$

Hence, applying expectations to both sides of (6.2.56) and using the independence assumption, we have

$$\underline{f}_{(j+1)N}=(I-\alpha_B R)\underline{f}_{jN}+\alpha_B \underline{g} \ . \tag{6.2.57}$$

As in Section 4.2 equation (4.2.9), we define a vector of expected filter coefficient errors: $\underline{u}_{jN}=E\{\underline{f}_{jN}\}-\underline{f}^*$. Then we may write (the analysis here is directly analogous to that of Section 4.2 for LMS, thus we abbreviate the algebra somewhat)

$$\underline{u}_{(j+1)N}=(I-\alpha_B R)\underline{u}_{jN} \ , \tag{6.2.58}$$

which is identical to the LMS mean error vector behavior (compare (6.2.58) with equation (4.2.10)), with the exception that the updates are indexed jN and $(j+1)N$, respectively. Hence we may apply the analysis of Section 4.2 directly, using the same orthogonal decomposition $R=Q\Lambda Q^t$, and ultimately, as in the LMS case, we can conclude that

$$\lim_{n\to\infty}\{E\{\underline{f}_n\}\}\to\underline{f}^* \ ,$$

subject to

$$0<\alpha_B<\frac{2}{\lambda_{\max}} \ , \tag{6.2.59}$$

where, as usual, $\lambda_{\max}$ is the largest eigenvalue of R.[9] The update

---

9. As with the LMS, convergence in the mean does not guarantee finite mean-squared error. Working with uncorrelated Gaussian inputs, Feuer [26] has obtained a more conservative limit which is necessary for convergence in mean-square. This result may be expressed as

$$0<\alpha_B\leq\frac{2}{(1+\frac{2}{L})tr\{R\}} \ .$$

for the orthogonalized mean error coefficients for the BLMS algorithm is identical to that for the LMS, that is

$$u'_{(j+1)N}(i)=(1-\alpha_B\lambda_i)u'_{jN}(i) \ ,\tag{6.2.60}$$

where $u'_n(i)$ is the i-th component of the decoupled error vector defined by

$$\underline{u}'_n=Q\underline{u}_n \ ,$$

which is identical to equation (4.2.17) for the usual LMS. In this case, however, the step relates to a block of N samples rather than a single data value. Consequently, in terms of the time-constants of Section 4.2, the BLMS time-constants are increased by a factor of N relative to LMS. That is

for LMS

$$t_i=\frac{1}{\alpha\lambda_i} \ ,\tag{4.2.27}$$

and for BLMS

$$t_{Bi}=\frac{N}{\alpha_B\lambda_i} \ .\tag{6.2.61}$$

Thus, the speed of convergence of each mode is reduced by a factor of N compared to that of LMS. On the other hand, this reduction is compensated by a reduction in excess mean-squared error due to the smoothing effect of the averaging operation. Clark *et al.* [25] give an expression for the steady-state mean-squared error for BLMS as (see Problem 6.24)

$$J=J_{\min}(1+\frac{\alpha_B}{2N}tr\{R\}) \ .\tag{6.2.62}$$

This compares with LMS (see Section 4.2.3, equation (4.2.52)) as

$$J=J_{\min}(1+\frac{\alpha}{2}tr\{R\}) \ .\tag{4.2.52}$$

Thus, the excess mean-squared error is reduced by a factor of N relative to LMS. Hence, subject to the stability considerations of (6.2.59), we may conclude that with stationary input data the choice

$$\alpha_B = N\alpha \ ,$$

gives equivalent convergence rates (time-constants) and steady-state behavior for the BLMS and LMS algorithms.[10]

6.2.10 LMS Volterra Filter

In recent years, interest in nonlinear systems, and particularly in nonlinear digital filtering, has steadily increased [20]. Much of this effort has focused on the well-known Volterra processor [30] which has been widely used in nonlinear system identification and filtering. While technological advances have increased available computational power, and thus improved the practicability of nonlinear processing schemes, most efforts have focused on second-order systems, for which the computation is most manageable. A digital second-order Volterra filter [31] with sampled input signals $x(n)$ produces an output $y(n)$ according to the relation

$$y(n) = \sum_{j=0}^{L-1} f^{(1)}(j)x(n-j) +$$
$$+ \sum_{j_1=0}^{L-1} \sum_{j_2=j_1}^{L-1} f^{(2)}(j_1,j_2)x(n-j_1)x(n-j_2) \ , \qquad (6.2.63)$$

where $f^{(1)}(j)$ are the usual linear filter coefficients, $f^{(2)}(j_1,j_2)$ are the $L(L+1)/2$ quadratic response coefficients, where the restriction of both linear and quadratic 'lengths' to L is for convenience, and where the limits j_1, j_2 have been chosen to avoid multiplicities in the quadratic components.[11]

---

10. More recent developments of the BLMS algorithm have included the use of a time-varying adaptation step-size $\alpha_B = \alpha_B(j)$ chosen at each block to optimize filter performance [27],[28], and a real-time implementation of the BLMS that has been undertaken by Young *et al.* [29].

11. Some implementations use the 'full' filter, without avoiding repeated terms in the quadratic kernel.

Given an input signal $x(n)$ and desired input $d(n)$ we may obtain a set of filter coefficients (linear and quadratic) which minimizes the mean-squared error

$$J = E\{e^2(n)\} = E\{[d(n) - y(n)]^2\} \ . \tag{6.2.64}$$

As with the linear filter, the solution may be obtained by differentiating J with respect to the individual coefficients and equating the results to zero (see Section 3.2). Differentiating with respect to the linear coefficients yields

$$\frac{\partial J}{\partial f^{(1)}(j)} = 0 = E\{e(n)x(n-j)\} \qquad ; 0 \le j \le L-1 \ , \tag{6.2.65}$$

which expresses the orthogonality of the error and the data, as occurs with the linear filter (see Section 3.2). Differentiating with respect to the second-order terms gives

$$\frac{\partial J}{\partial f^{(2)}(j_1,j_2)} = 0 = E\{e(n)x(n-j_1)x(n-j_2)\}$$

$$; 0 \le j_1 \le j_2 \le L-1 \ . \tag{6.2.66}$$

Equation (6.2.66) expresses the fact that for the quadratic minimum mean-squared error filter, a second condition holds; *orthogonality of the error and the data correlation.* For zero-mean stationary inputs,[12] the optimal filter is obtained from the solution of a set of normal equations which mirror the form of those for the linear filter:

$$R\underline{f} = \underline{g} \ , \tag{6.2.67}$$

but where for the second-order filter

$$\underline{f} = [\underline{f}^{(1)t}, \underline{f}^{(2)t}]^t,$$

with

$$\underline{f}^{(1)} = [f^{(1)}(0), f^{(1)}(1), ..., f^{(1)}(L-1)]^t \ ,$$
$$\underline{f}^{(2)} = [f^{(2)}(0,0), ..., f^{(2)}(0,L-1), f^{(2)}(1,1), ..., f^{(2)}(L-1,L-1)]^t \ .$$

---

12. In this case, we require stationarity of the moments to order four.

We also define an extended $[L+L(L+1)/2]\times 1$ data vector $\underline{x}_n$ as

$$\underline{x}_n=[\underline{x}_n^{(1)t},\underline{x}_n^{(2)t}]^t \;,$$

where

$$\underline{x}_n^{(1)}=[x(n),x(n-1),...,x(n-L+1)]^t \;,$$
$$\underline{x}_n^{(2)}=[x^2(n),...,x(n)x(n-L+1),x^2(n-1),...,x^2(n-L+1)]^t \;.$$

With these definitions, the filter output can be written in the usual form

$$y(n)=\underline{f}_n^t\underline{x}_n \;. \tag{6.2.63a}$$

In equation (6.2.67), R is the extended correlation matrix

$$R = \begin{bmatrix} R_2 & | & R_3 \\ - & | & - \\ R_3^t & | & R_4 \end{bmatrix} , \tag{6.2.68}$$

where

$$R_2=E\{\underline{x}_n^{(1)}\underline{x}_n^{(1)t}\} \;, \quad R_3=E\{\underline{x}_n^{(1)}\underline{x}_n^{(2)t}\} \;, \quad R_4=E\{\underline{x}_n^{(2)}\underline{x}_n^{(2)t}\} \;.$$

$\underline{g}$ is the extended cross-correlation vector given by

$$\underline{g}=[\underline{g}_2^t,\underline{g}_4^t]^t \;, \tag{6.2.69}$$

where

$$\underline{g}_2=E\{d(n)\underline{x}_n^{(1)}\} \;, \quad \underline{g}_4=E\{d(n)\underline{x}_n^{(2)}\} \;.$$

Here, R_2 is the usual $(L\times L)$ linear autocorrelation matrix. R_3 has dimension $L\times[L(L+1)/2]$ and R_4 is an $[L(L+1)/2]\times[L(L+1)/2]$ matrix. $\underline{g}_2$ is the $(L\times 1)$ vector of cross-correlation coefficients, and $\underline{g}_4$ is an $[L(L+1)/2]\times 1$ vector. Equations (6.2.67) are identical in structure to the usual normal equations (see Section 3.2). However, for the second-order case, we note that while R is symmetric it is not Toeplitz or block-Toeplitz. As in the linear case, R is positive semi-definite but not positive definite.

Note that if the third-order moments are zero, then R reduces to the block-diagonal form:

$$R= \begin{bmatrix} R_2 & | & 0 \\ - & | & - \\ 0 & | & R_4 \end{bmatrix} . \tag{6.2.70}$$

From this

$$R_2 \underline{f}^{(1)} = \underline{g}_2 \ ,$$

$$R_4 \underline{f}^{(2)} = \underline{g}_4 \ . \tag{6.2.71}$$

That is, the linear and quadratic components decouple. One particular case where this occurs is with Gaussian inputs. More generally, any linear process (see Section 2.4) generated using an iid input with samples drawn from a symmetric density will satisfy this condition. For simplicity we refer to such signals as **symmetric**. Note also that from equations (6.2.71), if $d(n)$, $x(n)$ are jointly Gaussian then $\underline{g}_4 = \underline{0}$ and hence $\underline{f}^{(2)} = \underline{0}$ (provided R_4 has full rank). This confirms the well-known result that for Gaussian signals, quadratic (or other nonlinear) terms offer no reduction in minimum mean-squared error compared to that obtained by the linear least-squares filter.

Coker and Simkins [32] defined an adaptive second-order Volterra filter, designed as a steepest descent iteration, updated using instantaneous estimates of the gradient vector. By direct analogy with the derivation of the LMS algorithm in Section 4.1, the algorithm has an update

$$\underline{f}_{n+1} = \underline{f}_n + \alpha e(n) \underline{x}_n \ , \tag{6.2.72}$$

with

$$y(n) = \underline{f}_n^t \underline{x}_n \ ,$$

$$e(n) = d(n) - y(n) \ ,$$

where α is the adaptation constant. The algorithm is therefore identical in form to the LMS update, although the nonlinear structure of $\underline{x}_n$ makes for rather different properties.

As with LMS, one may obtain conditions for convergence in the mean of $\underline{f}_n$ to $\underline{f}^*$. Rearranging (6.2.72) we have

$$\underline{f}_{n+1} = (I - \alpha \underline{x}_n \underline{x}_n^t) \underline{f}_n + \alpha d(n) \underline{x}_n \ .$$

Taking expectations of both sides and applying the usual independence assumption yields

$$E\{\underline{f}_{n+1}\} = (I - \alpha R) E\{\underline{f}_n\} + \alpha \underline{g} \ , \tag{6.2.73}$$

where R and g are given by equations (6.2.68) and (6.2.69) respectively, and where $g=Rf^*$. The form of (6.2.73) is identical to that for the usual LMS (see Section 4.2, equation (4.2.7)) and we may follow precisely the same analysis to conclude that

$$\lim_{n \to \infty} \{E\{\underline{f}_n\}\} \to \underline{f}^* \, ,$$

provided

$$0 < \alpha < \frac{2}{\lambda_{\max}} \, , \tag{6.2.74}$$

where $\lambda_{\max}$ is the largest eigenvalue of R.

In the case where $R_3=0$, that is where $x(n)$ is symmetric, substituting (6.2.70) into (6.2.73) we see that, in expectation at any rate, the linear and quadratic components of the filter may be decoupled to yield

$$\underline{f}_{n+1}^{(1)}=(I-\alpha R_2)\underline{f}_n^{(1)}+\alpha g_2 \, ,$$

$$\underline{f}_{n+1}^{(2)}=(I-\alpha R_4)\underline{f}_n^{(2)}+\alpha g_4 \, . \tag{6.2.75}$$

For this decoupled case we may conclude that the linear and quadratic components of the filter converge independently (again in expectation). Furthermore, the stability and relative convergence rates of the linear and quadratic modes will be determined by the eigenvalues of R_2 and R_4, respectively. When $x(n)$ is symmetric, it would seem expedient to use distinct adaptation constants, α_1 and α_2, say, for the linear and quadratic sections. If $\lambda_{\max 2}$, $\lambda_{\max 4}$ are the maximum eigenvalues of R_2 and R_4, respectively, then stability in the mean is assured provided

$$0 < \alpha_1 < \frac{2}{\lambda_{\max 2}} \, , \tag{6.2.74a}$$

for the linear updates, and

$$0 < \alpha_2 < \frac{2}{\lambda_{\max 4}} \, , \tag{6.2.74b}$$

for the quadratic.

Note that one feature that distinguishes the second-order Volterra LMS from the linear algorithm is that R is *not* diagonal even if $x(n)$ is iid. This in turn produces eigenvalue disparity, even with iid inputs. Consequently, unlike the linear filter, the quadratic LMS generally undergoes non-uniform convergence of the individual modes even when the input is white. This non-uniform behavior includes both distinct convergence factors and cross-coupling which distributes error across the filter coefficients. As a simple example consider $x(n)$ a zero-mean Gaussian iid sequence with variance σ^2 and $d(n)=x^2(n)$ with $L=2$. Defining the mean error vector $\underline{u}_n=E\{\underline{f}_n\}-\underline{f}^*$, then equation (6.2.73) can be written

$$\underline{u}_{n+1} = (I-\alpha R)\underline{u}_n ,\qquad(6.2.76)$$

with

$$R = \begin{bmatrix} \sigma^2 & 0 & | & 0 & 0 & 0 \\ 0 & \sigma^2 & | & 0 & 0 & 0 \\ - & - & | & - & - & - \\ 0 & 0 & | & 3\sigma^4 & 0 & \sigma^4 \\ 0 & 0 & | & 0 & \sigma^4 & 0 \\ 0 & 0 & | & \sigma^4 & 0 & 3\sigma^4 \end{bmatrix}.\qquad(6.2.77)$$

Equations (6.2.76) may be decoupled using the usual decomposition $R=Q\Lambda Q^t$ (see Section 4.2), where for this example a set of eigenvectors is given by:

$$Q = \begin{bmatrix} 1 & 0 & 0 & 0 & 0 \\ 0 & 1 & 0 & 0 & 0 \\ 0 & 0 & \dfrac{1}{\sqrt{2}} & 0 & \dfrac{1}{\sqrt{2}} \\ 0 & 0 & 0 & 1 & 0 \\ 0 & 0 & \dfrac{-1}{\sqrt{2}} & 0 & \dfrac{1}{\sqrt{2}} \end{bmatrix},\qquad(6.2.78)$$

and where Λ is the diagonal matrix of eigenvalues with elements $\lambda_1=\lambda_2=\sigma^2$, $\lambda_3=2\sigma^4$, $\lambda_4=\sigma^4$ and $\lambda_5=4\sigma^4$. If we assume that initially

$$\underline{f}_0=\underline{0} ,$$

then

$$\underline{u}_0^t = [0, 0, -1, 0, 0] \ .$$

Decoupling using (6.2.78), using repeated substitution, and then applying the inverse rotation to return to the original error coefficients (see Section 4.2), gives

$$u_n(5) = \frac{1}{2}(1 - \alpha\lambda_3)^n - \frac{1}{2}(1 - \alpha\lambda_5)^n \ . \tag{6.2.79}$$

Note that since $\lambda_3 \neq \lambda_5$ then $u_n(5) \neq 0$ for finite n. This says that even though no error is initially present in this filter component, $u_n(5)$ is non-zero for finite n. As we have seen in earlier sections, this type of behavior can arise with the linear LMS filter but *not* with white inputs.

As with the linear adaptive filter, the algorithm generally suffers from excess mean-squared error over and above the minimum mean-squared error $J_{\min}$. In steady-state we qualitatively describe this relationship via the equation

$$J_\infty = J_{\min} + excess\ mse \ , \tag{6.2.80}$$

By following similar analysis to that for the linear filter (see Section 4.2), and without invoking any further assumptions on the input, we may extend this relation to the second-order filter. The algebra involved follows exactly that for the linear filter and is left to the Problems section (see Problem 6.26). The only distinction is in the form of the correlation matrix which for the second-order case is given by equation (6.2.68). The expression for the *excess mse* becomes

$$excess\ mse \approx \sum_{j=0}^{L-1} \frac{\alpha_1}{2} J_{\min} \lambda_j^{(2)} + \sum_{k=0}^{L_2-1} \frac{\alpha_2}{2} J_{\min} \lambda_k^{(4)} \ , \tag{6.2.81}$$

where $L_2 = L(L+1)/2$ and where $\lambda_j^{(2)}$ are the eigenvalues of the matrix R_2 while $\lambda_k^{(4)}$ are the eigenvalues of the matrix R_4. Equivalently we may write

$$J_\infty \approx J_{\min}\left(1 + \frac{\alpha_1}{2} tr\{R_2\} + \frac{\alpha_2}{2} tr\{R_4\}\right) \ , \tag{6.2.82}$$

which is the direct extension of equation (4.2.52) for the conventional LMS.

6.3 Recursive Least-Squares Algorithms

6.3.1 Algorithms Derived From Newton's Method

In Chapter 4 we first considered iterative solutions to the normal equations. As we indicated, the main weaknesses of the steepest descent philosophy in general, and the LMS algorithm in particular, are slow and generally non-uniform convergence. We might add to this list the further disadvantage that LMS attempts to optimize a stochastic criterion using a local, single point estimate of the gradient at each iterative update. An alternative approach would be to solve for the optimum filter at time n, using all of the data available up to that point. In this section we will examine a far more powerful class of iterative solutions (and ultimately adaptive filters) which attempt to do just that. We shall derive these algorithms from consideration of Newton's method for obtaining the zeros of a function. However, as we shall see, this is not the only route the derivation can take.

Newton's method provides a technique for iteratively obtaining the zeros of a function $f(x)$. The method is known to converge rapidly for a wide class of functions. We will illustrate the algorithm in one dimension. Consider a function having a single zero $f(x)=0$, at $x=x^*$ as depicted in Figure 6.3.1. Newton's method starts with an initial guess x_0, as to the location of the zero, and its corresponding function value $f(x_0)$. A second estimate is then obtained by projecting down the tangent to the curve at x_0 to the point where the tangent crosses the x-axis. The equation of this tangent is the derivative of the function at x_0 and is given simply by

$$f'(x_0) = \frac{f(x_0)}{x_0 - x_1} \ , \tag{6.3.1}$$

rearranging (6.3.1), we have

$$x_1 = x_0 - \frac{f(x_0)}{f'(x_0)} \ ,$$

and the next estimate is then formed as $f(x_1)$. This formula is used as the basis of a general iteration:

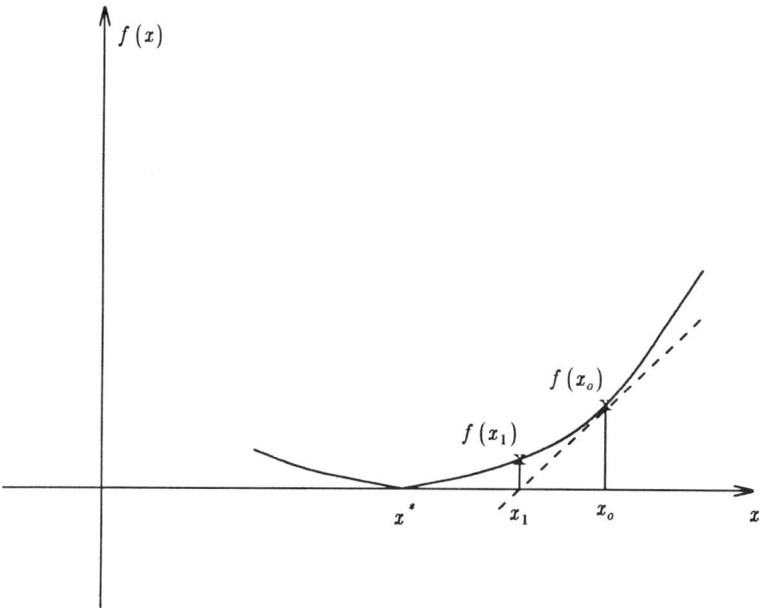

Figure 6.3.1 Finding the zeros of a function via Newton's iteration.

$$x_{n+1} = x_n - \frac{f(x_n)}{f'(x_n)} \,, \tag{6.3.2}$$

which, as is illustrated by the example in the diagram, converges rapidly for 'reasonably' behaved functions (the problem of multiple zeros is another matter). Of particular interest is the case where $f(x)$ is linear, since in this case the tangent is identical to the function itself, and hence convergence to the zero takes place in a single iterative step.

To apply this algorithm to the problem of least-squares minimization, we must find the zero of the gradient of the mean-squared error rather than of the function itself. Since the mean-squared error is quadratic, however, the gradient is linear, and consequently convergence takes place in a single step.

The adaptive filtering problem is more properly cast in a multidimensional framework. As usual, our concern is with the minimization of

$$J = E\{e^2(n)\} = E\{(d(n) - \underline{f}_n^t \underline{x}_n)^2\} = \sigma_d^2 - 2\underline{f}_n^t \underline{g} + \underline{f}_n^t R \underline{f}_n \,, \tag{6.3.3}$$

where $e(n) = d(n) - \underline{f}_n^t \underline{x}_n$. Recall from Section 4.2 that the derivative of J is

$$\nabla J = \frac{\partial J}{\partial \underline{f}_n} = 2R\underline{f}_n - 2\underline{g} , \qquad (6.3.4)$$

and that equating this to zero gives the optimum solution

$$\underline{f}^* = R^{-1}\underline{g} . \qquad (6.3.5)$$

J is quadratic and we anticipate that Newton's method should find the zero of the linear function ∇J in one step. For this multidimensional problem we may demonstrate that the Newton algorithm is given by

$$\underline{f}_{n+1} = \underline{f}_n - \frac{1}{2} R^{-1} \frac{\partial J}{\partial \underline{f}_n} . \qquad (6.3.6)$$

To verify this result we need only show that this update does indeed produce the result in one step. This may easily be done by substituting (6.3.4) for the gradient into the update (6.3.6), giving

$$\underline{f}_{n+1} = \underline{f}_n - \frac{1}{2} R^{-1} (2R\underline{f}_n - 2\underline{g}) .$$

Hence

$$\underline{f}_{n+1} = R^{-1}\underline{g} = \underline{f}^* ,$$

so that (6.3.6) is indeed the Newton iteration.

The immediate attractions of such a method must be tempered by the realization that the update equation requires both the autocorrelation matrix R and the gradient ∇J. Also the correlation matrix must be inverted, and the computation of the inverse is very expensive. In practice, it is necessary to use an estimate of the gradient. This might be provided by

$$\hat{\nabla} J = -2e(n)\underline{x}_n , \qquad (6.3.7)$$

that is, the LMS gradient estimate. The update of (6.3.6) then becomes

$$\underline{f}_{n+1} = \underline{f}_n + R^{-1} e(n)\underline{x}_n . \qquad (6.3.8)$$

With the introduction of the estimate for ∇J into the update, the filter coefficient vector will certainly be degraded by noise, and thus the result will produce excess mean-squared error. To allow a greater degree of control, a general constant α (constrained to be less than unity) is introduced, and the update becomes

$$\underline{f}_{n+1} = \underline{f}_n + \alpha R^{-1} e(n) \underline{x}_n \ . \tag{6.3.9}$$

In general, convergence to $\underline{f}^*$ will no longer occur in a single step. As we have indicated, this is not necessarily a disadvantage when excess mean-squared error is present. In fact for random inputs, the update is now driven by random quantities and we must think in terms of convergence in the mean as discussed in Section 4.2 in relation to the LMS algorithm.

Equation (6.3.9) represents an algorithm which is still 'idealized' due to the presence of the correlation matrix R. In practice this matrix, or more correctly its inverse, must be replaced by an estimate. Before addressing this issue, however, it is instructive to briefly consider the algorithm defined by (6.3.9) and in particular, to contrast this algorithm with the LMS [9].

Using the approach of Section 4.2 we may rewrite the update (6.3.9) as

$$\underline{f}_{n+1} = \underline{f}_n + \alpha R^{-1}(d(n) - \underline{f}_n^t \underline{x}_n) \underline{x}_n \ ,$$

$$= (I - \alpha R^{-1} \underline{x}_n \underline{x}_n^t) \underline{f}_n + \alpha R^{-1} d(n) \underline{x}_n \ . \tag{6.3.10}$$

Now, taking expectations and employing the usual independence assumption we have

$$E\{\underline{f}_{n+1}\} = (1-\alpha) E\{\underline{f}_n\} + \alpha R^{-1} \underline{g} \ . \tag{6.3.11}$$

We see that as expected if $\alpha = 1$, then

$$E\{\underline{f}_{n+1}\} = R^{-1} \underline{g} = \underline{f}^* \ ,$$

that is, mean convergence in a single step. For $\alpha < 1$, we may employ the usual mean error vector $\underline{u}_n = E\{\underline{f}_n\} - \underline{f}^*$, together with the solution (6.3.5) for $\underline{f}^*$. This yields

$$\underline{u}_{n+1} = (1-\alpha) \underline{u}_n \ .$$

Hence, using repeated substitution

$$\underline{u}_n = (1-\alpha)^n \underline{u}_0 \ . \tag{6.3.12}$$

From this equation we may observe the following:

i) The algorithm converges in the mean provided $|(1-\alpha)| < 1$. That is, if

$$0 < \alpha < 2 \ .$$

ii) Convergence proceeds exponentially, at a rate determined by $(1-\alpha)$.

iii) The convergence rate of each coefficient is identical, that is the convergence is independent of the eigenvalue spread of R.

This last point is a key advantage of this Newton algorithm and contrasts sharply with the LMS algorithm where, as we have seen, the convergence of different modes of the adaptive filter is dependent on the eigenvalue spread of R.

One may also develop expressions for the steady-state mean-squared error for the algorithm (6.3.8). The approach is similar to that employed in Section 4.2.3, for the LMS algorithm. For small α, the result for the excess mean-squared error is given by (see Problem 6.27)

$$excess \ mse \approx \frac{\alpha}{2} J_{\min} tr\{R\} \ . \tag{6.3.13}$$

It will be recalled that in Section 4.2.3 we obtained an identical result for the excess mean-squared error for the LMS algorithm (see equation (4.2.52a)). Thus, the advantages of the hybrid algorithm are obvious — it suffers no eigenvalue problems and has comparable misadjustment to the LMS for the same adaptation rate. However, we must keep in mind that these results relate to the idealized algorithm employing the true correlation matrix R. In practice we must estimate R using a time average. At time n we might estimate the i,j-th element of R by

$$R_n(i,j) = \sum_{l=0}^{n} x(l-i)x(l-j) \ , \tag{6.3.14a}$$

where we explicitly indicate the dependence on index n, and where the elements of R depend on both i and j because this time average does not produce a Toeplitz form for the matrix, and where we assume that $x(n)=0$ for $n<0$.[13] In vector form we may write

$$R_n=\sum_{l=0}^{n} \underline{x}_l \underline{x}_l^t \quad ; \, n=0,1,...$$ (6.3.14b)

The update (6.3.9) then becomes

$$\underline{f}_{n+1}=\underline{f}_n+\alpha R_n^{-1} e(n) \underline{x}_n \, .$$ (6.3.15)

There remains a considerable problem both in relation to computing R_n and in computing the inverse R_n^{-1}. R_n itself may be computed recursively using

$$R_{n+1}=R_n+\underline{x}_{n+1} \underline{x}_{n+1}^t \, .$$ (6.3.16)

The update for $\underline{f}_n$ still has a very high computational requirement, however, not least because R_n is not Toeplitz and therefore we require on the order of L^3 operations for each inversion of the matrix. This computational complexity can be reduced by using the well-known **matrix inversion lemma** [33], which provides a recursive update for the inverse of a structure $(A+\underline{u} \, \underline{u}^t)$ where A is a rectangular (non-singular) matrix, and $\underline{u}$ is an arbitrary $(N \times 1)$ vector. In view of the update for R_n of equation (6.3.16), this is precisely what we require. The lemma, stated here without proof, can be expressed as

$$(A+\underline{u} \, \underline{u}^t)^{-1}=A^{-1}-\frac{A^{-1} \underline{u} \, \underline{u}^t A^{-1}}{1+\underline{u}^t A^{-1} \underline{u}} \, .$$ (6.3.17)

Applying this to the update for the correlation matrix we have

---

13. Clearly the estimate (6.3.14a) differs from the time averaging for correlation coefficients discussed in Section 2.6. In those estimates a Toeplitz structure is assumed for the correlation matrix and averaging is performed using all the data in a window from $n=0$ to $N-1$. Here, no Toeplitz structure is assumed and the estimate is obtained by summing outer products of all available data vectors.

$$R_{n+1}^{-1}=(R_n+\underline{x}_{n+1}\underline{x}_{n+1}^t)^{-1}=R_n^{-1}-\frac{R_n^{-1}\underline{x}_{n+1}\underline{x}_{n+1}^t R_n^{-1}}{1+\underline{x}_{n+1}^t R_n^{-1}\underline{x}_{n+1}}. \qquad (6.3.18)$$

In spite of the apparent complexity of this equation, given R_n^{-1} no matrix inversions are required because $1+\underline{x}_{n+1}^t R_n^{-1}\underline{x}_{n+1}$ is a scalar. The computation is actually proportional to L^2 rather than L^3 as for the direct inversion. With this solution, R_n itself is not calculated at any stage in the iteration. The solution begins with an initial value for R_0^{-1} and simply updates the inverse using (6.3.18).

In summary, the operation of the overall algorithm proceeds as follows:

i) Initialize $\underline{f}_0$, R_{-1}^{-1}

ii) For $n=0,1,...$

$$R_n^{-1}=R_{n-1}^{-1}-\frac{R_{n-1}^{-1}\underline{x}_n\underline{x}_n^t R_{n-1}^{-1}}{1+\underline{x}_n^t R_{n-1}^{-1}\underline{x}_n}, \qquad (6.3.18)$$

$$e(n)=d(n)-\underline{f}_n^t\underline{x}_n,$$

$$\underline{f}_{n+1}=\underline{f}_n+\alpha R_n^{-1}e(n)\underline{x}_n. \qquad (6.3.15)$$

Treichler *et al.* [6] give two alternative procedures for the initial phase of computation for algorithms of this form:

i) R_n, g_n are built up using the initial input vectors until the rank of R_n^- is full. This requires at least L input vectors. R_n is then inverted.

ii) R_{-1}^{-1} is chosen as $\sigma^2 I$ where σ^2 is a small constant.

The first approach requires a one-off $O(L^3)$ calculation of the inverse, and implies a delay of at least L data points before any updates are performed. The second approach has the virtues of simplicity, and provides immediate adaptation but is obviously inaccurate initially in most cases. The use of a large constant for R is consistent with our uncertainty about the true correlation matrix. It has been reported [34],[35] that the algorithm works well with a wide range of choices for σ^2, and in practice this second procedure is most often employed.

6.3.2 Recursive Least-Squares

Derivation of the Recursive Least-Squares Algorithm

The algorithm specified by equations (6.3.15)-(6.3.18) is closely related to the **Recursive Least-Squares (RLS)** algorithm. This algorithm is derived from a quite different standpoint however; by considering the minimization of the deterministic measure:

$$J_n = \sum_{l=0}^{n} e^2(l) = \sum_{l=0}^{n} (d(l) - y(l))^2 , \qquad (6.3.19)$$

where $d(n)$, $y(n)$ are the desired signal and filter output, respectively. J_n is a recursive-in-time measure which changes at each point to reflect the arrival of a new data sample. The objective of the RLS algorithm is to maintain a solution which is optimal with respect to (6.3.19) at each iteration. Differentiating (6.3.19) and equating to zero leads, as usual, to a set of normal equations:

$$R_n \underline{f}_n = \underline{g}_n , \qquad (6.3.20)$$

with

$$R_n(i,j) = \sum_{l=0}^{n} x(l-i)x(l-j) , \qquad (6.3.21a)$$

and

$$g(i) = \sum_{l=0}^{n} d(l)x(l-i) . \qquad (6.3.21b)$$

Note that, as in (6.3.16)

$$R_n = R_{n-1} + \underline{x}_n \underline{x}_n^t . \qquad (6.3.22a)$$

Similarly, from (6.3.21b)

$$\underline{g}_n = \underline{g}_{n-1} + d(n)\underline{x}_n . \qquad (6.3.22b)$$

A recursive solution to the problem of minimizing J_n assumes that at time n we have R_n^{-1}, g_n and $\underline{f}_n$ available, where R_n, and g_n are defined as above, and where the optimal solution at the n-th iteration is

$$\underline{f}_n = R_n^{-1} \underline{g}_n \ . \tag{6.3.23}$$

Given this information at time n, we seek

$$\underline{f}_{n+1} = R_{n+1}^{-1} \underline{g}_{n+1} \ , \tag{6.3.24}$$

at time index $(n+1)$. We see that R_{n+1}^{-1} can be computed using the recursive formula (6.3.18) and that g_{n+1} can be obtained from (6.3.22b). The computation of $\underline{f}_{n+1}$ is achieved by substituting (6.3.18) and (6.3.22b) into (6.3.24) giving

$$\underline{f}_{n+1} = [R_n^{-1} - \frac{R_n^{-1} \underline{x}_{n+1} \underline{x}_{n+1}^t R_n^{-1}}{1 + \underline{x}_{n+1}^t R_n^{-1} \underline{x}_{n+1}}][\underline{g}_n + d(n+1)\underline{x}_{n+1}] \ . \tag{6.3.25}$$

Now, let $R_n^{-1} \underline{x}_{n+1} = \underline{z}$, say, and using (6.3.23) we may expand equation (6.3.25) as

$$\underline{f}_{n+1} = \underline{f}_n - \frac{\underline{z} \, \underline{x}_{n+1}^t \underline{f}_n}{1 + \underline{x}_{n+1}^t \underline{z}} + d(n+1)\underline{z} - \frac{d(n+1)\underline{z} \, \underline{x}_{n+1}^t \underline{z}}{1 + \underline{x}_{n+1}^t \underline{z}} \ . \tag{6.3.26}$$

Also, let $\underline{x}_{n+1}^t \underline{z} = k$, say. Then we may simplify (6.3.26) to

$$\underline{f}_{n+1} = \underline{f}_n - \left[\frac{\underline{x}_{n+1}^t \underline{f}_n - d(n+1)}{1+k} \right] \underline{z} \ ,$$

or

$$\underline{f}_{n+1} = \underline{f}_n - \left[\frac{\underline{x}_{n+1}^t \underline{f}_n - d(n+1)}{1+k} \right] R_n^{-1} \underline{x}_{n+1} \ . \tag{6.3.27}$$

We denote by $e(n+1/n)$, the quantity

$$e(n+1/n) = d(n+1) - \underline{f}_n^t \underline{x}_{n+1} \ . \tag{6.3.28}$$

This is often referred to as the **a priori error**, reflecting the fact that it is the error obtained using the old filter (that is prior to updating with the new data). This definition of *a priori* error also suggests the notion of **a posteriori error** − obtained after the filter coefficients have been updated. The prior error is actually the form used by the LMS − we calculate the error before updating the filter coefficients (although this is not obvious from the LMS update − the filters are indexed differently). Finally the RLS algorithm equation (6.3.27) may be written

$$\underline{f}_{n+1}=\underline{f}_n+\frac{1}{1+\underline{x}_{n+1}^t R_n^{-1}\underline{x}_{n+1}}e\,(n+1/n)R_n^{-1}\underline{x}_{n+1}\,.\qquad(6.3.29)$$

The RLS algorithm in summary is thus:

i) Initialize $\underline{f}_{-1}$, R_{-1}^{-1}

ii) For $n=0,1,\ldots$

$$e\,(n/n-1)=d\,(n)-\underline{f}_{n-1}^t\underline{x}_n\,,\qquad(6.3.28)$$

$$\alpha(n)=\frac{1}{1+\underline{x}_n^t R_{n-1}^{-1}\underline{x}_n}\,,\qquad(6.3.30)$$

$$\underline{f}_n=\underline{f}_{n-1}+\alpha(n)e\,(n/n-1)R_{n-1}^{-1}\underline{x}_n\,,\qquad(6.3.31)$$

$$R_n^{-1}=R_{n-1}^{-1}-\alpha(n)R_{n-1}^{-1}\underline{x}_n\underline{x}_n^t R_{n-1}^{-1}\,.\qquad(6.3.32)$$

Keeping in mind the fact that the LMS and Newton methods use the prior error, and the scalar nature of the gain $\alpha(n)$, we see the similarity of this RLS algorithm and the Newtonian algorithms of equations (6.3.15) $-$ (6.3.18). Note, however, that this algorithm may be expected to converge more quickly than that given in (6.3.15)-(6.3.18) due to the use of the step-size $\alpha(n)$.

Exponentially Weighted Recursive Least-Squares

A more general form for the RLS filter is obtained by replacing (6.3.19) by the modified functional

$$\tilde{J}_n=\sum_{l=0}^{n}\lambda^{n-l}e^2(l)\,,\qquad(6.3.33)$$

where now a weighting factor λ is included where

$$0<\lambda\leq 1\,.\qquad(6.3.34)$$

Values of $\lambda<1$ give more 'weight' (in $\tilde{J}_n$) to the most recent errors.[14] This is useful in nonstationary data where changes make the inclusion of old data less appropriate. We may employ a similar procedure to that for the unweighted RLS for the minimization of J_n. We obtain a similar set of normal equations (see Problem 6.28) with

$$R_n \underline{f}_n = \underline{g}_n , \qquad (6.3.20)$$

but where now

$$R_n = \sum_{l=0}^{n} \lambda^{n-l} \underline{x}_l \underline{x}_l^t , \qquad (6.3.35a)$$

$$\underline{g}_n = \sum_{l=0}^{n} \lambda^{n-l} d(l) \underline{x}_l . \qquad (6.3.35b)$$

Once again, we may write R_n, g_n in recursive form

$$R_n = \sum_{l=0}^{n-1} \lambda^{n-l} \underline{x}_l \underline{x}_l^t + \underline{x}_n \underline{x}_n^t ,$$

$$\underline{g}_n = \sum_{l=0}^{n-1} \lambda^{n-l} d(l) \underline{x}_l + d(n) \underline{x}_n ,$$

or

$$R_n = \lambda R_{n-1} + \underline{x}_n \underline{x}_n^t , \qquad (6.3.36a)$$

$$\underline{g}_n = \lambda \underline{g}_{n-1} + d(n) \underline{x}_n . \qquad (6.3.36b)$$

As before, given the solution to $R_{n-1} \underline{f}_{n-1} = \underline{g}_{n-1}$, we seek $\underline{f}_n$ which satisfies (6.3.20). Using a similar approach to that for the simple RLS we may obtain an **Exponentially Weighted RLS (EWRLS)** algorithm as:

i) Initialize $\underline{f}_{-1}$, R_{-1}^{-1}

ii) For $n = 1, 2, \dots$

$$e(n/n-1) = d(n) - \underline{f}_{n-1}^t \underline{x}_n , \qquad (6.3.37)$$

$$\alpha(n) = \frac{1}{\lambda + \underline{x}_n^t R_{n-1}^{-1} \underline{x}_n} , \qquad (6.3.38)$$

$$\underline{f}_n = \underline{f}_{n-1} + \alpha(n)e(n/n-1)R_{n-1}^{-1}\underline{x}_n \ , \tag{6.3.39}$$

$$R_n^{-1} = \frac{1}{\lambda}[R_{n-1}^{-1} - \alpha(n)R_{n-1}^{-1}\underline{x}_n\underline{x}_n^t R_{n-1}^{-1}] \ . \tag{6.3.40}$$

Comparison of these equations with (6.3.28)-(6.3.32) shows that, as expected, for $\lambda=1$ the algorithm reduces to the unweighted form. This choice for λ is obviously appropriate for stationary data. For other data $0.95<\lambda<0.9995$ has been suggested [36]. In this author's experience, values at the higher end of that range are more likely to prove satisfactory. The lower values being more likely to result in instability for the algorithm.

Notes on the RLS Algorithm

a) Computational Complexity

Computationally, RLS is considerably more expensive than simple LMS. As indicated in Section 4.1, LMS requires $O(L)$ operations per update (a total of $(2L+1)$ multiplications and $2L$ additions for the update and filtering combined). By contrast the requirement for RLS, even with the reductions obtained via the matrix inversion lemma, is $O(L^2)$. For the exponential RLS of equations (6.3.37) to (6.3.40), Alexander [36] gives a total of $(4L^2+4L)$ multiplications/divisions and $(3L^2+L-1)$ additions/subtractions. This total may be reduced somewhat by exploiting the symmetry of R_n^{-1} [37] but the total remains $O(L^2)$.[15]

b) Convergence and Eigenvalue Disparity

In principle, the rotation which occurs because of the pre-multiplication by R_n^{-1}, removes any eigenvalue dependence from the result (see equations (6.3.9)-(6.3.12)). In practice, however, as with the Newton algorithm discussed above, we usually start with an initial guess for $R=c^2I$, say. Clearly, the inverse of this matrix applies no rotation and therefore does not reduce the eigenvalue

---

14. More correctly we might write $e(i)$ as $e(i/n)=d(i)-\underline{f}_n^t\underline{x}_i$, since we minimize using all the data to produce $\underline{f}_n$.

15. Another problem [35] is that the update for R_n^{-1} for both RLS and EWRLS is not guaranteed to maintain positive definiteness for R_n, and numerical problems can result.

disparity. During convergence, if our estimate of the inverse correlation matrix tends to the true value, then the eigenvalue disparity disappears.

In fact, starting with $c^2 I$ and recursively computing an estimate of R_n denoted R_n', say, leads to

$$R_n' = \sum_{l=0}^{n} \underline{x}_l \underline{x}_l^t + c^2 I \quad ; \quad n = 0, 1, \ldots , \qquad (6.3.33)$$

instead of R_n as given by (6.3.14b). Clearly $E\{R_n'\} \neq E\{R_n\}$, and hence the effect of this initialization is to *bias* the solution away from $\underline{f}_n = R_n^{-1} g_n$ [35]. However, we see that as n increases, the effect of this initial inaccuracy diminishes and R_n' is asymptotically unbiased.

c) RLS versus LMS

We may illustrate the performance of the RLS and EWRLS algorithms in comparison with LMS via a simple example. Consider the system identification problem illustrated in Figure 6.3.2.

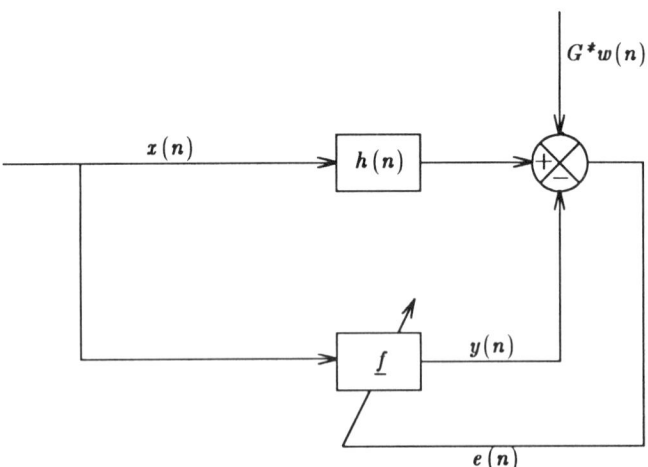

Figure 6.3.2 Adaptive filter model of the unknown system $h(0)=1$, $h(1)=0.5$.

An iid sequence $x(n)$ is input to a simple two-point impulse response $h(0)=1$, $h(1)=0.5$. The output is corrupted by measurement noise $w(n)$ with gain G. The measurement noise is assumed to be iid zero-mean Gaussian with unit variance. The adaptive filter is required to identify the unknown system using the input, and the noisy system output. Figure 6.3.3 shows a comparison of the RLS and LMS algorithms. In this trial, G was selected to give an SNR of 40dB. The diagram shows the norm of the coefficient errors $\underline{v}_k^t \underline{v}_k$ for $k=1,2,...,1000$ where $\underline{v}_k = \underline{f}_k - \underline{f}^*$. The results shown were obtained by averaging over an 'ensemble' of 100 trials. The averaging produces smooth coefficient error tracks. These results were obtained using $L=8$ and zero initial conditions for the filter coefficients for both algorithms. The adaptation constant for LMS was set as $\alpha=0.05$. The plot shows clear superiority for RLS in terms of both faster convergence, and lower steady-state mean-squared error. This increased convergence speed was obtained in spite of the fact that the adaptation step-size used for LMS was the largest value found that gave stable results. Similar demonstrations of RLS performance potential may be found in [35] and [38], for example. Moreover, this superiority was obtained with *white inputs*. For a colored input, the performance difference between the two methods should be even more dramatic since RLS should be relatively unaffected by the eigenvalue disparity. Figure 6.3.4 shows a similar performance comparison, obtained using identical parameters for both LMS and RLS, but where the input is an MA(1) process

$$x(n)=w(n)+0.8w(n-1) .$$

Comparing Figure 6.3.4 with Figure 6.3.3 confirms that for the colored input the LMS convergence is considerably slowed, whereas that for RLS is virtually unchanged.

Generally, Recursive Least-Squares algorithms have been employed extensively in parameter estimation and system identification problems. In signal processing applications, RLS has been less widely accepted. In part, this is a reflection of the higher computation and implementational complexity of RLS algorithms compared to LMS. It is also due to a perception that RLS is in some sense less robust than is LMS. The first of these problems has been addressed by the development of **Fast RLS** algorithms [39]. These are implementations of RLS and EWRLS algorithms that exploit the symmetry and redundancy in the computation of

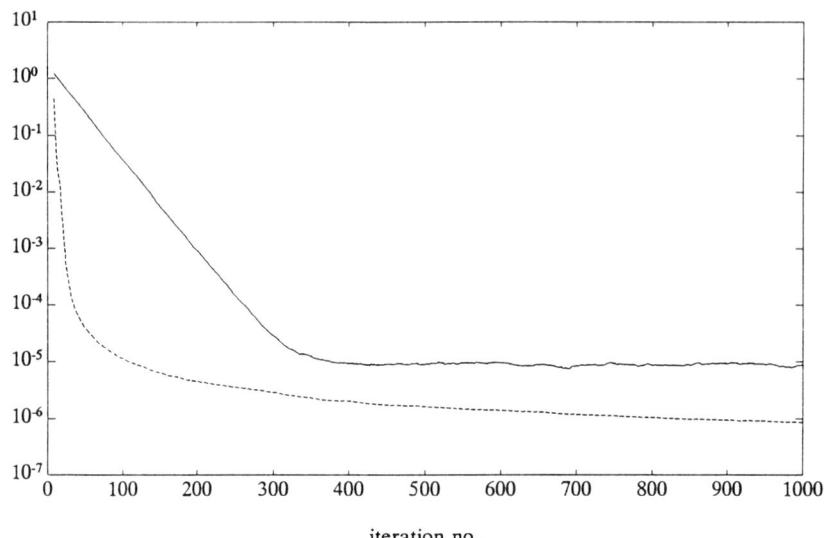

iteration no.

Figure 6.3.3 System modelling using LMS and RLS for an iid input. The error norm $\underline{v}_k^t \underline{v}_k$ is shown versus iteration number k. RLS (dashed line) converges faster, and to a lower steady-state value than LMS (solid line). The plots were obtained by averaging over 100 trials.

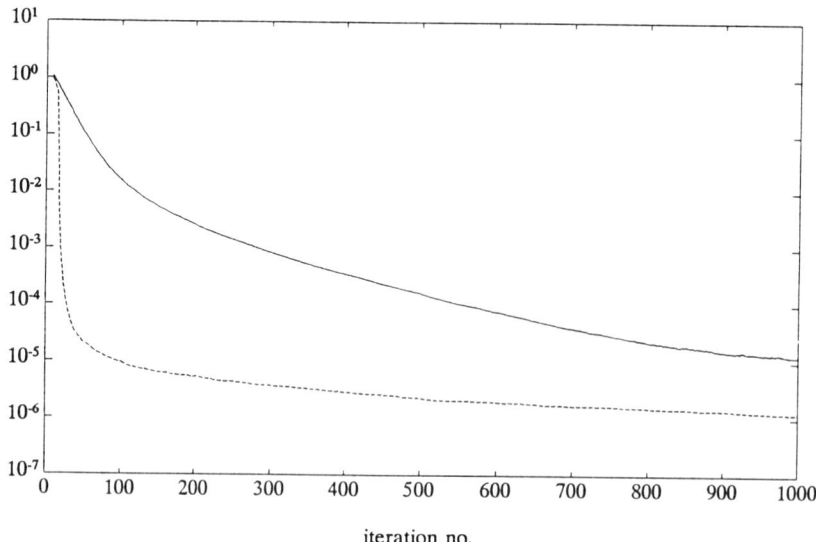

iteration no.

Figure 6.3.4 System modelling using LMS and RLS for an MA(1) input. The error norm $\underline{v}_k^t \underline{v}_k$ is shown versus iteration number k. RLS (dashed line) performance is comparable to that of Figure 6.3.3, LMS (solid line) is degraded. The plots were obtained by averaging over 100 trials.

the filter to reduce the computations to $O(L)$. The fastest such algorithm operates in about $7L$ multiplications/divisions per iteration. However, the numerical sensitivity of these algorithms remains problematic [38], and the development of stable fast algorithms remains an area of active research. We also note that in many practical problems, the initial RLS estimates may fluctuate substantially, and the LMS algorithm may in fact offer smoother convergence towards f^*, with no loss of speed [2]. Additionally, the superiority of RLS over LMS is often lost with nonstationary data. Without exponential weighting, RLS suffers poor tracking ability due to the equal weighting that is applied to all previous data in the correlation estimates. Exponential weighting can help tracking, but it is often not clear how λ should be chosen. Additionally, as we have indicated, $\lambda < 1$ tends to increase numerical problems in the algorithm.

Overall, we may conclude that the major advantage of RLS over LMS lies in faster convergence and reduced sensitivity to eigenvalue disparity for stationary inputs. On the other hand, the RLS algorithm is less useful in low eigenvalue disparity situations, in cases where unmodelled noise is high, and in tracking nonstationary data.

6.4 IIR Adaptive Filters

So far, we have concentrated exclusively on FIR filters and transversal implementations. In parallel with the development of adaptive filters with FIR structures, some work has been done on recursive implementations. These recursive forms generally give rise to IIR filters. The motivation for such developments are similar to those arising in ordinary filtering applications: An FIR filter may need very many coefficients to obtain a good response, and consequently may have an excessive computational requirement. A recursive structure can potentially produce comparable results with far fewer coefficients, and consequently with a much lower computational burden. This potential is not obtained easily, however, since the computational gain is offset by increased problems in guaranteeing stability and convergence. In particular, we note that IIR adaptive filters are prone to instability as a consequence of unbounded growth of the filter coefficients. Being recursive, and therefore having poles at locations other than the origin in the z-plane, the recursive adaptive filter may be unstable even if the adaptation is stable. Such instability results if the poles are positioned outside the unit circle in the z-plane.[16] Also, we note that for the IIR adaptive filter, unlike the FIR, the performance surface associated with the least-squares functional is not guaranteed to be unimodal [40]. Thus, apart from the additional stability worries, even if the algorithm converges, the convergence may be to a local rather than a global minimum. In spite of these drawbacks, the potential of IIR filters is such that interest in such structures remains high.

The basic form of the recursive filter is given by

$$y(n) = \sum_{j=1}^{N} h(j) y(n-j) + \sum_{j=0}^{L-1} f(j) x(n-j) . \qquad (6.4.1)$$

Here, we have extended the filter definition of previous sections by adding to the set of L non-recursive coefficients $f(i)$, a set of N recursive coefficients $h(i)$. At this stage we consider the fixed (non-adaptive) filter. We may extend the vector notation by defining a vector of filter coefficients

$$\underline{\theta} = [h(1), h(2), ..., h(N), f(0), f(1), ..., f(L-1)]^t ,$$

and by extending the data vector to include both input and output as

$$\underline{X}_n = [y(n-1), y(n-2), ..., y(n-N), x(n), x(n-1), ..., x(n-L+1)]^t .$$

The filter output is then written

$$y(n) = \underline{\theta}^t \underline{X}_n . \qquad (6.4.1\text{b})$$

As usual we minimize the mean-squared error:

$$J = E\{e^2(n)\} = E\{(d(n) - y(n))^2\} . \qquad (6.4.2)$$

The minimization is performed by differentiating and equating to

---

16. In referring to the z transform of the filter, we are essentially referring to the transform that would arise if the instantaneous coefficients were held fixed for all time.

zero, so that we write

$$\frac{\partial J}{\partial \theta(i)} = \frac{\partial E\{e^2(n)\}}{\partial \theta(i)} = 0 \quad ; i = 0,1,...,N+L-1 \; ,$$

or

$$2E\{e(n)\frac{\partial e(n)}{\partial \theta(i)}\} = 0 \quad ; i = 0,1,...,N+L-1 \; . \tag{6.4.3}$$

Since $e(n)$ is only dependent on θ through $y(n)$, then (6.4.3) becomes

$$E\{(d(n)-y(n))\frac{\partial y(n)}{\partial \theta(i)}\} = 0 \quad ; i = 0,1,...,N+L-1 \; . \tag{6.4.4}$$

Now,

$$\frac{\partial y(n)}{\partial \theta(i)} = \frac{\partial}{\partial \theta(i)} \left[\sum_{j=1}^{N} h(j)y(n-j) + \sum_{j=0}^{L-1} f(j)x(n-j) \right] \; . \tag{6.4.5}$$

This is more complex than the usual FIR minimization because $y(n-j)$, $j=1,2,...,N$ are also dependent on $\theta(i)$. We separate the derivatives into those with respect to h and those with respect to f. The derivative with respect to $h(i)$ is

$$\frac{\partial y(n)}{\partial h(i)} = \frac{\partial}{\partial h(i)} \left[\sum_{j=1}^{N} h(j)y(n-j) \right]$$

$$= y(n-i) + \sum_{j=1}^{N} h(j)\frac{\partial y(n-j)}{\partial h(i)} \quad ; 1 \le i \le N \; . \tag{6.4.6a}$$

Here, the first equality follows because f has no dependence on h, and the second because the derivative must be treated as a product. The derivative of (6.4.1) with respect to $f(i)$ is

$$\frac{\partial y(n)}{\partial f(i)} = x(n-i) + \sum_{j=1}^{N} h(j)\frac{\partial y(n-j)}{\partial f(i)} \quad ; 0 \le i \le L-1 \; . \tag{6.4.6b}$$

We may combine the last two equations to give

$$\frac{\partial y(n)}{\partial \underline{\theta}} = \underline{X}_n + \sum_{j=1}^{N} h(j)\frac{\partial y(n-j)}{\partial \underline{\theta}} \; , \tag{6.4.6}$$

which is thus a recursive update for the gradient. In total from
(6.4.6a) and (6.4.6b), equations (6.4.4) become

$$E\{e(n)\left[y(n-i)+\sum_{j=1}^{N}h(j)\frac{\partial y(n-j)}{\partial h(i)}\right]\}=0 \quad ; 1\leq i\leq N,$$

$$E\{e(n)\left[x(n-i)+\sum_{j=1}^{N}h(j)\frac{\partial y(n-j)}{\partial f(i)}\right]\}=0 \quad ; 0\leq i\leq L-1, \quad (6.4.7)$$

which are the recursive form for the normal equations of Section
3.2. Direct solution of these equations would yield the optimal
coefficient vector $\underline{\theta}^{*}=[\underline{f}^{*t},\underline{h}^{*t}]^{t}$. However, this is highly problematic
because the recursive nature of the derivatives produces equations
that are nonlinear in the filter coefficients. It is this nonlinearity
which results in the possibility of local minima.

As an alternative to direct solution, a steepest descent type of
iterative solution can be employed. By direct analogy with the
development of the FIR adaptive filter (see Section 4.1), this
update has the form

$$\underline{\theta}_{n+1}=\underline{\theta}_{n}-\frac{\alpha}{2}\nabla J. \quad (6.4.8)$$

From this it is simple to define an adaptive recursive filter. In line
with the LMS philosophy we replace $\nabla J=\nabla E\{e^{2}(n)\}$ by the
instantaneous version

$$\nabla e^{2}(n)=2e(n)\frac{\partial e(n)}{\partial\underline{\theta}_{n}}=-2e(n)\frac{\partial y(n)}{\partial\underline{\theta}_{n}},$$

so that

$$\underline{\theta}_{n+1}=\underline{\theta}_{n}+\alpha e(n)\frac{\partial y(n)}{\partial\underline{\theta}_{n}}. \quad (6.4.9)$$

As we have seen in (6.4.6), the derivative is itself recursive.
Replacing $\underline{\theta}$ in (6.4.6) by the time-varying $\underline{\theta}_{n}$, we have

$$\frac{\partial y(n)}{\partial\underline{\theta}_{n}}=\underline{X}_{n}+\sum_{i=1}^{N}h_{n}(i)\frac{\partial y(n-i)}{\partial\underline{\theta}_{n}}. \quad (6.4.10)$$

Equations (6.4.9) and (6.4.10), together with the filter output
equation (6.4.1b) represent the **Recursive LMS Algorithm (or**

IIRLMS) [41], summarized for convenience as:

$$y(n) = \underline{\theta}_n^t \underline{X}_n , \qquad (6.4.1b)$$

$$e(n) = d(n) - y(n) ,$$

$$\frac{\partial y(n)}{\partial \underline{\theta}_n} = \underline{X}_n + \sum_{i=1}^{N} h_n(i) \frac{\partial y(n-i)}{\partial \underline{\theta}_n} , \qquad (6.4.10)$$

$$\underline{\theta}_{n+1} = \underline{\theta}_n + \alpha e(n) \frac{\partial y(n)}{\partial \underline{\theta}_n} . \qquad (6.4.9)$$

Apart from the obvious approximation involved in using $\nabla e^2(n)$ rather than $\nabla E \{e^2(n)\}$, the recursive equation for the derivatives in (6.4.10) is also approximate. This is because $\underline{\theta}$ is not fixed so that in general we are differentiating with respect to $\underline{\theta}_{n-i}$, rather than $\underline{\theta}_n$. The basis for this approximation is the assumption that the change in the parameter vector $\underline{\theta}$ at each step is small. Obviously the validity of this approximation will depend on α, the input data, and N (the order of the recursive part of the filter).

It has been reported [9],[42], on the basis of numerous simulations, that this algorithm *generally* converges to a minimum of the mean-squared error performance surface for *sufficiently small* adaptation rates. However, IIR adaptive filters are known to experience slow convergence relative to their FIR counterparts. Moreover, convergence may not be to the global minimum, because as we have indicated, the surface may not have the simple quadratic structure of the FIR case and local minima can occur.[17]

Notes on the IIR Adaptive Filter

a) Computational Simplification

The computation of equation (6.4.10) requires a total of $N+L$, N-th order recursions. This computational burden can be reduced by employing the assumption of slowly-varying filter coefficients. Separating (6.4.10) into component derivatives with respect to h

---

17. Note that a global minimum can be assured, but only under highly restrictive conditions on both the filter input, and the model underlying the desired signal [41].

and f, for the adaptive case we have

$$\frac{\partial y(n)}{\partial h_n(i)} = y(n-i) + \sum_{j=1}^{N} h_n(j) \frac{\partial y(n-j)}{\partial h_{n-j}(i)} \quad ; 1,2,...,N \ , \qquad (6.4.10a)$$

$$\frac{\partial y(n)}{\partial f_n(i)} = x(n-i) + \sum_{j=1}^{N} h_n(j) \frac{\partial y(n-j)}{\partial f_{n-j}(i)} \quad ; i=0,1,...,L-1 \ . \quad (6.4.10b)$$

Now we can approximate these derivatives using $\gamma_1(n)$ and $\gamma_2(n)$, where

$$\gamma_1(n) = y(n) + \sum_{j=1}^{N} h_n(j)\gamma_1(n-j) \ , \qquad (6.4.11a)$$

and

$$\gamma_2(n) = x(n) + \sum_{j=1}^{N} h_n(j)\gamma_2(n-j) \ . \qquad (6.4.11b)$$

We may then define a vector of these approximate derivatives as

$$\underline{\phi} = [\gamma_1(n), \gamma_1(n-1), ..., \gamma_1(n-N), \gamma_2(n), \gamma_2(n-1), ..., \gamma_2(n-L+1)]^t \ .$$

The advantage of using γ_1, γ_2 instead of (6.4.10), is that the $N+L$ recursions are reduced to just two. We can illustrate the relation between γ_1, γ_2 and the recursive derivatives of (6.4.10). Consider, for example, the i-th element of γ_2

$$\gamma_2(n-i) = x(n-i) + \sum_{j=1}^{N} h_{n-i}(j)\gamma_2(n-i-j) \ .$$

We see that the difference between this and equation (6.4.10b) is $h_{n-i}(j)$, versus $h_n(j)$ in the i-th term. For sufficiently small α, this difference may be neglected.

An even simpler scheme has been proposed by Feintuch [44]. In this case the recursive structure of the gradient estimates of (6.4.10) is neglected entirely, and the gradient is approximated by simply using the first term of (6.4.10), that is the data vector. The modified equations are directly analogous to the FIRLMS algorithm and have the form

$$y(n) = \underline{\theta}_n^t \underline{X}_n \ , \qquad (6.4.12)$$

$$e(n)=d(n)-y(n) \ , \qquad\qquad (6.4.13)$$

$$\theta_{n+1}=\theta_n+\alpha e(n)\underline{X}_n \ . \qquad\qquad (6.4.14)$$

Of course, this is a less accurate approximation to the least-squares solution and has, in fact, been shown to converge to a solution which is biased from the least-squares minimum even when the IIRLMS converges to the minimum [45].

b) Stability of the IIR Filter

The stability of the IIR filter can be evaluated from the position of the poles of the transfer function relative to $|z|=1$, on any given iteration. Naturally, evaluating these poles (that is, the zeros of the denominator of the transfer function), constitutes a significant computational overhead on the algorithm. If α is small, this test may be conducted less frequently, though exactly how frequently will depend on how close the poles are to the unit circle, and therefore to instability. It should be noted that IIR adaptive filters may be most useful in situations where the system has lightly damped modes (sinusoidal or near sinusoidal components). This is because the FIR representation for such signals is likely to be require many coefficients. Unfortunately, such modes correspond to zeros close to $|z|=1$ and are thus most problematic for stability. Once potential instability is identified there are several possible courses of action [6]:

i) Ignore any update which would lead to unstable pole locations. This policy has the obvious disadvantage that the coefficient could become locked in a 'close-to-instability' condition with every update being rejected. This can be improved by successively reducing the step-size until a satisfactory update is obtained.

ii) Use an exponential weighting factor to push unstable or marginally stable roots inside $|z|=1$. That is, the coefficients are multiplied by a^i, where a is a constant with $0<a<1$. The weighting modifies the transform of the coefficients as

$$H(z)\rightarrow H(a^{-1}z) \ ,$$

and the zeros of $H(z)$ (that is the poles of the IIR filter) are pushed inwards along radial lines to the origin.

It is apparent that the formulation of the testing and projection schemes is essentially ad-hoc. More recently, several authors have suggested replacing the direct form implementation with **cascade** or **parallel** combinations of first or second-order sections [46],[47]. For these systems, it is simple to monitor stability for the first and second-order sections which make up the filter. These alternate realizations are also reported to have reduced sensitivity to round-off noise when implemented with finite precision arithmetic.

6.5 Lattice Structure Adaptive Filters

6.5.1 Forward and Backward Linear Prediction and the Lattice Structure

The lattice structure [48],[49] is an alternative to the usual direct form implementations of a digital filter. The lattice does not have the minimum number of multipliers and adders among all possible implementations of a given impulse response, but it does have some significant advantages. These include a simple test for filter stability, and a modular structure whereby lattice filters may be constructed from a cascade of identical sections. This can be an important property in hardware implementations. Lattice filters are also reported to have good performance in finite word length implementations and low sensitivity to quantization effects [50]. These advantages have led to the use of the lattice in a variety of applications including speech processing [51] and communications [52]. Adaptive filters implemented as lattice structures have other significant advantages. Principal among these is a greatly reduced sensitivity to the eigenvalue spread of the input signal. However, before discussing the adaptive filter in detail we will develop the basic lattice structure.

The simplest form of lattice can be derived from consideration of prediction error filters. Consider a zero-mean stationary input $x(n)$. Recall from Section 3.3.1 that we can define a p-th order single-step prediction of a signal $x(n)$ using a weighted sum of $x(n-1), x(n-2), ..., x(n-p)$. That is we form a prediction $\hat{x}(n)$ as

$$\hat{x}(n) = \sum_{i=1}^{p} c_i x(n-i) , \qquad (6.5.1)$$

where c_i for $i=1,2,...,p$ are the predictor coefficients associated with the p-th order predictor. The prediction error associated with this prediction is

$$e(n)=x(n)-\hat{x}(n) \; ,$$

or

$$e(n)=x(n)-c_1 x(n-1)-c_2 x(n-2)-....-c_p x(n-p) \; , \qquad (6.5.2)$$

and the operator $(1,-c_1,-c_2,...,-c_p)$ is a p-th order prediction error filter. As we have seen in Section 3.3, choosing the predictor coefficients c_i to minimize the mean-squared error $E\{e^2(n)\}$, leads to orthogonality of the prediction error and the input data. That is, for the optimal least-squares predictor

$$E\{e(n)x(n-j)\}=0 \quad ; j=1,2,...,p \; . \qquad (6.5.3)$$

Analogous to this forward predictor, it is also possible to define a **backwards predictor**, which for p-th order takes the form

$$\hat{x}(n-p)=\sum_{i=1}^{p} b_i x(n-p+i) \; , \qquad (6.5.4)$$

and thus predicts $x(n-p)$ using the p data values *ahead* of the point being predicted. This backwards predictor has prediction error

$$e^b(n)=x(n-p)-\hat{x}(n-p)$$
$$=x(n-p)-b_1 x(n-p+1)-b_2 x(n-p+2)-...-b_p x(n) \; , \quad (6.5.5)$$

where the superscript is used to distinguish $e^b(n)$ from the error associated with the **forward predictor** of (6.5.1), (6.5.2). To reinforce this distinction we shall henceforth denote the forward prediction error by $e^f(n)$. Note that we refer to the backwards predictor of $x(n-p)$ as $e^b(n)$ instead of $e^b(n-p)$. This is a notational convenience which reflects the use of data values $x(n-p+1)$, $x(n-p+2)$ through $x(n)$.

From equation (6.5.5) we can form a prediction error operator for the backwards predictor as $(1,-b_1,-b_2,...,-b_p)$, though we must keep in mind that this operator is applied to the data in the opposite direction to the forward prediction error filter. (That is, in the reverse order with increasing coefficient index applied to data with increasing index.)

It is interesting to examine the backwards predictor a little further. Consider the least-squares minimization

$$E\{[e^b(n)]^2\}=E\{[x(n-p)-b_1 x(n-p+1)-...-b_p x(n)]^2\} \ . \qquad (6.5.6)$$

Differentiating with respect to the b_j's and equating to zero as usual yields

$$E\{e^b(n)x(n-p+j)\}=0 \ ; \ \ j=1,2,...,p \ , \qquad (6.5.7)$$

which are the orthogonality conditions. Expanding this equation using (6.5.5) and taking expectations yields

$$\begin{bmatrix} r(0) & r(1) & . & r(p-1) \\ r(1) & r(0) & . & . \\ . & . & . & r(1) \\ r(p-1) & r(p-2) & . & r(0) \end{bmatrix} \begin{bmatrix} b_1 \\ . \\ . \\ b_p \end{bmatrix} = \begin{bmatrix} r(1) \\ . \\ . \\ r(p) \end{bmatrix} . \qquad (6.5.8)$$

Comparing this set of equations with the corresponding set for the forward linear predictor (see Section 3.3.1, equations (3.3.11))

$$\begin{bmatrix} r(0) & r(1) & . & r(p-1) \\ r(1) & r(0) & . & . \\ . & . & . & r(1) \\ r(p-1) & r(p-2) & . & r(0) \end{bmatrix} \begin{bmatrix} c_1 \\ . \\ . \\ c_p \end{bmatrix} = \begin{bmatrix} r(1) \\ . \\ . \\ r(p) \end{bmatrix} , \qquad (3.3.11)$$

and we see that the two sets are identical. Hence $c \equiv b$ and we reach the surprising conclusion that the forward and backward predictors are identical, although as noted above they are applied to the data in opposite directions. Note also that, as a consequence of the identical sets of predictor coefficients, the error energies associated with the minimum mean-squared error solutions for the forward and backwards predictors are identical. That is,

$$E\{[e(n)]^2\}=E\{[e^f(n)]^2\}=E\{[e^b(n)]^2\} \ , \qquad (6.5.9)$$

(See Appendix 6A). One other point to note at this stage, from equation (6.5.3) the forward prediction error $e^f(n)$ is orthogonal to

$x(n-1),...,x(n-p)$. From equation (6.5.7) the backwards prediction error is orthogonal to $x(n),...,x(n-p+1)$. On the other hand $e^b(n-1)$ is orthogonal to $x(n-1),...,x(n-p)$, the same data as for the forward predictor for time index n.

Now suppose that, given the forward and backward predictors of order p, it is desired to obtain the forward prediction error for the predictor of order $p+1$. One possibility would be to augment the normal equations (3.3.11), and solve the resulting $(p+1)$-th order system. Alternatively we could look for a set of coefficients which satisfy the $(p+1)$-th order orthogonality conditions:[18]

$$E\{e^f_{p+1}(n)x(n-j)\}=0 \quad ; j=1,2,...,p+1 . \qquad (6.5.10)$$

The pth order prediction error $e^f_p(n)$ satisfies this orthogonality requirement for $j=1,2,...,p$ as indicated by (6.5.3). To augment this we need to preserve these conditions and also to satisfy

$$E\{e^f_{p+1}(n)x(n-p-1)\}=0 . \qquad (6.5.11)$$

Suppose that we construct $e^f_{p+1}(n)$ by subtracting some multiple of $e^b_p(n-1)$ from $e^f_p(n)$, that is

$$e^f_{p+1}(n)=e^f_p(n)-\gamma_{p+1}e^b_p(n-1) , \qquad (6.5.12)$$

where γ_{p+1} is a constant. Then $e^f_{p+1}(n)$ is orthogonal to $x(n-1),x(n-2),...,x(n-p)$ because both $e^f_p(n)$ and $e^b_p(n-1)$ are orthogonal to this set of data. That is

$$E\{e^f_{p+1}(n)x(n-j)\}=E\{e^f_p(n)x(n-j)\}-$$
$$-\gamma_{p+1}E\{e^b_p(n-1)x(n-j)\}=0 \quad ; j=1,2,...,p .$$

Hence, all that is required is to force the orthogonality condition at $x(n-p-1)$. Substituting (6.5.12) into the desired condition (6.5.11) we see that the requirement is

$$E\{e^f_p(n)x(n-p-1)\}-\gamma_{p+1}E\{e^b_p(n-1)x(n-p-1)\}=0 ,$$

which upon rearrangement gives

---

18. Note that from this point onwards, as we shall be considering predictors of order p and $p+1$, it is necessary to use the subscript to distinguish the order.

$$\gamma_{p+1} = \frac{E\{e_p^f(n)x(n-p-1)\}}{E\{e_p^b(n-1)x(n-p-1)\}} \ . \tag{6.5.13}$$

As a final step we rearrange equation (6.5.5) at time $(n-1)$ to yield

$$x(n-p-1) = e_p^b(n-1) + \sum_{i=1}^{p} b_i x(n-p-1+i) \ .$$

Substituting into (6.5.13) and using the orthogonality of $e_p^b(n-1)$ and $e_p^f(n)$ with respect to $x(n-1),...,x(n-p)$ we obtain

$$\gamma_{p+1} = \frac{E\{e_p^f(n)e_p^b(n-1)\}}{E\{[e_p^b(n-1)]^2\}} \ . \tag{6.5.14}$$

Hence, if $e_{p+1}^f(n)$ is calculated according to (6.5.12) with γ_{p+1} selected according to (6.5.14), then the result is the desired $(p+1)$-th order least-squares forward prediction error.

To summarize, given the forward and backward prediction errors for order p we can obtain the forward prediction error for order $p+1$. This is achieved by forming a linear combination of forward and backward prediction errors of order p as indicated by equation (6.5.12). The required constant γ_{p+1} is given by equation (6.5.14). Similarly, and this is left as an exercise (see Problem 6.30), we can augment the order of the backward prediction error using

$$e_{p+1}^b(n) = e_p^b(n-1) - \gamma_{p+1}^b e_p^f(n) \ , \tag{6.5.15}$$

where

$$\gamma_{p+1}^b = \frac{E\{e_p^f(n)e_p^b(n-1)\}}{E\{[e_p^f(n)]^2\}} \ . \tag{6.5.16}$$

Taken together equations (6.5.12)-(6.5.16) give a procedure whereby the order of forward and backwards prediction error filters can be increased without solving the normal equations. In fact, we can construct a prediction error filter *solely* using this procedure starting with the zeroth order prediction error $e_0^f(n)$ and successively increasing the order. The initial values $e_0^f(n)$ and $e_0^b(n)$ are both equal to $x(n)$, since no prediction is actually made.

For convenience we now summarize the entire procedure:

$$e_0^f(n) = e_0^b(n) = x(n) \ . \tag{6.5.17}$$

For $i = 0, 1, \ldots, p$

$$\gamma_{i+1} = \frac{E\{e_i^f(n) e_i^b(n-1)\}}{E\{[e_i^b(n-1)]^2\}} \ , \tag{6.5.14}$$

$$e_{i+1}^f(n) = e_i^f(n) - \gamma_{i+1} e_i^b(n-1) \ , \tag{6.5.12}$$

$$\gamma_{i+1}^b = \frac{E\{e_i^f(n) e_i^b(n-1)\}}{E\{[e_i^f(n)]^2\}} \ , \tag{6.5.16}$$

$$e_{i+1}^b(n) = e_i^b(n-1) - \gamma_{i+1}^b e_i^f(n) \ . \tag{6.5.15}$$

This scheme is called a **lattice form prediction error filter** and is depicted in Figure 6.5.1.

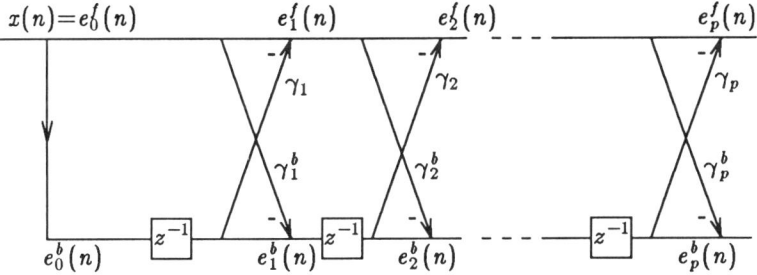

Figure 6.5.1 Lattice form p-th order prediction error filter.

The diagram clarifies the reason for the name. Note the equivalence of the individual sections of the lattice, illustrating the possibility of sharing hardware for different stages of the filter. For stationary inputs, a simplification of the equations is possible because the forward and backward prediction error energies at stage p are equal (see equation (6.5.9)). Thus, for stationary inputs, $\gamma_i^b = \gamma_i$ and the lattice equations have the simplified form:

$$e_0^f(n) = e_0^b(n) = x(n) \ . \tag{6.5.17}$$

For $i=0,1,...,p$

$$\gamma_{i+1}=\frac{E\{e_i^f(n)e_i^b(n-1)\}}{E\{[e_i(n-1)]^2\}} \; , \tag{6.5.18}$$

$$e_{i+1}^f(n)=e_i^f(n)-\gamma_{i+1}e_i^b(n-1) \; , \tag{6.5.19}$$

$$e_{i+1}^b(n)=e_i^b(n-1)-\gamma_{i+1}e_i^f(n) \; . \tag{6.5.20}$$

The γ_i's are known as **reflection coefficients** because of the parallel with transmission lines and with layered earth transmission in reflection seismology.

Properties of the Lattice Filter

a) Order Selection

As we have seen, in the lattice structure the predictor order can be modified without solving the normal equations separately for each increase in the order. Furthermore, because of the modular structure of the lattice, we may simply add or remove stages as required. Also, the forward and backward prediction errors for *all orders* are available during filter computation and may be used for decisions about the appropriate order for a given mean-squared error performance for the filter. In contrast, the predictor coefficients in the transversal filter must all be recalculated if the filter order changes.

b) Stability

It can be shown that the reflection coefficient values must satisfy

$$0\leq|\gamma_i|\leq1 \; , \tag{6.5.21}$$

for a stable analysis filter (see Appendix 6B). In fact, we may demonstrate that for stationary random variables, with γ_i calculated according to (6.5.18), this result is always obtained. Numerically, if estimates are used for γ_i the result may not hold. However, checking for stability is trivial in this case.

c) Orthogonality of Backward Prediction Errors

For stationary zero-mean inputs, the backward prediction errors at stages $i,j \leq p$ are mutually orthogonal with

$$E\{e_i^b(n)e_j^b(n)\}=0 \qquad ; i \neq j ,$$

$$=E\{[e_j^b(n)]^2\} \quad ; i=j . \qquad (6.5.22)$$

To see this, take $i \geq j$ for example, and substitute for $e_j^b(n)$ from (6.5.5) giving

$$E\{e_i^b(n)[x(n-j)-b_1x(n-j+1)-...-b_jx(n)]\} , \qquad (6.5.23)$$

but from (6.5.7)

$$E\{e_i^b(n)x(n-l)\}=0 \quad ; l=0,1,...,i-1 .$$

Thus, since $i \geq j$, all terms in (6.5.23) are zero. We may use a similar argument for $i<j$ (expanding $e_i^b(n)$ instead of $e_j^b(n)$) and conclude finally that

$$E\{e_i^b(n)e_j^b(n)\}=0 \quad ; i \neq j .$$

The lattice equations represent a stage-by-stage orthogonalization where each coefficient is determined separately. This is unlike the tapped delay line implementation of the prediction error filter where the predictor coefficients are interdependent. As we shall see, these properties are extremely useful in adaptive filtering problems.

d) Relation Between the Input Data and the Backward Prediction Errors

For each stage $i=0,1,...,p$ we may write out equation (6.5.5) as

$$e_0^b(n)=x(n)$$

$$e_1^b(n)=x(n-1)-b_1^{(1)}x(n)$$

$$\cdot \qquad \cdot \qquad \cdot$$

$$\cdot \qquad \cdot \qquad \cdot$$

$$e_p^b(n)=x(n-p)-b_1^{(p)}x(n-p+1)-...-b_p^{(p)}x(n) ,$$

where $b_i^{(j)}$ is the i-th backwards predictor coefficient for prediction order j. In vector notation these equations become

$$\underline{e}_n^b = L\underline{x}_n \ , \tag{6.5.24}$$

where $\underline{x}_n = [x(n), x(n-1), ..., x(n-p)]^t$,
$\quad\ \underline{e}_n^b = [e_0^b(n), e_1^b(n), ..., e_p^b(n)]^t$,

and where

$$L = \begin{bmatrix} 1 & 0 & 0 & . & . \\ -b_1^{(1)} & 1 & 0 & . & . \\ -b_2^{(2)} & -b_1^{(2)} & 1 & . & . \\ . & . & & . & . & 0 \\ -b_p^{(p)} & -b_{p-1}^{(p)} & . & . & 1 \end{bmatrix} . \tag{6.5.25}$$

L has eigenvalues $\lambda_1 = \lambda_2 = ... = \lambda_{p+1} = 1$, and hence it is non-singular. Also, the backwards prediction error correlation matrix may be expressed as

$$R_b = E\{\underline{e}_n^b \underline{e}_n^{bt}\} = E\{L\underline{x}_n \underline{x}_n^t L^t\} = LRL^t \ . \tag{6.5.26}$$

As a consequence of (6.5.22), this matrix has the diagonal form

$$R_b = diag\{\sigma_0^2, \sigma_1^2, ..., \sigma_p^2\} \ , \tag{6.5.27}$$

where $\sigma_i^2 = E\{[e_i^b(n)]^2\}$, and $\sigma_0^2 \geq \sigma_1^2 \geq ... \geq \sigma_p^2$ as a consequence of the monotonic behavior of the prediction error energy with predictor order (see Section 3.3.1).

e) Relation Between the Reflection Coefficients and the Predictor Coefficients

The lattice and transverse forms both produce least-squares prediction error filters for a p-th order predictor. It is apparent, therefore, that the prediction and reflection coefficients must be related. If we let $f_i^{(j)}$ represent the i-th coefficient of the forward prediction error filter for order j, then it can be shown (see Appendix 6B), that

$$f_j^{(j-1)}=0, \qquad f_0^{(j)}=1 , \tag{6.5.28}$$

$$f_i^{(j)}=f_i^{(j-1)}-\gamma_j f_{j-i}^{(j-1)} \quad ; i=1,2,...,j , \tag{6.5.29}$$

for $j=1,2,...,p$. Note that with this notation

$$\underline{f}^{(j)}=[f_0^{(j)},f_1^{(j)},...,f_j^{(j)}]^t=[1,-c_1^{(j)},...,-c_j^{(j)}]^t.$$

In applications where only the filter output (the prediction error) is required, then these relations are not required. For applications where the predictor structure is required (as, for example, in the AR spectrum estimation technique described in Section 7.4), the relations must be applied to the reflection coefficients in order to recover the predictor coefficients.

Estimation of the Reflection Coefficients

As with the transversal form least-squares filter of Section 3.2, the lattice equations given above are of limited practical utility because they depend on expected values. As usual, one possibility is to use a 'block-estimation' procedure, replacing the formal expectations in equations (6.5.14) and (6.5.16) with estimates obtained by time-averaging using the available samples. Let us assume that we observe the sequence $x(n)$ for $n=0,1,...,N-1$. Considering (6.5.14), for example, one possible estimator of γ_i is

$$\hat{\gamma}_i=\frac{\displaystyle\sum_{j=i}^{N-1} e_{i-1}^f(j)e_{i-1}^b(j-1)}{\displaystyle\sum_{j=i}^{N-1} [e_{i-1}^b(j-1)]^2} \quad ; i=1,2,...,p , \tag{6.5.30}$$

where the lower limit on the summation is determined by the dependence of $e_p(i)$ on data values $x(i),x(i-1),...,x(i-p)$. If it is desired to put more emphasis on recent data, a weighted estimate may be more appropriate. We may apply an exponential weighting to the time averaged quantities in an analogous manner to that used in Section 6.3 for RLS algorithms. In this case, equations (6.5.14) and (6.5.16) become

$$\hat{\gamma}_i=\frac{\displaystyle\sum_{j=i}^{N-1} c^{N-j-1}e_{i-1}^f(j)e_{i-1}^b(j-1)}{\displaystyle\sum_{j=i}^{N-1} c^{N-j-1}[e_{i-1}^b(j-1)]^2} \quad ; i=1,2,...,p , \tag{6.5.31}$$

and

$$\hat{\gamma}_i^b = \frac{\sum\limits_{j=i}^{N-1} c^{N-j-1} e_{i-1}^f(j) e_{i-1}^b(j-1)}{\sum\limits_{j=i}^{N-1} c^{N-j-1} [e_{i-1}^f(j)]^2} \quad ; \; i=1,2,...,p \; , \quad (6.5.32)$$

where $0<c<1$. Once again, if the data are stationary, then $\gamma_p^b = \gamma_p$. In this case several estimators for γ_p can be employed. For example

$$\hat{\gamma}_i = \frac{\sum\limits_{j=i}^{N-1} e_{i-1}^f(j) e_{i-1}^b(j-1)}{\left[\sum\limits_{j=i}^{N-1} [e_{i-1}^f(j)]^2 \sum\limits_{j=i}^{N-1} [e_{i-1}^b(j-1)]^2 \right]^{1/2}} \quad ; \; i=1,2,...,p \; , \quad (6.5.33)$$

or the computationally simpler:

$$\tilde{\gamma}_i = \frac{2 \sum\limits_{j=i}^{N-1} e_{i-1}^f(j) e_{i-1}^b(j-1)}{\left[\sum\limits_{j=i}^{N-1} [e_{i-1}^f(j)]^2 + [e_{i-1}^b(j-1)]^2 \right]} \quad ; \; i=1,2,...,p \; . \quad (6.5.34)$$

6.5.2 The Gradient Adaptive Lattice

The minimum mean-squared error solution for the lattice structure is equivalent to the normal equation solution for the transversal prediction error filter. The estimates of the filter parameters are obtained using a block approach; a segment of N data points drawn from a stationary process is used to compute each reflection coefficient estimate $\hat{\gamma}_i$. In this section, we develop a continuously adaptive update for the reflection coefficients, analogous to that employed in LMS for the transversal filter coefficients. The simplest algorithm of this type, known as the **Gradient Adaptive Lattice (GAL)** [53] is a steepest descent iteration designed to minimize the sum of forward and backwards prediction errors. That is, the reflection coefficients are chosen so as to minimize

$$J_i = E\{[e_i^f(n)]^2 + [e_i^b(n)]^2\} \quad ; i=1,2,...,p \; . \tag{6.5.35}$$

Analogous to the formula of Section 4.1, a steepest descent update for the i-th stage reflection coefficient is given by

$$\gamma_i(n+1) = \gamma_i(n) - \frac{\alpha}{2} \nabla J_i \quad ; i=1,2,...,p \; , \tag{6.5.36}$$

where $\gamma_i(n)$ represents the i-th stage reflection coefficient update for time index n,[19] and where, as usual, α is the adaptation constant. Note that in contrast to the steepest descent algorithm for the transversal filter, $\gamma_i(n)$ is a scalar, and hence we may develop the update for each reflection coefficient separately. As with LMS, we replace the function J_i of equation (6.5.35) by the instantaneous version:

$$\hat{J}_i = [e_i^f(n)]^2 + [e_i^b(n)]^2 \; . \tag{6.5.37}$$

Hence

$$\nabla \hat{J}_i = \frac{\partial \hat{J}_i}{\partial \gamma_i(n)} = 2[e_i^f(n) \frac{\partial e_i^f(n)}{\partial \gamma_i(n)} + e_i^b(n) \frac{\partial e_i^b(n)}{\partial \gamma_i(n)}] \; . \tag{6.5.38}$$

Now, from (6.5.19) and (6.5.20) for the time-varying reflection coefficient $\gamma_i(n)$ we have

$$e_i^f(n) = e_{i-1}^f(n) - \gamma_i(n) e_{i-1}^b(n-1) \; , \tag{6.5.19}$$

$$e_i^b(n) = e_{i-1}^b(n-1) - \gamma_i(n) e_{i-1}^f(n) \; . \tag{6.5.20}$$

Hence,

$$\frac{\partial e_i^f(n)}{\partial \gamma_i(n)} = -e_{i-1}^b(n-1), \quad \frac{\partial e_i^b(n)}{\partial \gamma_i(n)} = -e_{i-1}^f(n) \; . \tag{6.5.39}$$

Substituting into (6.5.38), we have

$$\nabla \hat{J}_i = 2[-e_i^f(n) e_{i-1}^b(n-1) - e_i^b(n) e_{i-1}^f(n)] \; . \tag{6.5.40}$$

Hence, using (6.5.37) the update is

---

19. Note that the use of a single reflection coefficient for both forward and backwards errors implicitly assumes stationarity of the input sequence.

$$\gamma_i(n+1)=\gamma_i(n)+\alpha[e_i^f(n)e_{i-1}^b(n-1)+e_i^b(n)e_{i-1}^f(n)] \ . \qquad (6.5.41)$$

This equation, together with the forward and backwards prediction error filters of (6.5.19) and (6.5.20) represents the basic GAL.

Notes on the GAL

a) Initialization

Initially we perform no prediction, and this corresponds to setting the reflection coefficient values to zero:

$$\gamma_i(0)=0 \quad i=1,2,...,p \ .$$

This results in an initial set of errors given by

$$e_i^f(0)=e_i^b(0)=x(0) \ .$$

b) Normalized Algorithm

As with LMS, it is common to replace the step-size α by a time-varying normalized parameter $\alpha(n)$. This normalization is conducted with respect to the error power. This has the effect of correcting for the error power reduction which occurs as both the time index and the stage increase. One approach is to normalize with respect to the sum of the squared forwards and backwards prediction errors [54]. A suitable estimator is given by

$$\alpha(n)=[1-k]\alpha(n-1)+k([e_i(n)]^2+[e_i^b(n)]^2) \ , \qquad (6.5.42)$$

where k is a weighting constant with $0<k<1$. With this choice, the update (6.5.41) becomes

$$\gamma_i(n+1)=\gamma_i(n)+\frac{c}{\alpha(n)}[e_i^f(n)e_{i-1}^b(n-1)+e_i^b(n)e_{i-1}^f(n)] \ , \ (6.5.43)$$

where c is a constant. This procedure is equivalent to normalizing with respect to the input power, and has the effect of equalizing convergence rate and misadjustment at each stage of the lattice.

c) Convergence

The convergence behavior of the GAL is rather complex because of a nonlinear interaction between the behavior of the i-th stage, and that of lower order stages [55]. Comparing GAL and LMS algorithms, the GAL algorithms generally enjoy faster convergence, at a rate which is largely independent of the eigenvalue spread of the input. On the other hand, the misadjustment has been shown to be higher than that for the transversal filter [56],[57]. The normalized GAL exhibits superior convergence behavior due to the removal of dependence on the input signal variance [58].

Example – Parameter Estimation for an AR Process

Assume that we have a sequence of samples $x(n)$ for $n=0,1,...,M-1$ drawn from a zero-mean stationary process of the form

$$x(n)=a_1 x(n-1)+a_2 x(n-2)+w(n) ,$$

where a_1, a_2 are unknown constants, and $w(n)$ is zero-mean iid. This signal is familiar as an AR(2) sequence. Figure 6.5.2 shows the reflection coefficients $\gamma_1(n)$, $\gamma_2(n)$ obtained using $x(n)$ as input to the GAL. In this experiment, the unnormalized form of the update (equation (6.5.41)) was used, and the step-size was fixed as $\alpha=0.005$. The results shown represent an average over an 'ensemble' of 100 trials. The simulations employed a unit variance Gaussian sequence for $w(n)$, and values for a_1 and a_2 of 0.75 and -0.125, respectively. The adaptive estimates of these coefficients are obtained from the reflection coefficients by using the relations (6.5.28), (6.5.29). (The details of this calculation, as well as the calculation of the exact values for γ_1, and γ_2 are considered in Problem 6.35.) Figure 6.5.3 shows the predictor coefficients calculated from the adaptive estimates for γ_1 and γ_2. As can be seen, the coefficients converge towards the exact values. We note that it is possible to calculate these predictor coefficients directly using the transversal filter. Moreover, it is clear that for a suitable choice of α, the transversal filter would also converge. A comparison between the two methods could not be based on convergence rate alone however, because similar values for α are not directly comparable for the two algorithms. Such a comparison would necessarily involve the relative convergence rates

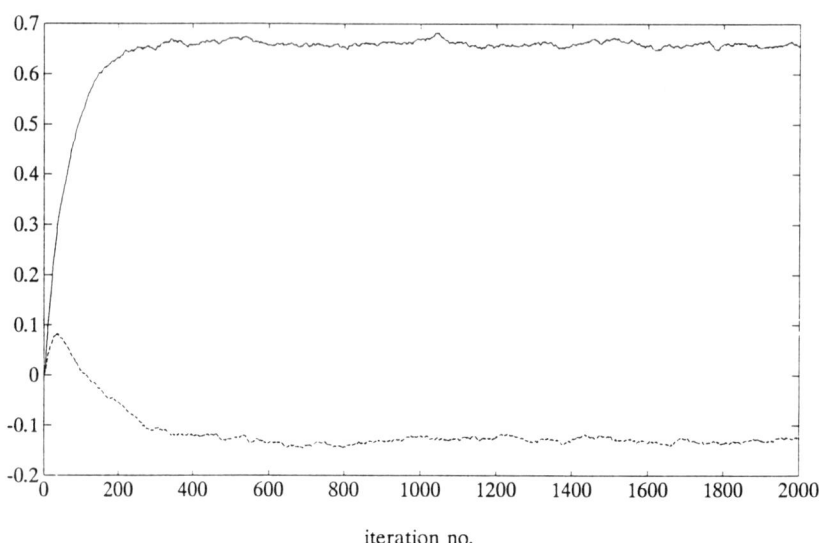

iteration no.

Figure 6.5.2 Reflection coefficients γ_1 (solid line), and γ_2 (dashed) for a GAL predictor with an AR(2) input with parameters $a_1=0.75$, and $a_2=-0.125$. The GAL system used a constant stpe-size $\alpha=0.005$. The results shown represent an average over 100 trials.

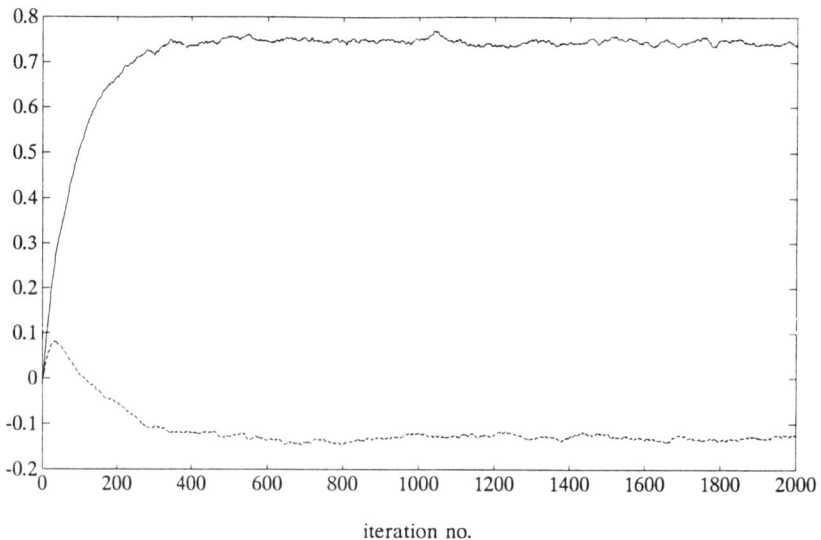

iteration no.

Figure 6.5.3 Predictor coefficients derived from the reflection coefficients in Figure 6.5.2. The results shown represent an average over 100 trials.

for an equivalent level of mean-squared error. This issue is explored further in Computer Problem 3.

6.5.3 Joint Process Estimator

The lattice structure of Figure 6.5.1 is equivalent to a transversal digital filter. However, the minimum mean-squared error solution for the reflection coefficients addresses a restricted class of problems — the formulation of prediction error filters. This, of course, represents only a small subset of digital and adaptive filtering problems. A general digital filtering implementation which exploits the attractive orthogonality properties of the lattice is the **joint process estimator** shown in Figure 6.5.4.

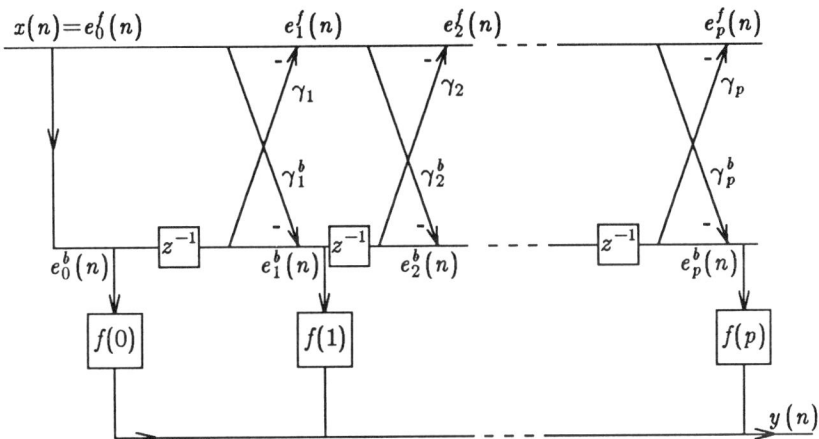

Figure 6.5.4 Lattice joint process estimator.

This filter has two distinct stages: a lattice predictor and a transverse or forward section. In the first stage, the inputs are processed by a lattice prediction error filter. This produces a set of forward and backwards prediction errors $e_i^f(n)$ and $e_i^b(n)$, respectively, according to the lattice equations (6.5.19) and (6.5.20). In the second stage, the lattice backwards prediction errors for stages $i=1,2,...,p$ provide the inputs to a transversal filter with coefficients $f(i)$; $i=1,2,...,p$. This produces an output $y(n)$. The structure can be viewed as a lattice preprocessor followed by a conventional filtering operation. The purpose of the preprocessor is to orthogonalize the inputs to the forward stage. There exists a one-to-one correspondence between the backwards prediction errors and the input data vector, as expressed by equation (6.5.24). Consequently, the joint process estimator is equivalent to a transversal digital filter. The structure is, however, more complex than a conventional transversal implementation.

The joint process estimator allows design of the $f(i)$'s so as to minimize the mean-squared error between the output $y(n)$ and a desired signal $d(n)$. The inputs to the processor are provided by the backwards prediction errors

$$e(n)=d(n)-y(n)=d(n)-\underline{f}^{\,t}\underline{e}_n^b \ , \qquad (6.5.44)$$

where $\underline{f}=[f(0),f(1),...,f(p)]^t$, $\underline{e}_n^b=[e_0^b(n),e_1^b(n),...,e_p^b(n)]^t$. Here, $e(n)$ is the overall system error rather than either the forward or backwards prediction errors. Minimization of

$$J=E\{e^2(n)\}=E\{[d(n)-\underline{f}^{\,t}\underline{e}_n^b]^2\} \ , \qquad (6.5.45)$$

gives[20]

$$R_b\underline{f}=\underline{g} \ , \qquad (6.5.46)$$

where $R_b=E\{\underline{e}_n^b\underline{e}_n^{bt}\}$, and $\underline{g}=E\{d(n)\underline{e}_n^b\}$. Referring to equation (6.5.27)

$$R_b=diag\{\sigma_0^2,\sigma_1^2,...,\sigma_p^2\} \ , \qquad (6.5.27)$$

where $\sigma_i^2=E\{[e_i^b(n)]^2\}$, and $\sigma_0^2\geq\sigma_1^2\geq...\geq\sigma_p^2$. The diagonal nature of the matrix R_b allows us to write the solution for each element of $f(i)$ independently as

$$f(i)=\frac{E\{d(n)e_i^b(n)\}}{\sigma_i^2} \quad ; i=0,1,...,p \ . \qquad (6.5.47)$$

The coefficients of the forward section can be adapted in the same manner as any transversal adaptive filter. Figure 6.5.5 shows a block diagram depicting the adaptive structure. The steepest

---

20. The minimization exactly parallels that for the transversal filter given in Section 3.2, with $\underline{e}_n^b$ replacing $\underline{x}_n$.

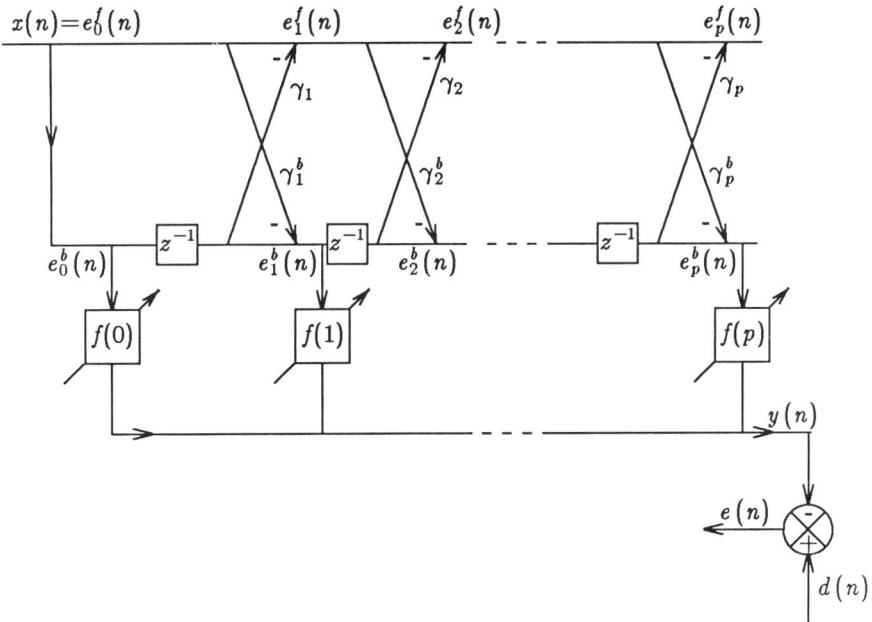

Figure 6.5.5 Adaptive Lattice joint process estimator.

descent update corresponding to J in (6.5.45) is obtained from

$$\underline{f}_{n+1} = \underline{f}_n - \frac{\alpha_f}{2} \nabla J \ , \tag{6.5.48}$$

where $\underline{f}_n$ represents the vector of forward section filter coefficients corresponding to time index n, and where α_f denotes the forward section adaptation constant. An LMS update is obtained by replacing the expectation in (6.5.48) by the instantaneous value. This leads to the update:

$$\underline{f}_{n+1} = \underline{f}_n + \alpha_f e(n) \underline{e}_n^b \ . \tag{6.5.49}$$

In the overall adaptive system, there are two sets of coefficients adapting in parallel: the reflection coefficients that are adapting to orthogonalize the input $x(n)$, and the forward section coefficients $\underline{f}_n$. The great advantage of the system compared to the transversal filter is that the lattice stage produces an orthogonal set of backwards prediction errors. Thus, in the lattice the transversal section operates on decoupled inputs, and hence the adaptive filter does not suffer from eigenvalue disparity problems. Of course, this behavior only occurs after the reflection coefficients have converged. In practice, there will be some dependence between backwards prediction errors during adaptation, and thus the forward section will not be entirely free of eigenvalue effects. Note however, that lattice reflection coefficient adaptation takes place only with respect to the input $x(n)$. Consequently, changes in the statistics of $d(n)$ will affect the forward coefficients $\underline{f}_n$, but not the reflection coefficients. For stationary $x(n)$, once the lattice predictor has converged, the backwards prediction errors will remain orthogonal even if $d(n)$ is nonstationary. Note finally that the joint process estimator reduces to the transversal filter when the inputs are uncorrelated, since in that case the reflection coefficients are zero and no orthogonalization takes place.

6.6 Frequency Domain Implementations of Adaptive Filters

6.6.1 Frequency Domain Adaptive Filter (FDAF)

In this section we return to the FIR transversal LMS filter, considering frequency domain implementations. A frequency domain implementation of the LMS algorithm can be obtained simply by transforming the input signals $x(n)$ and $d(n)$ using Discrete Fourier Transform (DFT) operations, filtering in the frequency domain, and then inverse transforming. There are two significant advantages to this approach. Firstly, as with conventional fixed coefficient filtering, the Fourier transforms can be computed via the Fast Fourier Transform (FFT), and this can greatly reduce the overall computational burden. Secondly, by judicious normalization it is possible to remove, or at least greatly reduce eigenvalue disparity problems. Against this, the frequency domain implementation is inherently a block-processing strategy, so that for on-line processing some delay is unavoidable while incoming data are stored in buffers pending a transform of a complete block.

A number of frequency domain implementations of adaptive filters have been proposed, the simplest approach [59], known as the **Frequency Domain Adaptive Filter (FDAF)** is depicted in Figure 6.6.1. Referring to the figure, the frequency domain

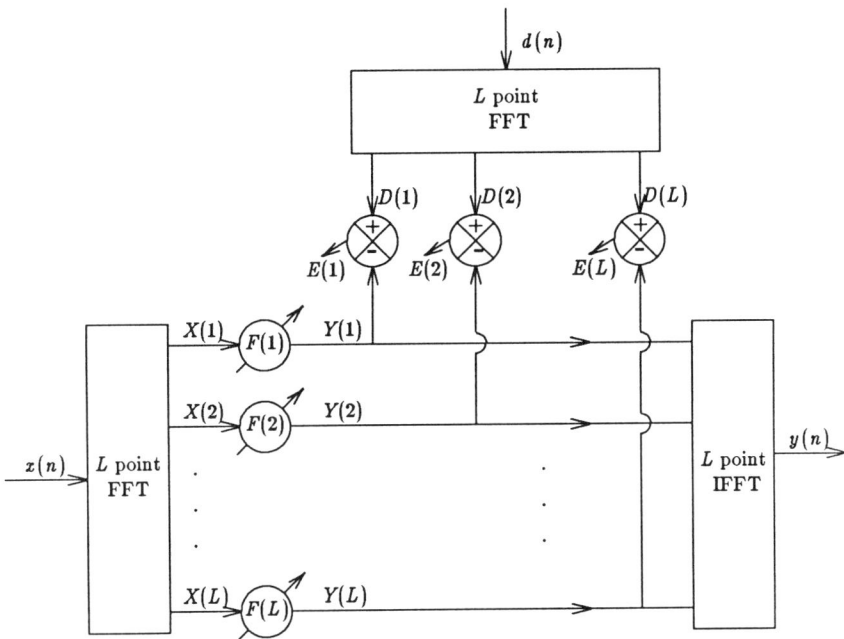

Figure 6.6.1 Frequency Domain Adaptive Filter (FDAF).

implementation provides a block-by-block approach to the processing. The inputs $x(n)$ and $d(n)$ are segmented into blocks of length L, say,

$$x_i(n)=x(iL+n), \quad d_i(n)=d(iL+n) \qquad ; n=0,1,...,L-1 , \qquad (6.6.1)$$
$$; i=0,1,...$$

Each block $x_i(n)$, $d_i(n)$ is transformed using an L point DFT to give $X_i(k)$ and $D_i(k)$, respectively, each of which is a set of L complex numbers corresponding to frequency indices ('bins') $k=0,1,...,L-1$

$$X_i(k)=DFT\{x_i(n)\}=\sum_{n=0}^{L-1} x_i(n)e^{-j\frac{2\pi nk}{L}} \qquad ; k=0,1,...,L-1 , \qquad (6.6.2a)$$

$$D_i(k)=DFT\{d_i(n)\}=\sum_{n=0}^{L-1} d_i(n)e^{-j\frac{2\pi nk}{L}} \qquad ; k=0,1,...,L-1 . \qquad (6.6.2b)$$

L complex adaptive filter coefficients are then applied to $X_i(k)$, one filter coefficient for each frequency bin. The adaptive filter is updated once per data block. The output in each frequency bin is computed as

$$Y_i(k)=F_i(k)X_i(k) \quad ; k=0,1,...,L-1 , \qquad (6.6.3)$$

where $F_i(k)$ is the i-th update (corresponding to the i-th block of data) for the filter coefficient associated with frequency component k. The error for the k-th frequency component is given by

$$E_i(k)=D_i(k)-Y_i(k) \quad ; k=0,1,...,L-1 . \qquad (6.6.4)$$

The system output is produced by inverse transforming

$$y_i(n)=y(iL+n)=IDFT\{Y_i(k)\} ,$$

$$=\frac{1}{L}\sum_{k=0}^{L-1} Y_i(k)e^{\frac{j2\pi nk}{L}} \quad ; n=0,1,...,L-1 . \qquad (6.6.5)$$

It remains to specify the update for the filter coefficients. Because of the complex nature of the data, the algorithm required is the complex LMS described in Section 6.2.8. By analogy with equation (6.2.46), for filter coefficient $F_{i+1}(k)$ we have

$$F_{i+1}(k)=F_i(k)+\alpha E_i(k)X_i^*(k) \quad ; k=0,1,...,L-1 , \qquad (6.6.6a)$$

where $*$ denotes complex conjugate and where, as usual, α is the adaptation constant. Taken together, equations (6.6.1)-(6.6.6) define the frequency domain LMS algorithm. Apart from the complex update, a key difference compared to the time domain LMS algorithm is that the update consists of L *independent* scalar update equations. We may write (6.6.6a) in vector form as

$$\underline{F}_{i+1}=\underline{F}_i+\alpha X_i^* \underline{E}_i , \qquad (6.6.6b)$$

where $\underline{F}_i=[F_i(0),F_i(1),...,F_i(L-1)]^t$, $\underline{E}_i=[E_i(0),E_i(1),...,E_i(L-1)]^t$, and

$$X_i=\begin{bmatrix} X_i(0) & 0 & 0 & . & & 0 \\ 0 & X_i(1) & . & . & & . \\ . & 0 & . & . & & . \\ . & . & . & X_i(L-2) & & . \\ 0 & . & . & 0 & X_i(L-1) \end{bmatrix}$$

$$= diag\{X_i(0), X_i(1), ..., X_i(L-1)\} \ .$$

The diagonal form of X_i emphasizes the decoupled nature of the update equations.

Stability, Convergence and Steady-State Mean-Squared Error of the FDAF

a) Convergence in the Mean

Consider $x(n)$, $d(n)$ jointly stationary. We may derive conditions for stability and convergence in the mean. As usual, we assume that the filter coefficients are initially zero

$$F_0(k) = 0 \qquad ; k = 0, 1, ..., L-1 \ . \tag{6.6.7}$$

Substituting (6.6.4) and (6.6.3) into (6.6.6a) we may write

$$F_{i+1}(k) = (1 - \alpha |X_i(k)|^2) F_i(k) + \alpha D_i(k) X_i^*(k) \ . \tag{6.6.8}$$

Now, taking expectations of both sides of this equation and assuming independence of $X_i(k)$ and $F_i(k)$, we have

$$E\{F_{i+1}(k)\} = [1 - \alpha R(k)] E\{F_i(k)\} + \alpha G(k) \ , \tag{6.6.9}$$

where

$$R(k) = E\{|X_i(k)|^2\} \ , \qquad G(k) = E\{D_i(k) X_i^*(k)\} \ ,$$

and where R and G are dependent on frequency bin k, but as a consequence of the stationarity of $x(n)$ and $d(n)$, not on index i. The independence assumption used to obtain equation (6.6.9) is directly analogous to that employed in Section 4.2.2 in the analysis of mean convergence for the time domain algorithm. Equation (6.6.9) is a recursive scalar difference equation which we may z transform to obtain

$$F^{(k)}(z) = [1 - \alpha R(k)] z^{-1} F^{(k)}(z) + \alpha \frac{G(k)}{(1 - z^{-1})} \ , \tag{6.6.10}$$

where

$$F^{(k)}(z) = \sum_{i=0}^{\infty} E\{F_i(k)\} z^{-i} \ ,$$

is the unilateral z transform. We may obtain the steady-state mean value $F_\infty(k)$ using the final value theorem for z transforms [60]:

$$E\{F_\infty(k)\}=\lim_{z\to 1}\{(z-1)F^{(k)}(z)\} \ . \tag{6.6.11}$$

Applying this result to equation (6.6.10), we have

$$E\{F_\infty(k)\}=\frac{G(k)}{R(k)}=\frac{E\{D_i(k)X_i^*(k)\}}{E\{|X_i(k)|^2\}} \ . \tag{6.6.12}$$

This gives the steady-state value for the FDAF coefficients, and is analogous to the normal equation solution for the least-squares filter in the time domain, although in general the two are not identical. We can obtain conditions for convergence to this value if we define a mean filter coefficient error $U_i(k)$ as

$$U_i(k)=E\{F_i(k)\}-E\{F_\infty(k)\} \ . \tag{6.6.13}$$

Then, substituting into (6.6.9) and using (6.6.12) we obtain

$$U_{i+1}(k)=[1-\alpha R(k)]U_i(k) \ .$$

Finally we observe

$$\lim_{i\to\infty}\{U_i(k)\}\to 0 \quad ; \ k=0,1,...,L-1 \ ,$$

provided

$$0<\alpha<\frac{2}{R(k)} \ , \tag{6.6.14a}$$

or

$$0<\alpha<\frac{2}{E\{|X_i(k)|^2\}} \ . \tag{6.6.14b}$$

As usual, the convergence and stability conditions for the algorithm are determined by the power of the input.

b) Non-Uniform Convergence

Consider the FDAF defined by the update equation (6.6.6). Equation (6.6.14b) indicates that the rate of convergence is determined by the power in each frequency band. In the vector form (6.6.6b), the convergence rate depends on the eigenvalues of the matrix

$$R = diag\{R(0), R(1), ..., R(L-1)\} \ . \tag{6.6.15}$$

In view of the diagonal nature of R, an important conclusion is that the convergence proceeds *independently* for each band k. This is quite different from the time domain filter in which the mean convergence of all the coefficients is coupled according to the structure of the correlation matrix (see Section 4.2). The convergence process can be represented through the use of time-constants. We recall from Section 4.2.2 that we may define a time-constant for the adaptation of an error coefficient as the time required for that component to decay by $1/e$. As in equation (4.2.27), we approximate the time-constant by

$$t_j \approx \frac{1}{\alpha \lambda_j} \ , \tag{4.2.27}$$

where, in this case

$$\lambda_j = R(j) = E\{|X_i(k)|^2\} \ .$$

Using this definition, each mode of the FDAF converges at a rate

$$t_{Bk} \approx \frac{1}{\alpha E\{|X_i(k)|^2\}} \ ,$$

relative to the once-per-block update. However, there are L samples per update, therefore

$$t_k \approx \frac{L}{\alpha E\{|X_i(k)|^2\}} \ . \tag{6.6.16}$$

For the example of a white input, the average power will be equal in each band, so that all coefficients will converge at the same rate, corresponding to equal time-constants. As we have seen, this is precisely what happens in the time domain case. In general, however, the power will differ between frequency bands and the FDAF will be subjected to a nonuniform convergence effect that is analogous to the eigenvalue disparity problems experienced by the time domain algorithm. However for the FDAF of (6.6.6), we may easily normalize the convergence in each band by employing the modified update:

$$F_{i+1}(k) = F_i(k) + \frac{\alpha E_i(k) X_i^*(k)}{E\{|X_i(k)|^2\}} \ , \qquad (6.6.17)$$

or, in practice we may employ an estimate of $E\{|X_i(k)|^2\}$ such as:

$$F_{i+1}(k) = F_i(k) + \alpha \frac{E_i(k) X_i^*(k)}{\dfrac{1}{N} \displaystyle\sum_{j=i-N+1}^{i} |X_j(k)|^2} \ . \qquad (6.6.18)$$

This is a relatively simple normalization, much simpler than that for the time domain algorithm where eigenvalue independent convergence requires the complexity of RLS or a lattice implementation.

c) Mean-Squared Error

Using the same independence assumptions, and following the analysis of Section 4.2.3, we can obtain an approximate expression for the algorithm excess mean-squared error as:

$$excess \ mse = \frac{\alpha L}{2} \ Power \ of \ the \ Input \ . \qquad (6.6.19)$$

This result can be obtained by directly following the analysis of Section 4.2.3, the details are left to Problem 6.37.

Notes on the FDAF Algorithm

a) Computation

The FDAF implementation provides a substantial reduction in computation when compared to the transversal time domain adaptive filter. In particular, processing a block of L data points in the time domain requires a total of $2L^2 + L$ multiplications [25]. The frequency domain algorithm requires a total of three L point FFT's and $2L$ complex multiplications. Ferrara [61], has evaluated the computational burden for the frequency domain implementation as $3L\log_2(L/2) + 4L$ real multiplications. This calculation includes the reduction in computation achieved by utilizing the symmetry in the transforms which occurs due to the real input data. Compared to a time domain implementation, the resulting computational gain is considerable, especially for long filters. For example, for $L=1000$ the number of multiplications required is reduced by more than a factor of 60.

b) Equivalence of Time-Domain and FDAF Implementations

The time and frequency domain implementations are not equivalent. There are two reasons for this disparity:

i) Circular Convolution Effect — The time domain (linear) convolution of the L point sequence $f(n)$ with L points of the input $x(n)$ produces an output $y(n)$ whose length is $2L-1$ points [60]. The calculation of this convolution using the DFT can only correspond to the linear convolution if each of the transforms is zero-padded to a minimum length of $2L-1$. In other cases, as in the adaptive configuration described here, the result is not equivalent to a linear convolution, and the DFT produces a circular convolution (see Appendix 6C for a discussion of circular convolution). As far as the adaptive system is concerned [61] the circular convolution effect can be minimized if L is chosen to be much greater than the length of the least-squares optimum $f^*(n)$. In this case, however, the resulting increase in computation renders the frequency domain algorithm less attractive.

ii) Block Processing — The frequency domain LMS algorithm operates by updating the filter coefficients only once per block of L points, thus the algorithm is a frequency domain parallel of the Block LMS (BLMS) algorithm, described in Section 6.2.9, though with the output computed using a circular convolution (see also Problem 6.36).

6.6.2 Fast LMS Algorithm

The Fast LMS (FLMS) algorithm is an alternative frequency domain implementation of the LMS algorithm designed to avoid circular convolution effects [62]. The algorithm is similar to the FDAF described in the previous section, except that care is taken to avoid circular convolution effects by using the **overlap-save** method of convolution [60]. Using this approach, the convolution

$$y_i(m) = \sum_{n=0}^{L-1} f_i(n) x_i(m-n) \quad ; \ m=0,1,...,L-1 \ , \qquad (6.6.20)$$

is implemented using $2L$ point transforms applied to input data vectors that are overlapped by L points. The $2L$ point response is defined as the transform of the L point time domain filter response padded with L zeros:

$$\underline{F}_i^t = DFT\{f_i(0), f_i(1), ..., f_i(L-1), 0, 0, ..., 0\} \ . \qquad (6.6.21)$$

We define the $2L$ point input data vector

$$\underline{\chi}_i^t = DFT\{x_{i-1}(0), ..., x_{i-1}(L-1), x_i(0), ..., x_i(L-1)\} \ . \qquad (6.6.22)$$

Using the overlap-save method (see Appendix 6C), the convolution is now obtained as

$$[y_i(0), y_i(1), ..., y_i(L-1)]^t = last \ L \ terms \ of \ IDFT\{\underline{F}_i \times \underline{\chi}_i\} \ , \quad (6.6.23)$$

where $\times$ denotes term-by-term multiplication. Overall, the system is intended to implement the BLMS algorithm with update

$$f_{i+1}(m) = f_i(m) + \alpha \nabla_i(m) \quad ; \ m=0,1,...,L-1 \ , \qquad (6.6.24)$$

where

$$\nabla_i(m) = \sum_{n=0}^{L-1} e_i(n) x_i(n-m) \quad ; \ m=0,1,...,L-1 \ , \qquad (6.6.25)$$

is the gradient estimate. This corresponds to the BLMS algorithm of Section 6.2.9. We implement an update

$$\underline{F}_{i+1} = \underline{F}_i + \alpha DFT\{\nabla_i(0), \nabla_i(1), ..., \nabla_i(L-1), 0, ..., 0\} \ . \qquad (6.6.26)$$

This equation effectively constrains the second set of L coefficients of the adaptive filter to zero (in the time domain that is). To see this, simply apply an inverse transform to each side of equation $(6.6.26)$.[21]

---

21. Note that to ensure that the constraint on the later half of the filter coefficients being zero, the initial condition must fulfill that requirement.

From (6.6.25), $\nabla_i(m)$ has the form of a sample correlation. We can view such an operation as the convolution of the two sequences, with one sequence time-reversed. This can be implemented using the FFT, provided we zero-pad the data. For the i-th block of errors $\{e_i(0), e_i(1), ..., e_i(L-1)\}$, we define a vector $\underline{\eta}_i$ as the transform of the time-reversed and zero-padded errors to length $2L$:

$$\underline{\eta}_i = DFT\{e_i(L-1), e_i(L-2), ..., e_i(0), 0, 0, ..., 0\} \ . \qquad (6.6.27)$$

Then using the overlap-save method the gradient is obtained as

$$\nabla_i(m) = (2L - m + 1)th \ term \ of \ IDFT \ \{\underline{\eta}_i \times \underline{\chi}_i^*\} \ , \qquad (6.6.28)$$

(see also Problem 6.36). This modified algorithm requires a total of five, $2L$ point transforms. For real inputs these transforms are symmetric and require computation of only the first $L+1$ terms. In spite of the increased computational complexity compared to the FDAF, the FLMS is computationally less demanding than the corresponding time domain implementation.

The FLMS is the precise frequency domain implementation of the BLMS algorithm of Section 6.2.9. Consequently, the convergence and mean-squared error behavior for the two algorithms is identical. Recall from Section 6.2.9 that the BLMS (and hence the FLMS) can be considered as a steepest descent iteration with block functional

$$J = \frac{1}{L} E\{\sum_{n=0}^{L-1} e_i^2(n)\} \ . \qquad (6.2.54)$$

For stationary random inputs we demonstrated that the BLMS is convergent to the usual least-squares solution $\underline{f}^* = R^{-1}\underline{g}$, subject to

$$0 < \alpha < \frac{2}{\lambda_{\max}} \ ,$$

where $\lambda_{\max}$ is the largest eigenvalue of R. This is precisely the same result as for the LMS. We recall further from Section 6.2.9, that BLMS (and hence FLMS) has time-constants which, when normalized to those for LMS, give the same excess mean-squared error as LMS (see equations (6.2.61) and (6.2.62)). Consequently, the performance characteristics of FLMS and LMS are essentially identical for stationary inputs.

Comparing the FLMS and FDAF algorithms, we note that the two have similar convergence requirements, time-constants and excess mean-squared error. However, unlike FDAF, FLMS avoids circular convolution effects, and thereby converges to the block optimum as defined by the functional (6.2.54). As we have indicated, for stationary inputs this is equivalent to the least-squares time domain solution. This gain is achieved at the expense of an increase in the computational requirements for the algorithm. Note, however, that for the FLMS the simple bin-by-bin independent normalization cannot be applied, because the update equations are coupled as a consequence of the constraint embodied by equation (6.6.26).

FLMS achieves linear convolution by using $2L$ point transforms and constraining L points of the filter response to zero. Mansour and Gray [63], proposed an alternative algorithm in which this constraint is removed. This removes the requirement for two of the five FFTs in the FLMS. The algorithm is otherwise identical to FLMS. Mansour and Gray were able to show that in addition to the reduced computational burden, this unconstrained frequency domain algorithm enjoys good convergence and simple normalization possibilities.

6.7 Applications Revisited

6.7.1 Adaptive Differential Pulse Code Modulation (ADPCM)

In Section 5.4.2, we described the elements of an Adaptive Differential Pulse Code Modulation (ADPCM) system. For a speech or data signal $s(n)$, the ADPCM system was defined by the equations:

$$\hat{s}(n/n-1) = \underline{f}_{n-1}^{t} \underline{e}_q(n-1) , \tag{5.4.3}$$

$$e(n) = s(n) - \hat{s}(n/n-1) , \tag{5.4.1}$$

$$e_q(n) = Q\{e(n)\} , \tag{5.4.2}$$

$$\hat{s}(n) = \hat{s}(n/n-1) + e_q(n) , \tag{5.4.4}$$

where $\hat{s}(n/n-1)$ is the prediction of $s(n)$, $e(n)$ is the prediction error, $\underline{e}_q=[e_q(n),e_q(n-1),...,e_q(n-L+1)]^t$ is the vector of quantized errors, $\underline{f}_n$ is the adaptive predictor coefficient vector, and $\hat{s}(n)$ is the reconstructed signal. In Section 5.4.2 we described an ADPCM system with adaptive predictor defined by a simple transversal LMS filter with update:

$$\underline{f}_n=\underline{f}_{n-1}+\alpha\underline{e}_{q(n-1)}e_q(n) , \qquad (5.4.5)$$

where $\underline{e}_{qn}=[e_q(n),e_q(n-1),...,e_q(n-L+1)]^t$, and where α is the adaptation constant.

The performance of the ADPCM system is directly related to the performance of the adaptive predictor. In fact, we can improve significantly over this simple algorithm by developing an adaptive predictor that incorporates some of the concepts introduced in earlier sections of this Chapter. A more general predictor for the ADPCM structure [64], incorporating several of these ideas, is provided by the updates

$$\underline{f}_n=\gamma_f\underline{f}_{n-1}+\alpha_f\left[\frac{\underline{e}_{q(n-1)}e_q(n)}{c+\underline{e}_q^t(n-1)\underline{e}_q(n-1)}\right] , \qquad (6.7.1a)$$

and

$$\underline{h}_n=\gamma_h\underline{h}_{n-1}+\alpha_h\left[\frac{\hat{\underline{s}}_{n-1}e_q(n)}{c+\hat{\underline{s}}_{n-1}^t\hat{\underline{s}}_{n-1}}\right] , \qquad (6.7.1b)$$

where $\underline{f}_n=[f_n(0),...,f_n(L-1)]^t$, and $\underline{h}_n=[h_n(0),...,h_n(M-1)]^t$, are vectors of transversal and recursive predictor coefficients, respectively. The predictor of (5.4.3) is replaced by

$$\hat{s}(n/n-1)=\underline{f}_{n-1}^t\underline{e}_{q(n-1)}+\underline{h}_{n-1}^t\hat{\underline{s}}_{n-1} , \qquad (6.7.2)$$

where $\hat{\underline{s}}_n=[\hat{s}(n),\hat{s}(n-1),...,\hat{s}(n-M+1)]^t$, and where the error $e(n)$ and quantized error $e_q(n)$ are defined as in equations (5.4.1) and (5.4.2), respectively. With this structure, the system of Figures 5.4.4a) and b) is replaced by that of Figures 6.7.1a) and b). The predictor of (6.7.1)-(6.7.2) differs from the update of equation (5.4.5) in several important respects. Firstly, this is a pole-zero adaptive algorithm with the recursive parameters $h_n(i)$ and transversal parameters $f_n(i)$ adapted separately. Values used for

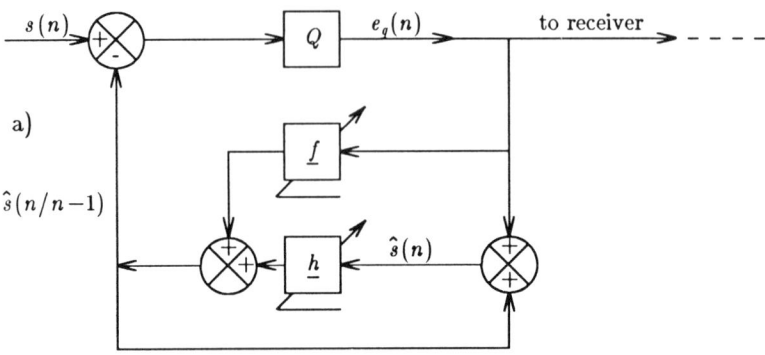

a)

$\hat{s}(n/n-1)$

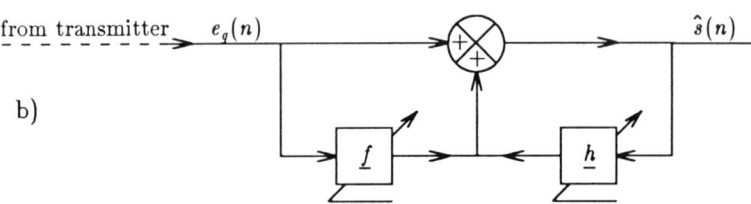

b)

Figure 6.7.1 ADPCM System with pole-zero predictor. Transversal coefficients $\underline{f}$ are driven by quantized error $e_q(n)$. Recursive coefficients $\underline{h}$ are driven by outputs $\hat{s}(n)$ (see equations (6.7.1)). Note that a single sample delay is implicit in the recursive filter.

the number of poles M and zeros L vary in different implementations, but are usually quite small. The CCITT international standard for 32 kbit/s ADPCM uses 2 poles and 6 zeros [13]. A low pole order is chosen because it facilitates control of stability of the predictor when transmission errors occur. Also, since speech signals exhibit significant fluctuations in power, both within speakers and between speakers, the filter employs a power normalization (see Section 6.2.1). The normalization uses an additive constant c ($c>0$) as protection against zero division in periods of silence or near silence in the input signal. The updates also employ leakage factors γ_f, γ_h with $0<\gamma\leq1$ in both cases (see Section 6.2.3). Leakage protects the filter coefficients from the impact of transmission channel errors, causing the effects of such errors to decay with time.[22] The leakage factors may be chosen separately for the poles and zeros, as may the adaptation rates α_f, α_h. As in the simpler algorithm discussed in Section 5.4.2, the update procedures and parameters are identical at both transmitter and receiver. (See also Computer Problem 6.4).

The adaptive predictor in an ADPCM system has been implemented in lattice form [66]. However, a comparison between LMS, the gradient lattice, and the more powerful least-squares lattice found the performance differences to be largely negligible. This was due, presumably, to the essentially secondary role of the prediction in ADPCM (as was discussed in Section 5.4.2). A more recent study found some performance improvement for the lattice and reduced sensitivity to channel errors [67]. Even so, the potential benefits of lattice implementation must be considered at best marginal in this application when the extra cost of implementation is taken into consideration.

6.7.2 Time-Delay Estimation

As we saw in Section 5.3, the LMS algorithm can be applied to the problem of time-delay estimation. Given measurements

$$x_1(t) = s(t) + v_1(t) ,$$
$$x_2(t) = s(t - D) + v_2(t) , \qquad (5.3.1)$$

we sample the measurements, and apply the LMS filter using $x_1(n) = x(n)$ and $x_2(n) = d(n)$ (see Figure 5.3.4). As in other applications, the advantages of the adaptive implementation are the very limited *a priori* knowledge required, the simplicity of the implementation, and the ability to track nonstationary signals. However, the tracking ability of LMS can be limited in this application. For example, suppose that we consider the received signals as noise-free, so that after sampling we have

$$x_1(n) = s(n) ,$$
$$x_2(n) = s(n - D) , \qquad (6.7.3)$$

where $s(n)$ is a white sequence with variance σ_w^2 and D is an

---

22. An alternative approach to protecting the system from the effects of transmission errors is provided by the Median LMS (MLMS) algorithm of Section 6.2.6 [65].

integer delay.[23] Assume further that we run the LMS filter with inputs $x(n)=x_1(n)$, $d(n)=x_2(n)$ and that by some iteration n_0 the algorithm has converged to the optimal solution $f^*(n)$ given by

$$f^*(i)=0 \quad ; i \neq D ,$$
$$f^*(i)=1 \quad ; i=D . \tag{6.7.4}$$

Suppose now that at index n_0, the delay between $x_1(n)$ and $x_2(n)$ suddenly changes to $D+1$ samples, say. The optimal solution now becomes

$$f^*(i)=0 \quad ; i \neq D+1 ,$$
$$f^*(i)=1 \quad ; i=D+1 . \tag{6.7.5}$$

In spite of the simple change in f^*, the adaptive filter cannot react quickly to the new situation and must 're-learn' the optimal solution. According to the theory of Section 4.2.2, the mean error components of LMS will propagate according to equation (4.2.10). For input $x_1(n)$ this takes the form

$$\underline{u}_{n+1}=(I-\alpha R_{11})\underline{u}_n , \tag{6.7.6}$$

where $\underline{u}_n=\underline{f}_n-\underline{f}^*$, and where R_{11} is the correlation matrix corresponding to the input $x(n)=x_1(n)$. The step change at time $n=n_0$ essentially corresponds to a new set of initial conditions with $\underline{f}^*$ given by (6.7.5), and with

$$f_{n_0}(i)=0 \quad ; i \neq D ,$$
$$f_{n_0}(i)=1 \quad ; i=D . \tag{6.7.7}$$

Hence,

$$u_{n_0}(i)=0 \quad ; i \neq D, D+1 ,$$
$$u_{n_0}(i)=1 \quad ; i=D ,$$
$$u_{n_0}(i)=-1 \quad ; i=D+1 . \tag{6.7.8}$$

Now, R is an $(L \times L)$ diagonal matrix with

---

23. The notation implies a sampling rate of unity. This restriction is for notational convenience only, and can easily be removed.

$$R = diag\{\sigma_w^2, \sigma_w^2, ..., \sigma_w^2\} \ ,$$

so that we may decouple equation (6.7.6) as

$$u_{n+1}(i) = (1 - \alpha\sigma_w^2) u_n(i) \quad ; \ n \geq n_0 \ ,$$

or

$$u_{n_0+m}(i) = (1 - \alpha\sigma_w^2)^m u_{n_0}(i) \quad ; \ m = 1, 2, ... \quad (6.7.9)$$

In particular,

$$u_n(D+1) = -(1 - \alpha\sigma_w^2)^{n-n_0} \ ,$$
$$u_n(D) = (1 - \alpha\sigma_w^2)^{n-n_0} \quad ; \ n \geq n_0 \ . \quad (6.7.10)$$

We see from this equation that the filter can only 'forget' the old optimal solution at the same exponential rate as it learns the new one.

Some Alternative Algorithms for TDE

a) Adaptive Delay Estimator [68]

A different approach to the TDE problem has been developed by Etter and Stearns [68] who replaced the LMS filter by a direct estimate of the delay (compare Figures 5.3.4 and 6.7.2). Their approach uses a gradient based update for the delay element of the form:

$$d_{n+1} = d_n - \frac{\alpha}{2} \hat{\nabla}_n \ , \quad (6.7.11)$$

where d_n is the delay estimate corresponding to update (index) n, and where $\hat{\nabla}_n$ is the gradient of the instantaneous squared error (as employed by the LMS algorithm). Now

$$\frac{\partial e^2(n)}{\partial d_n} = 2e(n) \frac{\partial e(n)}{\partial d_n} \ , \quad (6.7.12)$$

where from Figure 6.7.2

$$e(n) = [x_2(n) - x_1(n - d_n)] \ . \quad (6.7.13)$$

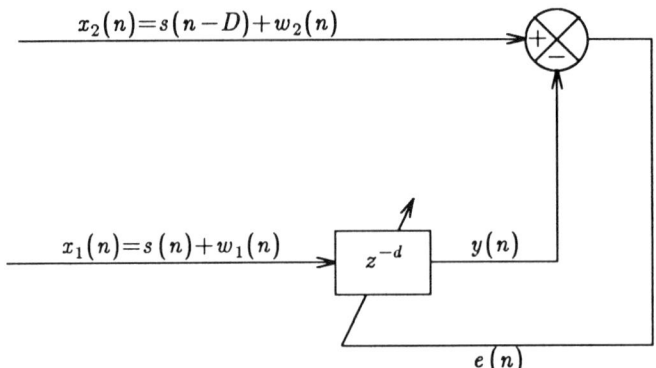

Figure 6.7.2 Etter and Stearns adaptive time-delay estimator. The delay estimate d_n is updated using a gradient formula (see equation (6.7.14)).

The derivative is approximated using a central difference formula, and the resulting update equation has the form:

$$d_{n+1}=d_n+\alpha e(n)[x_1(n-d_n-1)-x_1(n-d_n+1)] .\qquad (6.7.14)$$

A major advantage of this approach is the very low computational requirement as expressed by equation (6.7.14). However, Youn and Ahmed [69] have compared this algorithm with the usual LMSTDE, and concluded that the delay estimate may converge to a local minimum corresponding to a minimum of the cross-correlation of $x_1(n)$ and $x_2(n)$. Consequently, they concluded that the method is only effective if a sufficiently accurate initial estimate of delay is available to ensure convergence to the global minimum.

b) Smoothed Coherence Transform [70]

Scarborough *et al.* [70] implemented an algorithm which is intended as an adaptive implementation of the so-called **Smoothed Coherence Transform (SCOT)** [71]. The system configuration for this method is depicted in Figure 6.7.3. The system consists of two LMS adaptive filters $\underline{f}_1$, and $\underline{f}_2$. $\underline{f}_1$ operates with $x_1(n)$ as input and $x_2(n)$ as desired signal. For $\underline{f}_2$, the roles of input and desired signal are reversed. Using the technique of Section 4.2, the least-squares Wiener approximations for the two filters are given by:

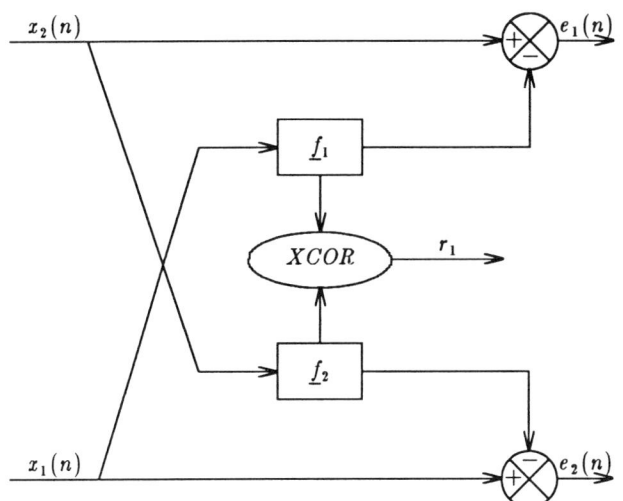

Figure 6.7.3 Adaptive estimation of magnitude squared coherence. $\underline{f}_1$, $\underline{f}_2$ are LMS adaptive filters, the adaptation mechanism is omitted for clarity.

$$F_1(e^{j\omega}) = \frac{R_{x_1 x_2}(e^{j\omega})}{R_{x_1 x_1}(e^{j\omega})} \ , \tag{6.7.15}$$

and

$$F_2(e^{j\omega}) = \frac{R_{x_2 x_1}(e^{j\omega})}{R_{x_2 x_2}(e^{j\omega})} \ . \tag{6.7.16}$$

Consequently, taking the product of the $F_1(e^{j\omega})$ and $F_2^*(e^{j\omega})$ gives

$$F_1(e^{j\omega}) F_2^*(e^{j\omega}) = \frac{|R_{x_1 x_2}(e^{j\omega})|}{R_{x_1 x_1}(e^{j\omega}) R_{x_2 x_2}(e^{j\omega})} = \gamma^2(\omega) \ . \tag{6.7.17}$$

$\gamma^2(\omega)$, is the **magnitude squared coherence**. This quantity is the square of the SCOT estimator. Now if $\gamma(\omega)$ is a complex quantity represented by $\gamma(\omega) = re^{-j\omega D}$, then $\gamma^2(\omega) = r^2 e^{-j2\omega D}$ and the delay can be estimated directly from the phase of $\gamma^2(\omega)$. In the time domain, the product in (6.7.17) is equivalent to the sample correlation of the two filter responses. Scarborough *et al.* reported results for this algorithm that compared favorably with LMSTDE.

c) A Gradient Adaptive Lattice Time-Delay Estimator (GALTDE)

A further approach to the usual LMSTDE system is depicted in Figure 6.7.4 [72].

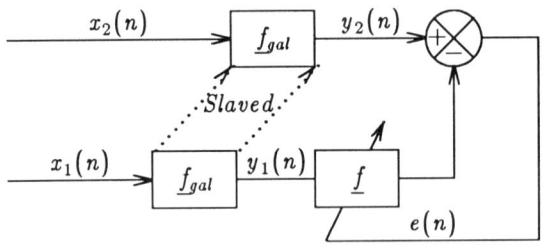

Figure 6.7.4 Gradient adaptive lattice time-delay estimator (GALTDE).

This system, which is referred to as the Gradient Adaptive Lattice Time-Delay Estimator (GALTDE), uses a pair of predictive lattice structures employed as prewhitening filters prior to the operation of an LMS filter. The lattice filter applied to $x_1(n)$ is defined by the equations:

$$e_i^f(n) = e_{i-1}^f(n) - \gamma_i(n)e_{i-1}^b(n-1) , \qquad (6.5.19)$$

$$e_i^b(n) = e_{i-1}^b(n-1) - \gamma_i(n)e_{i-1}^f(n) , \qquad (6.5.20)$$

$$e_0^f(n) = e_0^b(n) = x_1(n) , \qquad (6.5.17)$$

where $e_i^f(n)$ and $e_i^b(n)$ are the forward and backward prediction errors for stage i, and $\gamma_i(n)$ is the reflection coefficient for stage i. This is updated according to

$$\gamma_i(n+1) = \gamma_i(n) + \alpha[e_i^f(n)e_{i-1}^b(n-1) + e_i^b(n)e_{i-1}^f(n)]. \qquad (6.5.41)$$

where α is the adaptation constant. Recall from Section 6.5.2 that this represents the unnormalized form of the Gradient Adaptive Lattice (GAL) algorithm.

The lattice applied to $x_2(n)$ is a slaved version of that applied to $x_1(n)$. That is, no independent adaptation takes place in this second lattice. This is a reflection of our expectation that the spectra of $x_1(n)$ and $x_2(n)$ will be similar. The forward section of the lattice acts as an LMS filter operating on the backward prediction errors. That is, the update equation is

$$\underline{f}_{n+1}=\underline{f}_n+\alpha e(n)\underline{e}_n^b \ , \tag{6.5.49}$$

where $\underline{e}_n^b=[e_0^b(n),e_1^b(n),...,e_p^b(n)]^t$ is the vector of backward prediction errors at time n (see Section 6.5.3) and where

$$e(n)=x_2(n)-\underline{f}_n^t\underline{e}_n^b \ . \tag{6.7.18}$$

For the GALTDE processor, each predictive segment essentially acts as an inverse filter to the corresponding input. Assume $s(n)$ is a linear process (see Section 2.4) generated via

$$s(n)=g(n)^*w(n) \ , \tag{6.7.19}$$

where $w(n)$ is a zero-mean iid sequence, and where $g(n)$ is an arbitrary linear system. The lattice predictor acts to 'whiten' the input signals. In terms of the equivalent transversal representation for the predictive lattice (see Section 6.5), assuming that the measurement noise is zero, then the Wiener approximation to the steady-state behavior is given by

$$F_{gal}(e^{j\omega})=\frac{1}{G(e^{j\omega})} \ , \tag{6.7.20}$$

where $F_{gal}(e^{j\omega})$ is the equivalent transversal form for the predictive lattice.

From equation (6.7.20), each GAL can be interpreted as an (adaptive) prewhitening filter. Also, if the outputs (the backward prediction errors) from the lattices are denoted $y_1(n)$ and $y_2(n)$, where

$$y_1(n)=f_{gal}(n)^*x_1(n) \ , \tag{6.7.21}$$

and

$$y_2(n)=f_{gal}(n)^*x_2(n) \ , \tag{6.7.22}$$

then the Wiener approximation to the behavior of the LMS filter applied to $y_1(n)$ with $y_2(n)$ as desired input signal is

$$F(e^{j\omega})=\frac{R_{y_1y_2}(e^{j\omega})}{R_{y_1y_1}(e^{j\omega})}=\frac{|F_{gal}(e^{j\omega})|^2R_{x_1x_2}(e^{j\omega})}{|F_{gal}(e^{j\omega})|^2R_{x_1x_1}(e^{j\omega})}=\frac{R_{x_1x_2}(e^{j\omega})}{R_{x_1x_1}(e^{j\omega})}. \tag{6.7.23}$$

This is apparently identical to the direct least-squares processor. In practice, however, the solutions will differ because of the finite causal filters employed, and because the GALTDE will typically converge to the least-squares solution faster than the LMSTDE and other steepest descent based transversal filters. This latter point arises as a consequence of the non-uniform convergence that occurs due to the eigenvalue disparity associated with the inputs. As we have previously indicated, the ratio of the largest to smallest spectral components in the input signal $x_1(n)$, provides an upper bound for the ratio of the largest to smallest eigenvalues of R_{11}. Moreover, the eigenvalue disparity increases monotonically with filter length, and the disparity asymptotically approaches this bound as L increases (see Section 4.2.2). In particular, for bandlimited signals the ratio of $\lambda_{\max}/\lambda_{\min}$ is large even for short filters. This leads to highly non-uniform and ultimately slow convergence. The GALTDE has the potential to reduce this problem because the combined predictive lattice and LMS filter applied to $x_1(n)$ is less sensitive to eigenvalue problems.

To illustrate the differences between the GALTDE and LMSTDE, consider the mean error components for LMS as given by equation (6.7.6)

$$\underline{u}_{n+1}=(I-\alpha R_{11})\underline{u}_n , \qquad (6.7.6)$$

where R_{11} is the autocorrelation matrix of the input $x_1(n)$. Recall from Section 4.2.2, that these equations can be decoupled to yield

$$u'_{n+1}(j)=(1-\alpha\lambda_j)u'_n , \qquad (6.7.24)$$

where $u'_n(j)$ is the j-th element of the error vector at time n, rotated to orthogonal axes using the orthonormal matrix Q, whose columns are the eigenvectors of R_{11}, and where λ_j is the j-th eigenvalue of the matrix R_{11}. As we saw in Section 4.2, and again in Section 6.2.10, there is no requirement for each coefficient to converge directly to the solution. That is, according to (6.7.24) the error $u_n(j)$ at some intermediate point of convergence can actually be greater than $u_0(j)$, and thus, in a noisy environment the probability of a false delay estimate can be increased. This assumption can be substantiated by a simple example. Consider a signal MA(1) generating model

$$s(n)=w(n)+aw(n-1) \qquad ; 0<a<1, \qquad (6.7.25)$$

where $w(n)$ is a zero-mean, iid sequence with unit variance, and the delay $D=1$. The application of the LMSTDE system with filter length $L=3$ is sufficient to demonstrate the result. The fixed least-squares solution for this problem is easily seen to be

$$\underline{f}^* = [0,1,0]^t. \tag{6.7.26}$$

Hence, assuming $\underline{f}_0 = \underline{0}$, the error vector $\underline{u}_0$ is given by

$$\underline{u}_0 = [0,-1,0]^t. \tag{6.7.27}$$

Now, the first element $u_0(0)=0$, that is the initial solution for this element is correct. The solution $u_n(0)$ for $n=1,2,...$ is found by calculating the eigenvalues and eigenvectors of the matrix R_{11}, rotating to the orthogonal coordinates and thus obtaining the orthogonalized error vector. This difference equation is then solved, and $u_n(0)$ is obtained by applying the inverse transformation (multiplication by the transpose Q^t). The algebra is straightforward though rather tedious, and is left as an exercise. The final result can be expressed as

$$u_n(0) = \frac{-[1-\alpha((1+a^2)+\sqrt{2a})]^n + [1-\alpha((1+a^2)-\sqrt{2a}]^n}{2^{3/2}}, \tag{6.7.28}$$

which is generally non-zero for $1 < n < \infty$ (though $u_n(0) \to 0$ as $n \to \infty$ for $0 < \alpha < 2/(1+a^2)+\sqrt{2a}$). Consequently, the error associated with some coefficients of the response actually *increases* during convergence, before finally decaying to zero.

The performance of the gradient lattice is considerably more complex. The forward stage acts on

$$\underline{f}_{n+1} = (I - \alpha \underline{e}_n^b \underline{e}_n^{bt}) \underline{f}_n + \alpha e_i^b(n) \underline{e}_n^b , \tag{6.7.29}$$

For the GAL, we recall from Section 6.5 that after convergence of the predictive section, the backward prediction errors are orthogonal with

$$R_{GAL} = E\{\underline{e}_n^b \underline{e}_n^{bt}\} = diag\{\sigma_0^2, \sigma_1^2, ..., \sigma_{L-1}^2\} , \tag{6.5.27}$$

where $\sigma_i^2 = E\{[e_i^b(n)]^2\}$ is the backward prediction error energy for the i-th stage. As a consequence of the monotonic behavior of error energy with predictor length

$$\sigma_0^2 \geq \sigma_1^2 \geq \ldots \geq \sigma_{L-1}^2.$$

Hence, the mean error update equation corresponding to (6.7.6) is

$$\underline{u}_{n+1} = (I - \alpha R_{GAL})\underline{u}_n, \qquad (6.7.30)$$

and for the i-th coefficient

$$u_n(i) = (1 - \alpha \sigma_i^2)^n u_0(i). \qquad (6.7.31)$$

The GALTDE driven by equation (6.7.31) has two significant advantages compared to the LMSTDE:

i) As indicated by equation (6.7.31), the convergence of the modes of the forward section is independent (decoupled). Thus, the type of 'error spreading' which occurs with the LMSTDE (as, for example, in the 3 point problem discussed above) does not occur with the GALTDE.

ii) It can be shown [72] that the eigenvalue disparity for the forward section of the GALTDE driven by equation (6.7.31) is always less than or equal to that for LMS (see also Problem 6.34). Moreover, equality only occurs if the input is white or if the correlation matrix is singular.

In practice, the backward prediction errors are not orthogonal until the predictive section has converged, and the comments about GALTDE performance will only hold approximately during initial convergence. However, the GALTDE also has a tremendous advantage over LMSTDE in dealing with time-varying delays. To see this, consider the following scenario: assume that the adaptive filter has converged to the optimal solution by some time index n_0. Now assume that at some index $n_1 > n_0$ the time-delay between the signals changes. The filter now produces a non-zero error. As we saw in the simple example earlier, the LMSTDE must re-adapt using signals $x_1(n)$, $x_2(n)$, and this new adaptation will be subjected to precisely the same non-uniform convergence as during initialization. By contrast, the modes of the GALTDE will reconverge *completely independently*. This is because only the cross-correlation is affected by the new delay. The input autocorrelation is unaffected. Thus, the predictive component of the lattice (that which provides the decoupling and eigenvalue disparity reduction) remains fully adapted.

Appendix 6A — Forward and Backwards Prediction Error Energy

Consider a zero-mean stationary signal $x(n)$ as input to a p-th order predictive lattice. Referring to Section 6.5, equation (6.5.2), the forward prediction error for such a filter is

$$e_p^f(n) = x(n) - \sum_{i=1}^{p} c_i x(n-i) \ . \tag{A1}$$

Then

$$E\{[e_p^f(n)]^2\} = r(0) + \sum_{i=1}^{p} \sum_{j=1}^{p} c_i c_j r(i-j) - 2\sum_{i=1}^{p} c_i r(i) \ . \tag{A2}$$

Similarly, from equation (6.5.5) the backwards prediction error is

$$e_p^b(n) = x(n-p) - \sum_{i=1}^{p} b_i x(n-p+i) \ , \tag{A3}$$

and hence

$$E\{[e_p^b(n)]^2\} = r(0) + \sum_{i=1}^{p} \sum_{j=1}^{p} b_i b_j r(i-j) - 2\sum_{i=1}^{p} b_i r(i) \ . \tag{A4}$$

However, as we saw in (6.5.8), the least-squares filter coefficients satisfy $c_i = b_i$ for $i = 1, 2, ..., p$. Hence (A2) and (A4) are equivalent, and therefore the forward and backward prediction error energies are identical.

Appendix 6B — Lattice Reflection Coefficients and the Prediction Error Filter

For a zero-mean stationary signal $x(n)$, the reflection coefficients γ_i in the lattice structure are equivalent to the constants k_i in the Durbin recursion for the prediction error filter (see Appendix 3C). To verify this statement we first observe that the forward prediction error can be written as

$$e_p^f(n) = \sum_{i=0}^{p} f_i^{(p)} x(n-i) \ , \tag{B1}$$

where $f_i^{(p)}$ represents the i-th prediction error coefficient for the p-th order predictor. Here

$$\underline{f}=[f_0^{(p)},f_1^{(p)},...,f_p^{(p)}]^t=[1,-c_1^{(p)},...,-c_p^{(p)}]^t,$$

where the coefficients $c_i^{(p)}$ are the p-th order forward predictor coefficients of equation (6.5.2). From the Durbin's recursion for $\underline{f}$, (described in Appendix 3C), we have the following relations

$$f_i^{(j)}=f_i^{(j-1)}-k_j f_{j-i}^{(j-1)} \quad ; i=1,2,...,j ,$$
$$; j=1,2,...,p , \qquad (3.C16)$$

where

$$f_j^{(j-1)}=0 , \qquad f_0^{(j)}=1 \quad ; j=1,2,...,p . \qquad (3.C17)$$

Substituting (3.C16) into (B1), we have[24]

$$e_p^f(n)=\sum_{i=0}^{p}\left[f_i^{(p-1)}-k_p f_{p-i}^{(p-1)}\right]x(n-i) .$$

Hence combining with (B1) we have

$$e_p^f(n)=e_{p-1}^f(n)-k_p\sum_{i=1}^{p}f_{p-i}^{(p-1)}x(n-i) , \qquad (B2)$$

where we have used the fact that $f_p^{(p-1)}=0$ from (3.C17). Now

$$\sum_{i=1}^{p}f_{p-i}^{(p-1)}x(n-i)=[-c_{p-1}^{(p-1)}x(n-1)-c_{p-2}^{(p-1)}x(n-2)-$$
$$...-c_1^{(p-1)}x(n-p+1)+x(n-p)] ,$$

but in view of the equivalence of the forward and backward predictors as expressed by (6.5.8), we have

$$\sum_{i=1}^{p}f_{p-i}^{(p-1)}x(n-i)=[-b_{p-1}^{(p-1)}x(n-1)-b_{p-2}^{(p-1)}x(n-2)-$$
$$...-b_1^{(p-1)}x(n-p+1)+x(n-p)] ,$$

$$=e_{p-1}^b(n-1) , \qquad (B3)$$

---

24. This relation holds for $i=0$ as a consequence of (3.C17).

where the last equality follows from equation (6.5.5). Hence combining (B2) and (B3) we obtain

$$e_p^b(n)=e_{p-1}^f(n)-k_p\,e_{p-1}^b(n-1)\ . \tag{B4}$$

Comparing this equation with (6.5.12), we see that the constant k_p is equal to γ_p, the reflection coefficient. Furthermore, equation (3.C16) thereby provides a relation between the prediction error filter coefficients (transversal representation), and the reflection coefficients (lattice representation). As a consequence of the equality of k_p and γ_p, two further results may be inferred:

i)

$$E\{e_i^2(n)\}=(1-\gamma_i^2)E\{e_{i-1}^2(n)\}\qquad ;\ i=1,2,...,p\ , \tag{B5}$$

where $E\{e_i^2(n)\}=E\{[e_i^f(n)]^2\}=E\{[e_i^b(n)]^2\}$, because of the stationarity of the input. Equation (B5) follows directly from equation (3.C15) in Appendix 3C.

ii)

$$0\le|\gamma_i|\le1\ . \tag{B6}$$

This follows immediately from *i)* and the monotonic behavior of $E\{e_i^2(n)\}$ with i.

Appendix 6C — Linear and Circular Convolution

Consider the convolution of two sequences $x_1(n)$, $x_2(n)$, where to simplify the discussion we will assume both sequences are of length N. The result of this convolution $y(n)$, say, has a total length of $(2N-1)$ points [60]. The convolution

$$y(n)=x_1(n)\,{}^*x_2(n)\ , \tag{C1}$$

can be implemented in the frequency domain using the Discrete Fourier Transform (DFT). Since both input sequences are of length N points, N point transforms can be employed to produce $X_1(k)$ and $X_2(k)$

$$X_1(k)=DFT\{x_1(n)\}=\sum_{n=0}^{N-1} x_1(n)e^{-j\frac{2\pi nk}{N}} \, , \tag{C2}$$

$$X_2(k)=DFT\{x_2(n)\}=\sum_{n=0}^{N-1} x_2(n)e^{-j\frac{2\pi nk}{N}} \, . \tag{C3}$$

The product of the two transforms produces the N point sequence

$$X_3(k)=X_1(k)X_2(k) \, . \tag{C4}$$

Finally we may inverse transform, obtaining a sequence $y_c(n)$ as

$$y_c(n)=IDFT\{X_3(k)\}=\frac{1}{N}\sum_{k=0}^{N-1} X_3(k)e^{j\frac{2\pi nk}{N}} \, , \tag{C5}$$

where *IDFT* denotes the inverse discrete Fourier transform. The output $y(n)$ resulting from the time-domain convolution is, however, $(2N-1)$ points in length. Clearly, therefore, $y(n)$ and $y_c(n)$ are not the same. Let us examine $y_c(n)$ more carefully. Substituting for $X_1(k)$, $X_2(k)$ into (C5) we have

$$y_c(n)=IDFT\{X_1(k)X_2(k)\} \, .$$

Hence, using (C2), (C3)

$$y_c(n)=\frac{1}{N}\sum_{m=0}^{N-1} x_1(m)\sum_{r=0}^{N-1} x_2(r)\left[\sum_{k=0}^{N-1} e^{j\frac{2\pi}{N}(n-m-r)k}\right] \, , \tag{C6}$$

but

$$\sum_{k=0}^{N-1} e^{j\frac{2\pi}{N}(n-m-r)k} = N \quad ; \ n-m-r=...-2N,-N,0,N,2N,... \tag{C7}$$

$$= 0 \quad ; \ otherwise \, .$$

Thus, (C6) collapses to a single summation with $r=n-m$. $y_c(n)$ can then be written

$$y_c(n)=\sum_{m=0}^{N-1} x_1(m)x_2(n-m)_{mod \ N} \quad ; \ n=0,1,...,N-1 \, , \tag{C8}$$

where *mod N* denotes the sequence taken **modulo** N. Thus the product of the DFTs corresponds to a convolution as indicated by equation (C1), but a convolution which employs the sequence $x_2(n)_{mod\ N}$. One way to view this operation is to think of equation (C8) as the result of a convolution of the periodically extended $x_2(n)$ viewed through a window from 0 to $N-1$. As the convolution proceeds, the periodic sequence is shifted through the window. Taking the example of $N=4$, say, we may explicitly evaluate the convolution:

$$y_c(n) = \sum_{m=0}^{3} x_1(m)x_2(n-m)_{mod\ 4} \qquad ; n=0,1,2,3 .$$

The individual values are given by:

$$y_c(0) = \sum_{m=0}^{3} x_1(m)x_2(-m)_{mod\ 4} ,$$

or

$$y_c(0) = \mathbf{x_1(0)x_2(0)} + x_1(1)x_2(3) + x_1(2)x_2(2) + x_1(3)x_2(1) .$$

Note that only the term in bold occurs in the corresponding linear convolution. Similarly

$$y_c(1) = \mathbf{x_1(0)x_2(1)} + \mathbf{x_1(1)x_2(0)} + x_1(2)x_2(3) + x_1(3)x_2(2) ,$$
$$y_c(2) = \mathbf{x_1(0)x_2(2)} + \mathbf{x_1(1)x_2(1)} + \mathbf{x_1(2)x_2(0)} + x_1(3)x_2(3) ,$$
$$y_c(3) = \mathbf{x_1(0)x_2(3)} + \mathbf{x_1(1)x_2(2)} + \mathbf{x_1(2)x_2(1)} + \mathbf{x_1(3)x_2(0)} ,$$

beyond this point the sequence repeats. Note that of the four values produced, only $y_c(3)$ is equal to the linear convolution. All of the others contain additional terms. Examining the sequence of $x_2(n)$ for each of the $y_c(n)$ values, we see that the sequence circulates. That is $x_2(n)$ goes through $x_2(n), x_2(n-1), .., x_2(0)$ and then returns to $x_2(N-1), x_2(N-2)...$ For this reason the result $y_c(n)$ is called the **circular convolution** of the two sequences.

Convolution performed using the DFT always produces a circular result. However, for the convolution of two N point sequences, the circular convolution is identical to the linear result providing all three transforms, that is both forward transforms and the inverse DFT, are zero-extended to at least $(2N-1)$ points in length.

For the Frequency Domain Adaptive Filter (FDAF) implementation of Section 6.6.1, the convolution of the filter and the data is a circular convolution. Moreover, in that case the input data is much longer than the filter. We obviously cannot wait until all the data are available before transforming. The filtering can be performed in blocks of N, provided it is recognized that the outputs from successive blocks will overlap, and that steps must be taken to ensure that successive blocks are put together correctly. There are two principal methods for implementing such a convolution known as the **overlap-add** and **overlap-save** methods [60]. Both methods can be used to produce linear convolution in the FDAF, but in practice the overlap-save is used as it is computationally cheaper. The overlap-save method is implemented by sectioning the input into blocks of $2N$ data points which overlap by 50%, that is by N points. The output for each block is now computed using $2N$ point circular convolution, producing an output $y'(n)$, say. The key to the overlap-save method is that part of the $2N$ point sequence $y'(n)$ is identical to the corresponding linear convolution. In particular the first N points of $y'(n)$ differ from $y(n)$, while the latter N points are the same. Thus, the result of the convolution is identical to the linear result providing the transforms are of length $2N$, and the first N points are discarded. Hence the need for the N point overlap between successive blocks.

Problems

6.1 Consider the LMF update [1]

$$\underline{f}_{n+1} = \underline{f}_n + \alpha N e^{2N-1}(n)\underline{x}_n \ ,$$

with desired input

$$d(n) = \underline{f}^{*t}\underline{x}_n + w(n) \ ,$$

where $\underline{f}^* = [f^*(0), f^*(1), ..., f^*(L-1)]^t$, and where $w(n)$ is zero-mean, iid Gaussian with variance σ^2, and is independent of $x(n)$. Define an error vector $\underline{v}_n = \underline{f}_n - \underline{f}^*$. By neglecting powers of $v_n(i)$ higher than one (that is, by assuming $\underline{f}_n$ is close to the minimum mean-squared error solution), show that

$$\lim_{n\to\infty} \{E\{\underline{v}_n\}\}\to\underline{0} ,$$

provided

$$0<\alpha<\frac{2}{N(2N-1)\sigma^{2N-2}\lambda_{\max}} ,$$

where $\lambda_{\max}$ is the largest eigenvalue of the autocorrelation matrix R. State any further assumptions you require to obtain this result.

Note: For a zero-mean Gaussian random variable u, with variance σ^2

$$E\{u^n\}=0 \qquad\qquad ; n \ odd ,$$
$$=1.3...(n-1)\sigma^n \ ; n \ even .$$

6.2 An alternative update procedure for a gradient-based adaptive algorithm is given by:

$$\underline{f}_{n+1}=\underline{f}_n+\alpha\hat{e}(n)\underline{x}_n ,$$

where

$$\hat{e}(n)=d(n)-\underline{f}_{n+1}^t\underline{x}_n .$$

Show that this procedure results in an update which is equivalent to the normalized algorithm

$$\underline{f}_{n+1}=\underline{f}_n+\frac{e(n)\underline{x}_n}{\underline{x}_n^t\underline{x}_n+1/\alpha} .$$

where $e(n)=d(n)-\underline{f}^t\underline{x}_n$ is the usual error.

6.3 A form of 'variable step' LMS algorithm has update equation:

$$\underline{f}_{n+1}=\underline{f}_n+\alpha_n e(n)\underline{x}_n .$$

a) Assuming $E\{\underline{x}_n\underline{x}_n^t\underline{f}_n\}=E\{\underline{x}_n\underline{x}_n^t\}E\{\underline{f}_n\}$, and making use of the decomposition $R=Q\Lambda Q^t$, show that for stationary inputs,

$E\{\underline{f}_n\}\rightarrow\underline{f}^*$ as $n\rightarrow\infty$ provided

$$0<\alpha_n<\frac{2}{\lambda_j} \ , \ \text{for all } j,n \ .$$

b) Find

$$\lim_{n\rightarrow\infty}\{E\{\underline{f}_n\}\} \ ,$$

if

$$\alpha_n>\frac{2}{\lambda_{\max}} \qquad ; \ n=n_1 \ ,$$

that is, if α_n exceeds the limit at one time instant n_1.

c) Discuss the advantages and disadvantages of the choice

$$\alpha_n=\frac{1}{n} \quad ; \ n\geq 1 \ ,$$

according to the nature of the input signals.

6.4 Consider the variable step algorithm

$$\underline{f}_{n+1}=\underline{f}_n+e(n)M_n\underline{x}_n \ ,$$

where M_n is an $(L\times L)$ diagonal matrix of adaptation factors $\alpha_i(n)$:

$$M_n=diag\{\alpha_0(n),\alpha_1(n),...,\alpha_{L-1}(n)\} \ .$$

For zero-mean stationary inputs, use the independence assumption to analyze the mean behavior of the algorithm. Find upper and lower limits on $\alpha_i(n)$, for mean convergence to occur.

6.5 Assuming $x(n)$, $d(n)$ are zero-mean stationary inputs, use the independence assumption to show that the leaky LMS algorithm with update

$$\underline{f}_{n+1}=\gamma\underline{f}_n+\alpha e(n)\underline{x}_n \ , \tag{6.2.8}$$

where $0<\alpha<1$, converges as

$$\lim_{n\to\infty}\{\underline{f}_n\}\to\left[R+\frac{(1-\gamma)I}{\alpha}\right]^{-1}\underline{g} \ , \tag{6.2.11}$$

provided

$$0<\alpha<\frac{2}{\lambda_{\max}+\dfrac{(1-\gamma)}{\alpha}} \ , \tag{6.2.12}$$

where $\lambda_{\max}$ is the largest eigenvalue of $R=E\{\underline{x}_n\underline{x}_n^t\}$.

6.6 Consider the leaky update:

$$\underline{f}_{n+1}=(1-\alpha\gamma)\underline{f}_n+\alpha e(n)\underline{x}_n \ .$$

Show that using the independence assumption, and subject to a stability constraint on α,

$$\lim_{n\to\infty}\{E\{\underline{f}_n\}\}\to(R+\gamma I)^{-1}\underline{g} \ ,$$

where R, g, have their usual definitions, and find the limits on α for convergence to occur.

6.7 Show that minimization of the modified functional [10]

$$J=e^2(n)+a\underline{f}_n^t\underline{f}_n \ , \tag{6.2.15}$$

using an LMS type gradient descent algorithm, leads to the leaky update

$$\underline{f}_{n+1}=(1-\alpha a)\underline{f}_n+\alpha e(n)\underline{x}_n \ . \tag{6.2.16}$$

6.8 Show that minimization of the modified functional

$$J=E\{|e(n)|\} \ ,$$

using an LMS-type instantaneous gradient descent algorithm leads to the signed error algorithm

$$\underline{f}_{n+1}=\underline{f}_n+\alpha sgn[e(n)]\underline{x}_n \ . \tag{6.2.18}$$

6.9 [14] Consider the homogeneous problem $d(n)=\underline{f}^{*t}\underline{x}_n$, where

$\underline{f}^{*t}=[c,c,c]$; $c>0$, and where $x(n)$ is the '3-periodic' sequence $\underline{x}_n^t=[3,-1,-1,3,-1,-1,...]$. Given $\underline{f}_0\equiv\underline{0}$:

a) Show that the Sign-Sign algorithm

$$\underline{f}_{n+1}=\underline{f}_n+\alpha sgn[e(n)]sgn[\underline{x}_n] ,$$

is always divergent in the sense that

$$\lim_{n\to\infty}\{\underline{f}_n\}\to\infty \quad ;\alpha>0 .$$

b) Find limits on α for the conventional LMS algorithm to converge.

6.10 Using the signal inputs of **6.9**, consider the Signed Regressor algorithm:

$$\underline{f}_{n+1}=\underline{f}_n+\alpha e(n)sgn[\underline{x}_n] .$$

Is this algorithm convergent for the configuration described?

6.11 [11] Repeat **6.10** for the periodic inputs:

   i) $x_n=\{...1,1,1,-3,1,1,1,-3,...\}$.
   ii) $x_n=\{...1,8,1,-5,-5,1,8,1,-5,-5,...\}$.

6.12 Consider the Momentum LMS algorithm

$$\underline{f}_{n+1}=\underline{f}_n+k[\underline{f}_n-\underline{f}_{n-1}]+\alpha e(n)\underline{x}_n \quad ; |k|<1 ,$$

where α, k are constants:

a) By assuming that $\underline{f}_\infty\to\underline{c}$ a constant vector, show that the algorithm converges in the mean to the usual least-squares solution.

b) By decoupling using $R=Q\Lambda Q^t$, and applying the final value theorem for z transforms, confirm the result of part a) and show that the convergence occurs subject to:

$$0<\alpha<\frac{2(1+k)}{\lambda_{max}} .$$

6.13 Consider the third-order LMS algorithm:

$$\underline{D}_n=\frac{1}{1-\delta}\left[e(n)\underline{x}_n-\delta e(n-1)\underline{x}_{n-1}\right],$$

$$\underline{C}_n=\underline{C}_{n-1}+\gamma(\underline{D}_n-\underline{C}_{n-1}),$$

$$\underline{B}_n=\underline{B}_{n-1}+\beta(\underline{C}_n-\underline{B}_{n-1}),$$

$$\underline{f}_{n+1}=\underline{f}_n+\alpha\underline{B}_n,$$

where α, β, γ, and δ are constants. Show that this update can be written as

$$\underline{f}_{n+1}=\gamma_1\underline{f}_n+\gamma_2\underline{f}_{n-1}+\gamma_3\underline{f}_{n-2}+\gamma_4 e(n)\underline{x}_n+\gamma_5 e(n-1)\underline{x}_{n-1},$$

and find expressions for γ_1, $\gamma_2,...,\gamma_5$, in terms of α, β, γ and δ.

6.14 Calculate the output from an $N=3$ point median filter for each of the following input sequences:

i) $x=\{0,0,1,0,-20,-15,-10,-5,0\}$.

ii) $x=\{0,0,-1/2,0,10,7.5,5,2.5,0\}$.

Comparing *i)* and *ii)* what general property can you infer about the median filtering operation?

Note: Assume zero values for x outside the specified range.

6.15 Calculate and sketch the outputs from a 3 point median filter, and from a 3 point weighted average linear filter with coefficients $(1/4,1/2,1/4)$, with inputs:

i) $x=\{0,0,1,1,12,0,0,1,1\}$.

ii) $x=\{1,2,2,8,8,8\}$.

What general properties of median filters might be inferred from these results?

Note: Assume zero values for x outside the specified range.

6.16 Using two-sample constant extension at the beginning and end points, apply the 5 point **trimmed-mean filter** obtained by ranking the input data within an N point moving window, and then applying a linear weighting $\underline{a}=[a(0),a(1),...,a(N-1)]^t$, where $N=5$, and $\underline{a}=[0,1,1,1,0]^t$ to each of the following sequences:

  i) $x=\{2, -3, 4, 5, -7\}$.

  ii) $x=\{-5, 7.5, -10, -12.5, 17.5\}$.

  iii) $x=\{2, 3, 7, 9, 11\}$.

What general properties of the filter can you infer from these examples?

6.17 Repeat the previous question applying a 5 point **outer-mean** filter $(1,0,0,0,1)$ to each of the sequences *i)*, *ii)* and *iii)*. Can you infer similar properties for this filter? Repeat the example with the **maximum filter** $(0,0,0,0,1)$. What properties do you now infer?

6.18 For the signal conditions given in **6.9**, and assuming that changes in the filter coefficients occur sufficiently slowly as to be neglected over any N point window, examine the behavior of the median LMS algorithm

$$\underline{f}_{n+1}=\underline{f}_n+\alpha Med\{e(n)\underline{x}_n\}_N , \qquad (6.2.31)$$

with $N=3$.

6.19 The Median LMS (MLMS) algorithm has an update:

$$\underline{f}_{n+1}=\underline{f}_n+\alpha Med\{e(n)\underline{x}_n\}_N . \qquad (6.2.31)$$

Assuming a zero-mean iid input $x(n)$, that the off-diagonal terms of $\underline{x}_n\underline{x}_n^t$ can be neglected in the median operation, and using the independence assumption, find conditions for convergence in the mean to $\underline{f}^*$, and find the corresponding time-constants.

6.20 Consider the LMS algorithm subjected to input signals:

$$d(n)=\underline{f}^{*t}\underline{w}_n+\eta(n) ,$$

$$x(n)=w(n)+\varsigma(n) ,$$

where $w(n)$, $\eta(n)$, $\varsigma(n)$ are zero-mean mutually uncorrelated iid sequences with variances σ_w^2, σ_1^2, and σ_2^2, respectively, and where $\underline{w}=[w(n),w(n-1),...,w(n-L+1)]^t$. Show that for the LMS algorithm, $E\{\underline{f}_n(i)\}$ does *not* converge to $f^*(i)$.

6.21 Consider the complex LMS algorithm

$$\underline{f}_{n+1}=\underline{f}_n+\alpha e(n)\underline{x}^*(n) , \qquad (6.2.46)$$

where $*$ denotes complex conjugate. If $x(n)$ and $d(n)$ are zero-mean complex stationary input sequences (see Section 2.1.6), find *i)* the least-squares solution for the filter, and *ii)* conditions for convergence in the mean for the algorithm.

6.22 For the complex LMS algorithm with update

$$\underline{f}_{n+1}=\underline{f}_n+\alpha e(n)\underline{x}^*(n) , \qquad (6.2.46)$$

if the input $x(n)$ has the form of a complex sinusoid

$$x(n)=e^{j\omega_0 n} ,$$

use the method of Section 4.2.4 to find an equivalent transfer function relating $D(z)$ to $E(z)$. State carefully any assumptions needed in the derivation.

6.23 Consider a modified Block Least-Squares functional which for the j-th block has the form

$$J=\frac{1}{N}E\{\sum_{i=0}^{N-1} e^2(jN+i)\} \quad ; j=0,1,...,$$

where

$$e(jN+i)=d(jN+i)-\underline{f}^t\underline{x}_{jN+i} ,$$

with $\underline{x}_{jN+i}=[x(jN+i),x(jN+i-1),...,x(jN+i-L+1)]^t$. Given zero-mean stationary inputs $x(n)$, $d(n)$:

a) Show that direct minimization of J gives rise to the usual normal equations:

$$R\underline{f}=\underline{g} \; ,$$

where R, g are the autocorrelation matrix and cross-correlation vector, respectively.

b) Show that adopting a steepest descent minimization, and replacing $E\{\ \}$ with the instantaneous value leads to the Block LMS (BLMS) algorithm

$$\underline{f}_{(j+1)N}=\underline{f}_{jN}+\frac{\alpha_B}{N}\sum_{i=0}^{N-1} e\,(jN+i)\underline{x}_{jN+i} \quad ; \; j=0,1,... \qquad (6.2.47)$$

c) Assuming that changes in the filter coefficient vector occur sufficiently slowly as to be negligible over N iterations, show that subject to the choice of a suitable normalizing factor for the adaptation constant, the mean behavior of the algorithm is identical to that of the Average LMS (ALMS) (see equation (6.2.22)) with averaging interval M.

6.24 Use the analysis method of Section 4.2.3 to show that the excess mean-squared error for the Block LMS algorithm with update (6.2.47) can be approximated by

$$J_{exc}=\frac{\alpha_B}{2N}tr\{R\} \; .$$

State any additional assumptions required to complete the derivation.

6.25 For a zero-mean iid sequence with samples that are uniform on $[-1,1]$, find the extended correlation matrix of equation (6.2.68) for a filter length $L=2$. Find the eigenvalues of this matrix and hence obtain stability limits for the quadratic Volterra filter corresponding to equations (4.2.20) and (4.2.23).

6.26 Follow the analysis of Section 4.2.3 to obtain an approximate expression for the steady-state mean-squared error for the quadratic Volterra LMS filter of Section 6.2.10. State carefully any assumptions required in your derivation, and contrast these with the corresponding assumptions for the conventional LMS.

6.27 Apply the analysis of Section 4.2.3 to the hybrid update:

$$\underline{f}_{n+1} = \underline{f}_n + \alpha R^{-1} e(n) \underline{x}_n ,$$

to show that the excess mean-squared error for the algorithm can be approximated by:

$$excess\ mse \approx \frac{\alpha}{2} J_{\min} tr\{R\} . \qquad (6.3.13)$$

6.28 Show that the direct minimization of

$$\tilde{J}_n = \sum_{i=0}^{n} \lambda^{n-i} e^2(i) ,$$

where $e(i) = d(i) - \underline{f}_n^t \underline{x}_i$, leads to

$$R_n \underline{f}_n = \underline{g}_n , \qquad (6.3.20)$$

where

$$R_n = \sum_{i=0}^{n} \lambda^{n-i} \underline{x}_i \underline{x}_i^t , \qquad (6.3.35a)$$

$$\underline{g}_n = \sum_{i=0}^{n} \lambda^{n-i} d(i) \underline{x}_i . \qquad (6.3.35b)$$

6.29 Consider the second-order IIR structure:

$$H(z) = \frac{\gamma + \eta z^{-1}}{1 + \alpha z^{-1} + \beta z^{-2}} .$$

Derive IIRLMS type updates for the parameters γ, η, α and β, and corresponding recursive updates for the derivatives. (Use the slowly-varying parameters assumption to obtain the derivative updates.)

6.30 Follow the analysis of Section 6.5 to show that given the p-th order forward and backwards prediction errors $e^p(n)$, $e_b^p(n)$, we may augment the backwards predictor as

$$e_{p+1}^b(n) = e_p^b(n-1) - \gamma_{p+1}^b e_p^f(n) , \qquad (6.5.15)$$

where

$$\gamma_{p+1}^b = \frac{E\{e_p^f(n)e_p^b(n-1)\}}{E\{[e_p^f(n)]^2\}} . \qquad (6.5.16)$$

6.31 Consider the lattice equations

$$e_i^f(n) = e_{i-1}^f(n) - \gamma_i e_{i-1}^b(n-1) , \qquad (6.5.19)$$

$$e_i^b(n) = e_{i-1}^b(n-1) - \gamma_i e_{i-1}^f(n) , \qquad (6.5.20)$$

with the modified reflection coefficient

$$\gamma_i = \frac{E\{e_{i-1}^f(n)e_{i-1}^b(n-1)\}}{\frac{1}{2}\left[E\{[e_{i-1}^b(n-1)]^2\} + E\{[e_{i-1}^f(n)]^2\}\right]} ,$$

where

$$e_i^f(n) = x(n) - c_1^{(i)}x(n-1) - \ldots - c_i^{(i)}x(n-i) ,$$

$$e_i^b(n) = x(n-i) - b_1^{(i)}x(n-i+1) - \ldots - b_i^{(i)}x(n) ,$$

where for the *j*-th order predictor, $c_i^{(j)}$, $b_i^{(j)}$ are the *i*-th forward and backwards prediction errors, respectively. Show that for a stationary zero-mean input, the lattice forward prediction error $e_2^f(n)$ is equivalent to the prediction error derived from the normal equations.

6.32 a) Using equations (6.5.19) and (6.5.20), show that for a stationary random input, minimization of

$$J = E\{[e_p^f(n)]^2 + [e_p^b(n)]^2\} ,$$

with respect to γ_p, leads to the modified reflection coefficient γ_i in **6.31**.

b) Show that the resulting reflection coefficients γ_p satisfy $|\gamma_p| \leq 1$.

6.33 Consider the two point stochastic gradient adaptive lattice joint process operator:

$$\underline{f}_n=[f_n(0),f_n(1)]^t .$$

Assuming that the input $x(n)$ is zero mean, stationary with unit variance, and assuming that the lattice section has fully converged, show that:

a) The filter input (that is, the backward prediction errors) are orthogonal.

b) The eigenvalue disparity $\lambda_{max}^{(jp)}/\lambda_{min}^{(jp)}$ associated with this joint process operator satisfies

$$\frac{\lambda_{max}^{(jp)}}{\lambda_{min}^{(jp)}} \leq \frac{\lambda_{max}}{\lambda_{min}} ,$$

where $\lambda_{max}/\lambda_{min}$ is the eigenvalue disparity of a direct (transversal) 2 point predictive filter applied to $x(n)$.

6.34 For a zero-mean stationary random input $x(n)$, show that the lattice joint process operator of Figure 6.5.4, designed to minimize the mean-squared error between the process output $y(n)$ and the desired signal $d(n)$, is related to the minimum mean-squared error transversal filter for the problem via:

$$\underline{f}_L=(L^{-1})^t\underline{f}^* ,$$

where $\underline{f}_L$ and $\underline{f}^*$ are the minimum mean-squared error filters for the joint process and transversal filters, respectively, and where

$$L=\begin{bmatrix} 1 & 0 & 0 & . & . \\ -b_1^{(1)} & 1 & 0 & . & . \\ -b_2^{(2)} & -b_1^{(2)} & 1 & . & . \\ . & . & & . & . & 0 \\ -b_p^{(p)} & -b_{p-1}^{(p)} & . & . & 1 \end{bmatrix} ,$$

$$(6.5.25)$$

where $b_i^{(j)}$ is the i-th minimum mean-squared error backwards predictor coefficient for predictor order j.

6.35 Consider the AR(2) signal

$$x(n)=a_1 x(n-1)+a_2 x(n-2)+w(n) \; ,$$

where $w(n)$ is a zero-mean Gaussian white sequence, and a_1, a_2 are constants. In Section 6.5.1, a signal generated according to this model was used to illustrate the performance of the GAL algorithm. For the minimum mean-squared error lattice of order 2, find the relations between the predictor coefficients c_1, c_2 and the reflection coefficients γ_1, γ_2. Given $a_1=0.75$, $a_2=-0.125$, find explicit values for γ_1, γ_2. Compare your results with Figure 6.5.2.

6.36 Consider the FDAF algorithm defined by update equation (6.6.6) where the vector $\underline{X}_i=[X_i(0),X_i(1),...,X_i(L-1)]^t$ corresponds to the Discrete Fourier Transform (DFT) for the i-th input data block with elements $(x_i(0),x_i(1),...,x_i(L-1))$, where $x_i(j)=x(iL+j)$, and where $\underline{D}_i$ is defined similarly for the desired input. The DFT can be expressed as a matrix operator P with elements

$$P(n,k)=e^{-j\frac{2\pi nk}{L}} \quad ; \; n,k=0,1,...,L-1 \; ,$$

with inverse DFT given by

$$P^{-1}=\frac{1}{L}P^{*} \; ,$$

where * denotes complex conjugate. Define the $(L\times L)$ diagonal matrix $X_D=diag\{X_i(0),X_i(1),...,X_i(L-1)\}$.

a) Find the form of the matrix

$$C=P^{-1}X_D P \; .$$

b) Show that the time domain equivalent of the FDAF update can be written

$$\underline{f}_{i+1}=\underline{f}_i+\alpha \sum_{m=0}^{L-1} e_i(m)\underline{c}_i(m) \; ,$$

where $\underline{f}_i=P^{-1}\underline{F}_i$, $\underline{c}_i(m)$ is the m-th row of the matrix C, and

$$e_i(m) = mth \ element \ of \ \{\underline{d}_i - C\underline{f}_i\} \ ,$$

where $\underline{d}_i = P^{-1}\underline{D}_i$.

6.37 Consider the FDAF update in vector form as

$$\underline{F}_{i+1} = \underline{F}_i + \alpha \underline{E}_i X_i^* \ ,$$

where $\underline{F}_i = [F_i(0), ..., F_i(L-1)]^t$, $\underline{E}_i = [E_i(0), ..., E_i(L-1)]^t$ and where $X_i = Diag\{X_i(0), X_i(1), ..., X_i(L-1)\}$. For zero-mean stationary inputs $x(n)$ and $d(n)$, using the analysis of Section 4.2.3, and noting any independence assumptions required, find an expression for the the excess mean-squared error for the algorithm.

6.38 Let $x(n) = \cos(\omega_0 nT)$. Apply the analysis of Section 4.2.4 to derive an equivalent transfer function for the FDAF. (Assume that the frequency resolution of the FFT is equal to the fundamental frequency f_0.)

6.39 Consider an adaptive filtering configuration with inputs of the form $x(n) = d(n) = w(n)$, where $w(n)$ is a zero-mean iid sequence with variance σ_w^2. Find

$$\lim_{i \to \infty} \{E\{F_i(k)\}\} \ ,$$

for *i)* The FDAF algorithm and *ii)* the FLMS algorithm.

Computer Problem 1 – LMS versus Variable Step LMS

Implement single input adaptive filter systems with LMS and Variable Step LMS (VS-LMS) updates (see Section 6.2.2). A single input adaptive filter can be obtained by using a desired input $d(n) = s(n)$, and delaying the input $x(n) = s(n - \Delta)$ (see also Section 5.1). For the LMS, we implement such a filter using the update equation:

$$\underline{f}_{n+1} = \underline{f}_n + \alpha e(n)\underline{x}_n \ ,$$

where $\qquad e(n) = x(n) - y(n),$

and $\qquad\qquad y(n)=\underline{f}_n^t\underline{x}_n$,

where $\underline{f}_n=[f_n(0),f_n(1),...,f_n(L-1)]^t$,
$\qquad\underline{x}_n=[x(n),x(n-1),...,x(n-L+1)]^t$,

α is the adaptation constant, and Δ is an integer delay. For the VS-LMS algorithm, the update equation is

$$\underline{f}_{n+1}=\underline{f}_n+e(n)M_n\underline{x}_n ,$$

where M_n is an $(L\times L)$ diagonal matrix with

$$M_n=diag\{\alpha_0(n),\alpha_1(n),...,\alpha_{L-1}(n)\} ,$$

where the $\alpha_i(n)$'s are adaptation factors. The $\alpha_i(n)$'s are adapted according to the following rule:

For $n=0$:

$$\alpha_i(n)=\alpha_{\max} \quad ; i=0,1,...,L-1 .$$

For $n>0$:

If $e(n)x(n-i)$ undergoes N_1 successive sign changes, then

$$\alpha_i(n)=\alpha_i(n)*c_1 ,$$

where $c_1<1$. If $e(n)x(n-i)$ does not change sign for N_2 successive updates, then

$$\alpha_i(n)=\alpha_i(n)*c_2 ,$$

where $c_2<1$.
In all cases $\alpha_i(n)$ is constrained such that

$$\alpha_{\min}\leq\alpha_i(n)\leq\alpha_{\max} .$$

Assume $\underline{f}_0=\underline{0}$ throughout. The inputs to both algorithms should be:

i) Data file $x(n)$ of length M.
ii) Delay value Δ.
iii) Filter length L.

In addition, for LMS the adaptation constant α is required. Additional inputs for VS-LMS are α_{min}, α_{max}, c_1, c_2, N_1, and N_2. For both algorithms, the outputs should be:

i) Data file of error values.
ii) Data file of filter output values.

Test the program using $G=0$, with $x(n)$ comprising 5000 points of zero-mean Gaussian white noise with unit variance. $\Delta=0$, filter length $L=32$, $\alpha=0.01$. Plot the error and confirm that values for the input parameters can be found such that the error decays as n increases. Consider inputs

$$d(n)=w(n)+Gv(n) ,$$

$$x(n)=w(n-\Delta) ,$$

where $w(n)$, $v(n)$ are mutually uncorrelated zero-mean iid Gaussian sequences, with $M=5000$ samples, and where G is a constant gain. Compare the algorithms by calculating the total squared error

$$J=\sum_{n=0}^{M-1} e^2(n) .$$

a) Set $G=1$ and $L=10$. For LMS use an empirical search procedure to find the α that minimizes J.

b) Fix $N_1=N_2=3$, $\alpha_{min}=0$, $\alpha_{max}=0.5$. Can you find values for c_1, c_2 for which the value for total squared error for VS-LMS is lower than that for LMS?

Computer Problem 2 – LMS versus Median LMS

Part 1 – Implement the LMS adaptive filter given by the update equation:

$$\underline{f}_{n+1}=\underline{f}_n+\alpha e(n)\underline{x}_n ,$$

where $\qquad e(n)=d(n)-y(n),$

and $\qquad\qquad y(n)=\underline{f}_n^t\underline{x}_n,$

with $\underline{f}_n=[f_n(0),f_n(1),....,f_n(L-1)]^t$,
$\underline{x}_n=[x(n),x(n-1),...,x(n-L+1)]^t$.

Assume $\underline{f}_0=\underline{0}$ in all trials. The inputs to this program should be:

    i) Data files $d(n)$ and $x(n)$.
    ii) Filter length, L.
    iii) Adaptation constant, α.

Part 2 – Generate 5000 samples of each of the following types of test data using $d(n)=x(n)$ in each case:

    i) iid samples, uniformly distributed over [-1,1].
    ii) iid samples, Gaussianly distributed with zero-mean and unit variance.
    iii) Sinusoid with digital frequency $\pi/5$ and phase angle zero.

Part 3 – The main focus of this assignment is identifying the structure of a simple system in the presence of output measurement noise (see diagram)

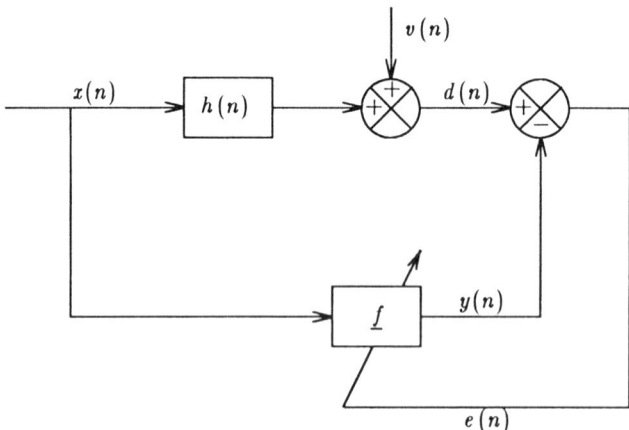

In all cases we consider $x(n)=w(n)$ as zero-mean unit variance Gaussian white noise, and the system $h(n)$ to have the form:

$h(0)=1.0,\ h(1)=0.5,\ h(2)=0.25,\ h(3)=0.125,\ h(4)=0.0625$.

We are concerned here with the performance of the identifier in the presence of various types of uncorrelated noise $v(n)$. For this problem, we know that

$$\underline{f}^* = [f^*(0), f^*(1), ..., f^*(4)]^t = [1, 0.5, 0.25, 0.125, 0.0625]^t,$$

so we set $L=5$. To further constrain the problem, we fix $\alpha=0.005$. Consider the steady-state error measure:

$$E = \left[\sum_{i=0}^{L-1} [f_M(i) - f^*(i)]^2 \right]^{1/2},$$

where $M=5000$. Evaluate and tabulate values of E for the following cases:

i) $v(n)$ iid Gaussian for cases $\sigma_v^2 = 1, 2, 4,$ and 8.
ii) $v(n)$ uniform $[-1,1]$.
iii) $v(n)$ sparse impulsive with

$$prob(v(n)=0) = 1-p ,$$
$$prob(v(n) \neq 0) = p ,$$

where $p \ll 1$, and where non-zero values for $v(n)$ have Gaussian amplitude distribution with zero-mean and standard deviation 10.

Part 4 – Implement two modified LMS algorithms:

i) Average LMS (ALMS) for which

$$\underline{f}_{n+1} = \underline{f}_n + \frac{\alpha}{N} \sum_{i=0}^{N-1} e(n-i)\underline{x}_{n-i} .$$

ii) Median LMS (MLMS) for which the j-th component has update

$$f_{n+1}(j) = f_n(j) + \frac{\alpha}{N} Med\{e(n)x(n-j),$$
$$, ..., e(n-N+1)x(n-j-N+1)\} ,$$

where in both cases $\underline{x}_n$, $d(n)$, $y(n)$ and $e(n)$ have the usual definitions. Set $N=3$, calculate E for the same examples as used in **Part 3** and tabulate your results. What conclusions (if any) can you reach about the relative performance of the three algorithms?

Computer Problem 3 – LMS versus GAL

The purpose of this assignment is to investigate the relation between the transversal LMS and the normalized GAL algorithms in a linear prediction problem.

Part 1 – Implement the LMS algorithm as a single step predictor using the equations

$$\underline{f}_{n+1}=\underline{f}_n+\alpha e(n)\underline{x}_n \ ,$$

and $$y(n)=\underline{f}_n^t \underline{x}_n,$$

where $$e(n)=x(n+1)-y(n),$$

and $\underline{f}_n=[f_n(0),f_n(1),....,f_n(L-1)]^t$,
$\underline{x}_n=[x(n),x(n-1),...,x(n-L+1)]^t$.

Assume $\underline{f}_0=\underline{0}$ throughout.

Part 2 – Implement the unnormalized GAL using the update equation

$$\gamma_i(n+1)=\gamma_i(n)+\alpha[e_i^f(n)e_{i-1}^b(n-1)+e_i^b(n)e_{i-1}^f(n)] \ , \quad (6.5.41)$$

where
$$e_i^f(n)=e_{i-1}^f(n)-\gamma_i(n)e_{i-1}^b(n-1) \ , \qquad\qquad (6.5.19)$$
$$e_i^b(n)=e_{i-1}^b(n-1)-\gamma_i(n)e_{i-1}^f(n) \ , \qquad\qquad (6.5.20)$$

and where initially,

$$\gamma_i(0)=0 \ \ ; \ i=1,2,...,p \ .$$

Part 3 – Test the LMS and GAL algorithms by calculating the updates for $k=1,2,...,2000$. Use $L=2$ for the LMS filter, and $p=2$ for the lattice. Try values of $\alpha=0.005$ for both algorithms. Use sequences of 2000 points of data derived from the AR(2) model

$$x(n)=a_1x(n-1)+a_2x(n-2)+w(n) \, ,$$

where a_1, a_2 are unknown constants, and $w(n)$ is a zero-mean iid sequence. Use an ensemble of 100 trials and average the prediction coefficient values. (For the GAL it will be necessary to calculate the predictor coefficients from the reflection coefficients using equations (6.5.28), (6.5.29).) Compare your results with those obtained in the trial described in Section 6.5.1. Calculate the steady-state mean-squared error for both algorithms by averaging the final squared error value over 100 trials.

Part 4 – Repeat the trial of Part 3, varying α for LMS until the steady-state mean-squared errors are approximately equal. Which algorithm converges faster now?

Computer Problem 4 – ADPCM

Many forms of ADPCM system are possible depending on the quantization law used, the predictor structure, update rule, and so on. In this assignment, we shall use the system described by Koo and Gibson [64] (we use the all-pole version only and assume an ideal channel). Write a computer program to simulate the system and then test system performance for a range of parameters as described below. We restrict attention to data sampled at 8000 *Hz*. With this sample rate, for a 32 kbit/s system the algorithm uses a 16-level symmetric quantizer with adaptive step-size $\Delta(n)$. Use a uniform quantizer, with step-size adaptation chosen via the one-word memory backward adaptive algorithm

$$\Delta(n+1)=\Delta^\beta(n).M(|I(n)|) \, , \tag{1}$$

where $M(\ )$ is a time-invariant multiplier depending on the magnitude corresponding to the transmitted code-word at time n, (denoted $|I(n)|$), and $\beta<1$ is a damping factor. The values should be chosen as:

M_1	0.9
M_2	0.9
M_3	0.9
M_4	0.9

$$M_5 \quad 1.2$$
$$M_6 \quad 1.6$$
$$M_7 \quad 2.0$$
$$M_8 \quad 2.4$$

Limits of $\Delta_{min}=8$ and $\Delta_{max}=2048$ should be imposed (set $\Delta(0)=\Delta_{min}$) and β should be fixed at 63/64 in all trials. The predictor has the form:

$$\hat{s}(n/n-1)=\underline{a}_{n-1}^t\underline{\hat{s}}_{n-1}, \tag{2}$$

where $\underline{\hat{s}}_n=[\hat{s}(n),\hat{s}(n-1),...,\hat{s}(n-L+1)]^t$,
$\quad \underline{a}_n=[a_n(0),a_n(1),...,a_n(L-1)]^t$

is an $(L\times1)$ vector of predictor coefficients. The error $e(n)$ is

$$e(n)=s(n)-\hat{s}(n/n-1) , \tag{3}$$

and $e_q(n)=Q\{e(n)\}$ is the quantized error. The update equation for $\underline{a}_n$ is

$$\underline{a}_n=\lambda\underline{a}_{n-1}+\alpha(n)e_q(n)\underline{\hat{s}}_{n-1} , \tag{4}$$

where

$$\alpha(n)=\frac{\alpha}{100+\underline{\hat{s}}_{n-1}^t\underline{\hat{s}}_{n-1}} , \tag{5}$$

and where λ is a leakage factor. Set $\underline{a}_0=\underline{0}$. At the receiver, the quantized error and the prediction are combined to produce a reconstructed signal $\hat{s}(n)$ as

$$\hat{s}(n)=\hat{s}(n/n-1)+e_q(n) , \tag{6}$$

where $\hat{s}(n/n-1)$ is generated using an identical update to that employed at the transmitter. Design suitable test procedures to ensure correct operation of each stage of your program. Evaluate your system using *i)* white inputs, *ii)* AR(2) inputs generated as:

$$x(n)=x(n-1)+0.25x(n-2)+w(n) . \tag{7}$$

If you have digitized speech data available, evaluate the performance achievable for such data using one or two sentences. Two objective criteria should be employed in the evaluation:

i) Signal-to-Prediction Error Ratio (SPER)

$$SPER = 10\log_{10}\left[\frac{E\{s^2(n)\}}{E\{(s(n)-\hat{s}(n/n-1))^2\}}\right]. \tag{8}$$

ii) Signal-to-Noise Ratio (SNR)

$$SNR = 10\log_{10}\left[\frac{E\{s^2(n)\}}{E\{(s(n)-\hat{s}(n))^2\}}\right]. \tag{9}$$

SPER is used to evaluate the prediction quality at the transmitter and SNR is used to evaluate the overall reconstruction at the receiver. Note that in both cases the expectations are, in practice, replaced by time averages taken over the entire data segment. Evaluate and tabulate results for both measures for $L=4,8,10$, with $\lambda=127/128, 255/256$ and 1 (a total of nine experiments). Use $\alpha=0.03$ throughout.

References

---

1. E.Walach and B.Widrow, "The least mean fourth (LMF) adaptive algorithm and its family," *IEEE Trans. Inform. Theory*, vol. IT-30, pp. 275-283, 1984.

2. C.R.Johnson, Jr., *Lectures on Adaptive Parameter Estimation*, Prentice-Hall, 1988.

3. N.J.Bershad, "Analysis of the normalized LMS algorithm with Gaussian inputs," *IEEE Trans. Acoust., Speech, Signal Processing*, vol. ASSP-34, pp. 793-806, 1986.

4. D.Ristow and B.Kosbahn, "Time-varying prediction filtering by means of updating," *Geophysical Prospecting*, vol. 27, pp. 40-61, 1979.

5. R.W.Harris, D.M.Chabries and F.A.Bishop, "A variable step (VS) adaptive filter algorithm," *IEEE Trans. Acoust., Speech, Signal Processing*, vol. ASSP-34, pp. 309-316, 1986.

6.   J.R.Treichler, C.R.Johnson Jr. and M.G.Larimore, *Theory and Design of Adaptive Filters*, Wiley, 1987.

7.   J.D.Gibson, "Adaptive prediction in speech differential encoding systems," *Proc. IEEE*, vol. 68, pp. 488-525, 1980.

8.   R.D.Gitlin, H.C.Meadors, Jr., and S.B.Weinstein, "The tap-leakage algorithm: An algorithm for the stable operation of a digitally implemented, fractionally spaced adaptive equalizer," *Bell Syst. Tech. J.*, vol. 61, pp. 1817-1839, 1982.

9.   B.Widrow and S.D.Stearns, *Adaptive Signal Processing*, Prentice-Hall, 1985.

10.   P.Darlington and G.Xu, "Equivalent transfer functions of minimum output variance mean-square estimators," *IEEE Trans. Signal Processing*, vol. SP-39, pp. 1674-1677, 1991.

11.   W.A.Sethares, I.M.Y.Mareels, B.D.O.Anderson, C.R.Johnson and R.R.Bitmead, "Excitation conditions for signed regressor least mean-squares adaptation," *IEEE Trans. Circuits Systs.*, vol. CAS-35, pp. 613-624, 1988.

12.   C.F.Cowan and P.M.Grant (Eds.) *Adaptive Filters*, Prentice-Hall, 1985.

13.   "32 kbit/s Adaptive differential pulse code modulation (ADPCM)," *CCITT Standard Recommendation G.271*, Melbourne, 1988.

14.   S.Dasgupta and C.R.Johnson, Jr., "Some comments on the behavior of sign-sign adaptive identifiers," *Systems and Control Letters*, vol. 7, pp. 75-82, 1986.

15.   J.R.Kim and L.D.Davisson, "Adaptive linear estimation for stationary $M-$dependent processes," *IEEE Trans. Inform. Theory*, vol. IT-21, pp. 23-31, 1975.

16.   J.G.Proakis, "Channel identification for high-speed digital communications," *IEEE Trans. Automat. Contr.*, vol. AC-19, pp. 916-922, 1974.

17.   J.R.Glover, "High order algorithms for adaptive filters," *IEEE Trans. Commun.*, vol. COM-27, pp. 216-221, 1979.

18.   S.Roy and J.J.Shynk, "Analysis of the momentum LMS algorithm," *IEEE Trans. Acoust., Speech, Signal Processing*, vol. ASSP-38, pp. 2088-2098, 1990.

19.   P.M.Clarkson and T.I.Haweel, "A median LMS algorithm," *Proc. IEE Electr. Letts.*, vol. 25, pp. 520-522, 1989.

20. I.Pitas and A.Venetsanopoulos, *Nonlinear Digital Filters: Principles and Applications*, Kluwer Academic, 1990.

21. T.I.Haweel and P.M.Clarkson, "Analysis and generalization of a median adaptive filter," *Proc. IEEE Int. Conf. Acoust., Speech, Signal Processing*, pp. 1269-1272, 1990.

22. G.A.Williamson, P.M.Clarkson and W.A.Sethares, "Performance characteristics of the median LMS algorithm," *IEEE Trans. Signal Processing*, vol. SP-41, pp. tbd, 1993.

23. T.I.Haweel and P.M.Clarkson, "A class of order statistic LMS filters," *IEEE Trans. Signal Processing*, vol. SP-40, pp. 44-53, 1992.

24. B.Widrow, J.M.McCool and M.Ball, "The complex LMS algorithm," *Proc. IEEE*, vol. 63, pp. 719-720, 1975.

25. G.A.Clark, S.K.Mitra and S.R.Parker, "Block implementation of adaptive digital filters," *IEEE Trans. Circuits Systs.*, vol. CAS-28, pp. 584-592, 1981.

26. A.Feuer, "Performance analysis of the block least mean squares algorithm," *IEEE Trans. Circuits Systs.*, vol. CAS-32, pp. 960-963, 1985.

27. W.B.Mikhael and F.H.Wu, "Fast algorithms for block FIR adaptive digital filtering," *IEEE Trans. Circuits Systs.*, vol. CAS-34, pp. 1152-1160, 1987.

28. W.B.Mikhael and F.H.Wu, "A fast block FIR adaptive digital filtering algorithm with individual adaptation of parameters," *IEEE Trans. Circuits Systs.*, vol. CAS-36, pp. 1-10, 1989.

29. M.C.S.Young, P.M.Grant and C.F.N.Cowan, "Block LMS adaptive equalizer design for digital radio," *Signal Processing IV: Theories and Applications (Proc. of Eurasip Conference)*, pp. 1349-1352, 1988.

30. M.Schetzen, *The Volterra and Wiener Theories of Nonlinear Systems*, Wiley, 1980.

31. T.Koh and E.J.Powers, "Second-order Volterra filtering and its application to nonlinear system identification," *IEEE Trans. Acoust., Speech, Signal Processing*, vol. ASSP-33, pp. 1445-1455, 1985.

32.   M.J.Coker and D.N.Simkins, "A nonlinear adaptive noise canceller," *Proc. IEEE Int. Conf. Acoust., Speech, Signal Processing*, pp. 470-473, 1980.

33.   R.A.Horn and C.R.Johnson, *Matrix Analysis*, Cambridge University Press, 1985.

34.   D.D.Falconer and L.Ljung, "Application of fast Kalman estimation to adaptive equalization," *IEEE Trans. Commun.*, vol. COM-26, pp. 1439-1445, 1978.

35.   S.Haykin, *Adaptive Filter Theory*, Prentice-Hall, 2nd Ed, 1991.

36.   S.T.Alexander, *Adaptive Signal Processing Theory and Applications*, Springer-Verlag, 1986.

37.   B.Friedlander, "Adaptive algorithms for finite impulse response filters," in *Adaptive Filters*, C.F.Cowan and P.M.Grant (Eds), Prentice-Hall, 1985.

38.   C.F.N.Cowan, "Performance comparisons of finite linear adaptive filters," *IEE Proceedings F*, vol. 134, pp. 211-216, 1987.

39.   J.M.Cioffi and T.Kailath, "Fast recursive least squares transversal filters for adaptive filtering," *IEEE Trans. Acoust., Speech, Signal Processing*, vol. ASSP-32, pp. 304-337, 1984.

40.   S.Stearns, "Error surfaces of recursive adaptive filters," *IEEE Trans. Acoust., Speech, Signal Processing*, vol. ASSP-29, pp. 763-766, 1981.

41.   S.A.White, "An adaptive recursive digital filter," *Proc. 9th Asilomar Conf. Circuits, Systs., Comp.*, pp. 21-25, 1975.

42.   J.R.Treichler, "Adaptive algorithms for infinite impulse response filters," in *Adaptive Filters*, C.F.Cowan and P.M.Grant (Eds), Prentice-Hall, 1985.

43.   H.Fan and M.Nayeri, "On error surfaces of sufficient order adaptive IIR filters: proofs and counterexamples to a unimodality conjecture," *IEEE Trans. Acoust., Speech, Signal Processing*, vol. ASSP-37, pp. 1436-1442, 1989.

44.   P.L.Feintuch, "An adaptive recursive LMS filter," *Proc. IEEE*, vol. 64, pp. 1622-1624, 1976.

45. C.R.Johnson, Jr., and M.G.Larimore, "Comments on and additions to 'An adaptive recursive LMS filter'," *Proc. IEEE*, vol. 65, pp. 1399-1401, 1977.

46. M.Nayeri and W.K.Jenknis, "Alternate realizations to adaptive IIR filters and properties of their performance surfaces," *IEEE Trans. Circuits Systs.*, vol. CAS-36, pp. 485-496, 1989.

47. J.J.Shynk, "Adaptive IIR filtering using parallel form realizations," *IEEE Trans. Acoust., Speech, Signal Processing*, vol. ASSP-37, pp. 519-533, 1989.

48. F.Itakura and S.Saito, "Digital filtering techniques for speech analysis and synthesis," *Proc. 7th Int. Conf. on Acoust.*, pp. 261-264, 1971.

49. A.H.Gray and J.D.Markel, "Digital lattice and ladder filter synthesis," *IEEE Trans. Audio Electroacoust.*, vol. AU-21, pp. 491-500, 1973.

50. R.E.Crochiere, "Digital ladder structures and coefficient sensitivity," *IEEE Trans. Audio Electroacoust.*, vol. AU-20, pp. 240-246, 1972.

51. T.W.Parsons, *Voice and Speech Processing*, McGraw-Hill, 1987.

52. E.H.Satorius and S.T.Alexander, "Channel equalization using adaptive lattice algorithms," *IEEE Trans. Commun.*, vol. COM-27, pp. 899-905, 1979.

53. L.J.Griffiths, "A continuously adaptive filter implemented as a lattice structure," *Proc. IEEE Int. Conf. Acoust., Speech, Signal Processing*, pp. 683-686, 1977.

54. L.J.Griffiths, "An adaptive lattice structure for noise cancelling applications," *Proc. IEEE Int. Conf. Acoust., Speech, Signal Processing*, pp. 87-90, 1978.

55. M.L.Honig and D.G.Messerschmitt, "Convergence properties of an adaptive digital lattice filter," *IEEE Trans. Acoust., Speech, Signal Processing*, vol. ASSP-29, pp. 642-653, 1981.

56. G.R.L.Sohie and L.H.Sibul, "Stochastic convergence properties of the adaptive gradient lattice," *IEEE Trans. Acoust., Speech, Signal Processing*, vol. ASSP-32, pp. 102-107, 1984.

57. M.J.Rutter, P.M.Grant, D.Renshaw and P.B.Denyer, "Design and realization of adaptive lattice filters," *Proc. IEEE Int. Conf. Acoust., Speech, Signal Processing*, pp. 21-24, 1983.

58. M.L.Honig and D.G.Messerschmitt, *Adaptive Filters: Structures, Algorithms, and Applications*, Kluwer Academic, 1984.

59. M.Dentino, J.McCool and B.Widrow, "Adaptive filtering in the frequency domain," *Proc. IEEE*, vol. 66, pp. 1658-1659, 1978.

60. A.V.Oppenheim and R.W.Schafer, *Discrete-Time Signal Processing*, Prentice-Hall, 1989.

61. E.R.Ferrara, "Frequency domain adaptive filtering," in *Adaptive Filters*, C.F.Cowan and P.M.Grant (Eds.), Prentice-Hall, 1985.

62. E.R.Ferrara, "Fast implementation of LMS adaptive filters," *IEEE Trans. Acoust., Speech, Signal Processing*, vol. ASSP-28, pp. 474-475, 1980.

63. D.Mansour and A.H.Gray, "Unconstrained frequency-domain adaptive filters," *IEEE Trans. Acoust., Speech, Signal Processing*, vol. ASSP-30, pp. 726-734, 1982.

64. B.Koo and J.D.Gibson, "Experimental comparison of all-pole, all-zero, and pole-zero predictors for ADPCM speech coding," *IEEE Trans. Commun.*, vol. COM-34, pp. 285-290, 1986.

65. M.Givens and P.M.Clarkson, "The application of the median LMS algorithm to ADPCM systems," *Proc. IEEE Int. Conf. Acoust., Speech, Signal Processing*, pp. 3665-3668, 1991.

66. M.L.Honig and D.G.Messerschmitt, "Comparison of adaptive linear prediction algorithms in ADPCM," *IEEE Trans. Commun.*, vol. COM-30, pp. 1775-1785, 1982.

67. R.C.Reininger and J.D.Gibson, "Backward adaptive lattice and transversal predictors in ADPCM," *IEEE Trans. Commun.*, vol. COM-33, pp. 74-82, 1985.

68. D.M.Etter and S.D.Stearns, "Adaptive estimation of time-delays in sampled data systems," *IEEE Trans. Acoust., Speech, Signal Processing*, vol. ASSP-29, pp. 582-587, 1981.

69. D.H.Youn and N.Ahmed, "Comparison of two adaptive methods for time-delay estimation ," *Proc. IEEE Int. Conf. Acoust., Speech, Signal Processing*, pp. 883-886, 1983.

70. K.Scarborough, N.Ahmed, D.H.Youn and G.C.Carter, "On the SCOT and ROTH algorithms for time-delay estimation," *Proc. IEEE Int. Conf. Acoust., Speech, Signal Processing*, pp. 371-374, 1982.

71. C.H.Knapp and G.C.Carter, "The generalized correlation method for the estimation of time-delay," *IEEE Trans. Acoust., Speech, Signal Processing*, vol. ASSP-24, pp. 320-327, 1976.

72. M.V.Dokic and P.M.Clarkson, "Real-time adaptive filters for time-delay estimation," *Mechanical Systems and Signal Processing*, vol. 6, pp. 403-418, 1992.

chapter seven

Methods of Spectral Estimation

7.1 Introduction

It will be recalled from Chapter 2 that the power spectrum of a zero-mean stationary signal $x(n)$ with autocorrelation $r(n)$ is defined by

$$R(e^{j\omega}) = \sum_{n=-\infty}^{\infty} r(n)e^{-j\omega n} . \qquad (7.1.1)$$

As we indicated in Chapter 2, the autocorrelation sequence $r(n)$ is usually not available in practical applications, and must instead be estimated from a finite set of data. It is clear that under these conditions $R(e^{j\omega})$ must also be estimated. The concern is to estimate the power spectrum from a single realization of $x(n)$ observed over a finite time interval. Any such estimator, derived from the stationary random sequence $x(n)$ will itself be random. We should be careful to distinguish between estimates of discrete parameters, as in Section 3.1, and estimation of the spectrum which is a continuous function of frequency. Nevertheless, certain properties, useful in assessing parameter estimation procedures, can be applied to spectral estimators. For example if the signal is stationary, then for a consistent spectral estimator, the longer the observation interval, the lower the variance (at each frequency) and hence the better the estimate we obtain. Practically, however, the data may only be stationary in some local sense, and this reduction in variance must be traded against 'smearing' caused by averaging together nonstationary data. The objectives of power spectrum estimation are thus to find the best possible estimate using the shortest possible data record.

The subject of spectral estimation has a long history (see, for example, Robinson [1], for a broad perspective), and has been the subject of several texts [2]-[4] and of a special issue of the *IEEE Proceedings* [5]. A good review article is that by Kay and Marple [6]. Additionally, spectral estimation is discussed in more or less detail in numerous texts on time-series analysis, and in many standard works on digital signal processing.

Methods for estimating spectra are often divided into two groups: **classical**, and **modern**. Classical methods do not attempt to impose any model or structure on the data. They operate purely as nonparametric estimators. Within this class, given a data sequence $x(n)$ for $n=0,1,...,M-1$, we may estimate the spectrum (7.1.1) either using the data directly, or by first computing an estimate of the autocorrelation and then transforming this estimate. Both of these approaches will be considered in the sequel. Modern methods, by contrast, assume a model structure for the observed data, estimate the parameters of the model, then use those estimates to form an estimate of the spectrum. We consider each approach to the problem separately in the following sections. We begin, however, by noting that equation (7.1.1) is not, in fact, the only possible definition for the spectrum. A 'nearly equivalent definition' [3] is given by

$$R(e^{j\omega})= \lim_{M\to\infty}\left\{\frac{1}{2M+1}E\{|\sum_{n=-M}^{M} x(n)e^{-j\omega n}|^2\}\right\}, \qquad (7.1.2a)$$

(see Appendix 7A). Equation (7.1.2a) may be expressed as:

$$R(e^{j\omega})= \lim_{M\to\infty}\left\{\frac{1}{2M+1}E\{|X_M(e^{j\omega})|^2\}\right\}, \qquad (7.1.2b)$$

where we have used a definition which is sufficiently general to include a two-sided sequence $x(n)$,[1] and where

$$X_M(e^{j\omega})= \sum_{n=-M}^{M} x(n)e^{-j\omega n} .$$

This interpretation is useful because it shows $R(e^{j\omega})$ explicitly as the *average power of the signal decomposed into constituent frequencies*.

(7.1.2) and (7.1.1) give two alternate definitions of $R(e^{j\omega})$, one **direct** (from the data), and one **indirect** (from the correlation function). Spectral estimation methods derived from the direct

---

1. The single-sided equivalent can be recovered by simply setting $x(n)=0$ for $n<0$.

definitions are known as **periodograms**, while those obtained via the indirect definition are sometimes called **correlograms**. We discuss periodogram methods in Section 7.2, and the correlogram in Section 7.3.

7.2 Periodogram Methods of Spectral Estimation

7.2.1 The Periodogram

The periodogram estimate of the power spectrum is defined as

$$R_p(e^{j\omega}) = \frac{1}{M} |\sum_{n=0}^{M-1} x(n) e^{-j\omega n}|^2 , \qquad (7.2.1)$$

or

$$R_p(e^{j\omega}) = \frac{1}{M} |X(e^{j\omega})|^2 ,$$

where

$$X(e^{j\omega}) = \sum_{n=0}^{M-1} x(n) e^{-j\omega n} .$$

Comparing (7.2.1) with (7.1.2), we see that relative to the true spectrum, the periodogram is obtained by replacing the expectation operator by the raw data, and by transforming using only a finite sequence of M samples. The periodogram is thus obtained simply by Fourier transforming $x(n)$, squaring the absolute value, and scaling. We observe that like the true spectrum, for real $x(n)$ the periodogram is a real and even function of frequency. From the definition (7.2.1), $R_p(e^{j\omega})$ also enjoys the non-negative property of the true spectrum $R(e^{j\omega})$.

We may expand the expression (7.2.1) for the periodogram as

$$R_p(e^{j\omega}) = \frac{1}{M} \left[\sum_{n_1=0}^{M-1} \sum_{n_2=0}^{M-1} x(n_1) x(n_2) e^{-j\omega n_1} e^{+j\omega n_2} \right] ,$$

$$R_p(e^{j\omega}) = \frac{1}{M} \left[x^2(0) + x(1)x(0) e^{-j\omega} + x(2)x(0) e^{-j2\omega} + \right.$$

$$...+ x(M-1)x(0) e^{-j(M-1)\omega} + x(0)x(1) e^{j\omega} +$$

$$\left. ...+ x(0)x(M-1) e^{j(M-1)\omega} + ... + x^2(M-1) \right] ,$$

or

$$R_p(e^{j\omega})=\frac{1}{M}\left[\sum_{n=0}^{M-1}x^2(n)+\sum_{n=0}^{M-2}x(n)x(n+1)e^{-j\omega}+\right.$$

$$...+x(0)x(M-1)e^{-j(M-1)\omega}+\sum_{n=0}^{M-2}x(n)x(n+1)e^{j\omega}+$$

$$\left.+\sum_{n=0}^{M-3}x(n)x(n+2)e^{j2\omega}+...+x(0)x(M-1)e^{j(M-1)\omega}\right].$$

Recall from Section 2.6, equation (2.6.18), that the biased correlation estimator $r^{'}(m)$ for a sample of length M is given by

$$r^{'}(m)=\frac{1}{M}\sum_{n=0}^{M-|m|-1}x(n)x(n+|m|)\ .\qquad(2.6.18)$$

Hence we may write $R_p(e^{j\omega})$ as

$$R_p(e^{j\omega})=r^{'}(0)+r^{'}(1)e^{-j\omega}+...+r^{'}(M-1)e^{-j\omega(M-1)}+r^{'}(-1)e^{j\omega}+$$

$$...+r^{'}(1-M)e^{j\omega(M-1)}\ ,$$

or

$$R_p(e^{j\omega})=\sum_{m=-(M-1)}^{M-1}r^{'}(m)e^{-j\omega m}\ .\qquad(7.2.2)$$

Hence *the periodogram is the Fourier transform of the biased correlation estimate $r^{'}(m)$.*

Properties of the Periodogram

a) Bias

From equation (7.2.2)

$$E\{R_p(e^{j\omega})\}=E\{\sum_{m=-(M-1)}^{M-1}r^{'}(m)e^{-j\omega m}\}\ ,$$

but it will be recalled from Section 2.6 that $r^{'}(m)$ is itself biased with

$$E\{r^{'}(m)\}=\left[\frac{M-|m|}{M}\right]r(m)\ ,\qquad(2.6.19)$$

so that $R_p(e^{j\omega})$ is also biased in general, even if $r(m){=}0$ for $|m|{\geq}M$. Note that an exception occurs when $x(n)$ is white. In that case no bias results since the correlation is non-zero only for $m{=}0$. More generally, we observe that the periodogram estimate is asymptotically unbiased because as $M{\rightarrow}\infty$

$$\frac{M-|m|}{M}{\rightarrow}1 \ .$$

b) Variance and Consistency

Perhaps the most surprising property of the periodogram is that it is *not* a consistent estimator. Hence, increasing the number of data samples does *not* decrease the variance of the estimator at any one frequency. As a numerical example we consider the periodogram estimates for a first order AR process

$$x(n){=}0.9x(n-1){+}w(n) \ . \tag{7.2.3}$$

Figure 7.2.1 shows a sample of 1024 points from a single realization of $x(n)$. The data shown are samples $w(n)$ from a zero-mean iid, unit variance Gaussian sequence. Figure 7.2.2 shows the raw periodograms derived from this data for lengths $M{=}32$, 128 and 1024. It is apparent that at any particular frequency, the variance of the estimate does not reduce as M increases. Note that although the spectra are shown as continuous lines, in practice, computation of the periodogram involves sampling in frequency using the Discrete Fourier Transform (DFT). That is, $R_p(e^{j\omega})$ in (7.2.1) is sampled at discrete frequencies given by

$$\omega{=}\frac{2\pi k}{N} \quad ; \ k{=}0,1,...,N{-}1 \ . \tag{7.2.4}$$

For the examples in Figure 7.2.2, in all cases the transform length was $N{=}1024$, and the sample rate was set at 1 *kHz*. The plots show the spectrum from *dc* to the folding frequency. Issues related to the practical implementation and the resultant sampling in frequency are discussed below.

Analytically, evaluation of the variance of the periodogram is difficult. We can illustrate the lack of consistency, however, using the relatively simple case resulting when $x(n)$ is a zero-mean iid Gaussian sequence. The variance of the periodogram is given by

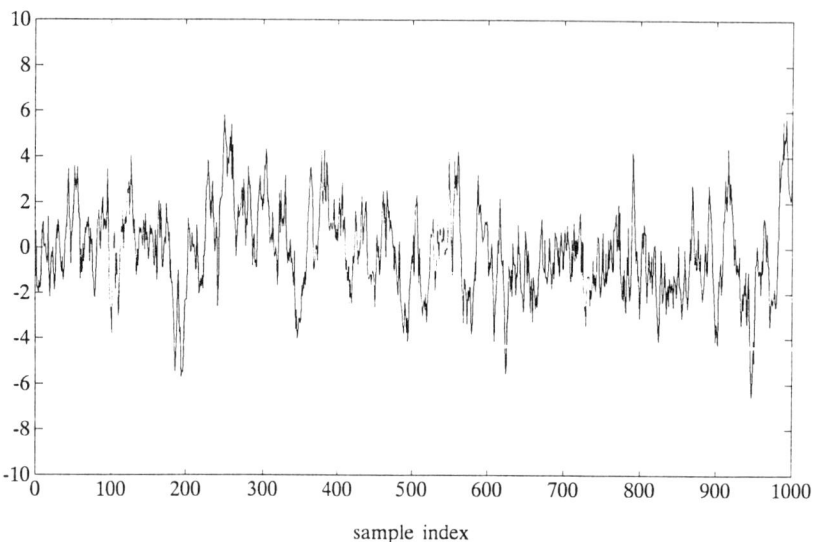

Figure 7.2.1 1024 samples of the AR(1) process (7.2.3) with $N(0,1)$ iid excitation.

$$Var\{R_p(e^{j\omega})\} = E\{R_p^2(e^{j\omega})\} - E\{R_p(e^{j\omega})\}^2 \ . \qquad (7.2.5)$$

Evaluation of $E\{R_p(e^{j\omega})\}$ is straightforward from equation (7.2.1)

$$E\{R_p(e^{j\omega})\} = E\{\frac{1}{M} \mid \sum_{n=0}^{M-1} x(n)e^{-j\omega n} \mid^2\} \ ,$$

$$= \sigma_w^2 \ , \qquad (7.2.6)$$

because $E\{x(n)x(m)\} = \sigma_w^2 \delta(n-m)$, where σ_w^2 is the variance of $w(n)$. Now, expanding $R_p^2(e^{j\omega})$ we have

$$E\{R_p^2(e^{j\omega})\} = \frac{1}{M^2} \sum_{n_1=0}^{M-1} \sum_{n_2=0}^{M-1} \sum_{n_3=0}^{M-1} \sum_{n_4=0}^{M-1} E\{x(n_1)x(n_2)x(n_3)x(n_4)\} \times$$

$$\times e^{-j\omega(n_1-n_2+n_3-n_4)} \ .$$

For a general input this is difficult to evaluate because of the need to compute the fourth moments of $x(n)$. For an iid Gaussian sequence, however, [3]

a)

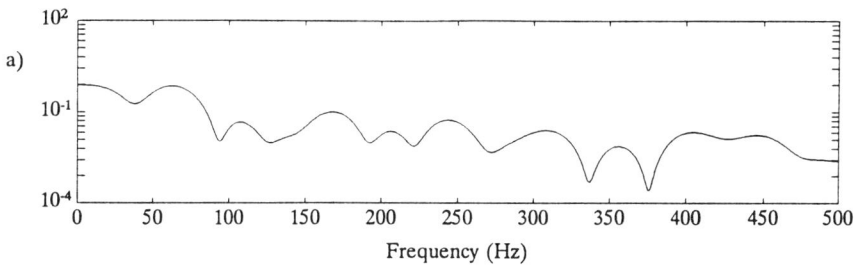

b)

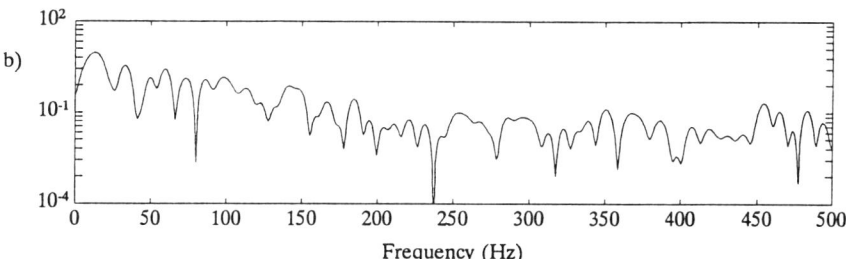

c)

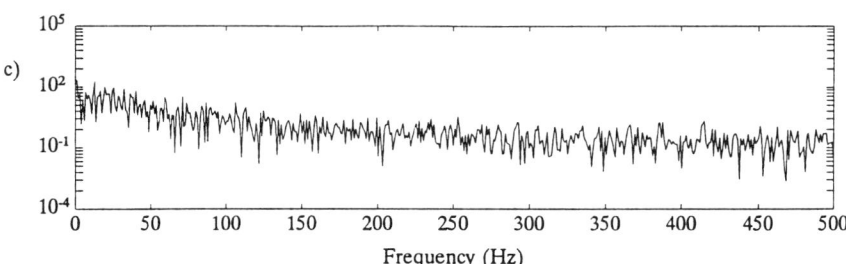

Figure 7.2.2 Raw periodograms for the data of Figure 7.2.1; a) M=32, b) M=128, c) M=1024.

$$E\{x(n_1)x(n_2)x(n_3)x(n_4)\}=E\{x(n_1)x(n_2)\}E\{x(n_3)x(n_4)\}+$$
$$+E\{x(n_1)x(n_3)\}E\{x(n_2)x(n_4)\}+$$
$$+E\{x(n_1)x(n_4)\}E\{x(n_2)x(n_3)\}\ ,$$

or

$$E\{x(n_1)x(n_2)x(n_3)x(n_4)\}=\sigma_w^4 \qquad ; \; n_1=n_2, \; n_3=n_4 \; ,$$
$$; \; n_1=n_3, \quad n_2=n_4 \; ,$$
$$; \; n_1=n_4, \quad n_2=n_3 \; ,$$
$$=0 \qquad ; \; otherwise \; . \qquad (7.2.7)$$

Hence, $E\{R_p^2(e^{j\omega})\}$ becomes

$$E\{R_p^2(e^{j\omega})\}=\frac{\sigma_w^4}{M^2}\left\{2M^2+\sum_{n_1=0}^{M-1}\sum_{n_2=0}^{M-1}e^{-j2\omega(n_1-n_2)}\right\}, \qquad (7.2.8)$$

where the first term arises from $n_1=n_2$, $n_3=n_4$ and from $n_1=n_4$, $n_2=n_3$. The second term arises from $n_1=n_3$, $n_2=n_4$, with a total of M^2 terms in each case. Also,

$$\sum_{n_1=0}^{M-1}\sum_{n_2=0}^{M-1}e^{-j2\omega(n_1-n_2)}=\sum_{n_1=0}^{M-1}e^{-j2\omega n_1}\sum_{n_2=0}^{M-1}e^{j2\omega n_2} \; ,$$

$$=\left[\frac{1-e^{-j2\omega M}}{1-e^{-j2\omega}}\right]\left[\frac{1-e^{j2\omega M}}{1-e^{j2\omega}}\right]=\left(\frac{\sin(\omega M)}{\sin\omega}\right)^2 \; .$$

Hence, from (7.2.8)

$$E\{R_p^2(e^{j\omega})\}=\sigma_w^4\left\{2+\left[\frac{1}{M}\frac{\sin(\omega M)}{\sin\omega}\right]^2\right\} \; . \qquad (7.2.9)$$

Finally combining (7.2.5), (7.2.6) and (7.2.9), the variance is

$$Var\{R_p(e^{j\omega})\}=\sigma_w^4\left\{1+\left[\frac{1}{M}\frac{\sin(\omega M)}{\sin\omega}\right]^2\right\} \; . \qquad (7.2.10)$$

Note that the minimum value for the variance occurs when the second term of (7.2.10) is zero, so that $Var\{R_p(e^{j\omega})\}=\sigma_w^4$. Also, this is the asymptotic value as $M\rightarrow\infty$. That is

$$\lim_{M\rightarrow\infty}\left\{Var\{R_p(e^{j\omega})\}\right\}=\sigma_w^4 \; .$$

Hence, the variance does *not* decay to zero as $M\rightarrow\infty$, and so by counterexample we have demonstrated that $R_p(e^{j\omega})$ is not a consistent estimator.[2]

As a generalization of these results we may calculate the covariance of the periodogram at frequencies ω_1, ω_2 defined by

$$cov\{R_p(e^{j\omega_1})R_p(e^{j\omega_2})\}=E\{R_p(e^{j\omega_1})R_p(e^{j\omega_2})\}-$$
$$-E\{R_p(e^{j\omega_1})\}E\{R_p(e^{j\omega_2})\} . \qquad (7.2.11)$$

For a zero-mean Gaussian white input with variance σ_w^2, a similar development to that used for the variance yields

$$cov\{R_p(e^{j\omega_1})R_p(e^{j\omega_2})\}=\sigma_w^4\left\{\left[\frac{\sin[(\omega_1+\omega_2)M/2]}{M\sin[(\omega_1+\omega_2)/2]}\right]^2+\right.$$
$$\left.+\left[\frac{\sin[(\omega_1-\omega_2)M/2]}{M\sin[(\omega_1-\omega_2)/2]}\right]^2\right\}, \qquad (7.2.12)$$

(see Problem 7.2). When the periodogram is computed using the DFT (with consequent sampling in frequency), with samples occurring at $\omega_1=2\pi k_1/M$ and $\omega_2=2\pi k_2/M$, for k_1, k_2 integer, then equation (7.2.12) becomes

$$cov\{R_p(k_1)R_p(k_2)\}=\sigma_w^4\left\{\left[\frac{\sin[\pi(k_1+k_2)]}{M\sin[\pi(k_1+k_2)/M]}\right]^2+\right.$$
$$\left.+\left[\frac{\sin[\pi(k_1-k_2)]}{M\sin[\pi(k_1-k_2)/M]}\right]^2\right\} . \qquad (7.2.13)$$

We see that if $k_1 \neq k_2$, then in general both terms of this equation are zero. This says that *for white inputs the covariance between samples of the periodogram that are adjacent in frequency is zero.* As M increases, the resolution of the periodogram increases but the correlation between these more closely spaced samples remains zero, resulting in an increasingly ragged spectral estimate. This is apparent from the example of Figure 7.2.2.

Although the variance and covariance expressions given above are restricted to Gaussian white noise, they can be used as the basis for approximate expressions for the variance and covariance

2. As we saw in Section 2.6, the estimator $r'(m)$ is consistent. In view of (7.2.2) we may make the interesting observation that the Fourier transform of a consistent estimator need not be consistent.

for a general input. In the interests of simplicity we will only consider the extension for the variance, the covariance expressions are similar. Consider a general zero-mean stationary input $x(n)$ with periodogram [3]

$$R_{px}(e^{j\omega}) = \frac{1}{M} |X(e^{j\omega})|^2 .$$

Assume that the process $x(n)$ is generated by passing a zero-mean iid sequence $w(n)$ through a linear time invariant system, $h(n)$. We know from Section 2.3 that for such a signal

$$R_{xx}(e^{j\omega}) = |H(e^{j\omega})|^2 R_{ww}(e^{j\omega}) = |H(e^{j\omega})|^2 \sigma_w^2 . \qquad (7.2.14)$$

We may obtain approximate expressions for the variance and covariance of the periodogram if we assume that an analogous result holds for the periodogram,[4] *viz*

$$R_{px}(e^{j\omega}) = |H(e^{j\omega})|^2 R_{pw}(e^{j\omega}) , \qquad (7.2.15)$$

we then write

$$Var\{R_{px}(e^{j\omega})\} = E\{\left[|H(e^{j\omega})|^2 R_{pw}(e^{j\omega})\right]^2\} -$$

$$- \left[E\{|H(e^{j\omega})|^2 R_{pw}(e^{j\omega})\}\right]^2 ,$$

$$= \left(|H(e^{j\omega})|^2\right)^2 Var\{R_{pw}(e^{j\omega})\} . \qquad (7.2.16)$$

Hence, if we make the further approximation that the white noise generating the process $x(n)$ is Gaussian, then substituting (7.2.10) into (7.2.16) we have

$$Var\{R_{px}(e^{j\omega})\} = \sigma_w^4 \left(|H(e^{j\omega})|^2\right)^2 \left[1 + \left(\frac{1}{M} \frac{\sin(\omega M)}{\sin\omega}\right)^2\right] ,$$

---

3. The double subscript is used here to distinguish the fact that this is a periodogram estimate (subscript p) for signal $x(n)$ (subscript x).

4. We note that this expression can only be approximately correct because the periodograms are not obtained as expectations as are the true spectra — they are instead raw transforms of finite length data segments.

which, using (7.2.14), gives

$$Var\{R_{pz}(e^{j\omega})\}=R_{xx}^2(e^{j\omega})\left[1+\left[\frac{1}{M}\frac{\sin(\omega M)}{\sin\omega}\right]^2\right] . \qquad (7.2.17)$$

We see that, subject to the approximation (7.2.15), the variance is asymptotically

$$\lim_{M\to\infty}\left\{Var\{R_{pz}(e^{j\omega})\}\right\}=R_{xx}^2(e^{j\omega}) ,$$

that is, *the variance of the periodogram estimate is proportional to the square of the true spectrum.* Once again we observe that the variance does not decay with increasing M, confirming that the periodogram is not a consistent estimator.

c) The Periodogram and the Bartlett Window

We may obtain another interpretation of the periodogram from equation (7.2.2)

$$R_p(e^{j\omega})=\sum_{m=-(M-1)}^{M-1} r^{'}(m)e^{-j\omega m} . \qquad (7.2.2)$$

Recall that this is the transform of the biased estimator $r^{'}(m)$ which has the property

$$E\{r^{'}(m)\}=\left[\frac{M-|m|}{M}\right]r(m) . \qquad (2.6.19)$$

Taking the expectation of (7.2.2) and using (2.6.19) we have

$$E\{R_p(e^{j\omega})\}=\sum_{m=-(M-1)}^{M-1}\left[\frac{M-|m|}{M}\right]r(m)e^{-j\omega m} .$$

We can now think of $E\{R_p(e^{j\omega})\}$ as the Fourier transform of $r(m)$ multiplied by a window $w_b(m)$. That is,

$$E\{R_p(e^{j\omega})\}=\sum_{m=-\infty}^{\infty}[r(m)w_b(m)]e^{-j\omega m} , \qquad (7.2.18)$$

where

$$w_b(m) = \frac{M - |m|}{M} \qquad ; |m| < M ,$$

$$= 0 \qquad\qquad ; otherwise . \qquad (7.2.19)$$

This window has extent from $m = -(M-1)$ to $(M-1)$ and is called a **triangular** or **Bartlett** window. The window can be obtained by forming the sample autocorrelation of a rectangular window.

Equation (7.2.18) can be viewed as the Fourier transform of the product $[r(m)w_b(m)]$. Multiplication in the time domain becomes convolution in the frequency domain, so that (7.2.18) may equivalently be written

$$E\{R_p(e^{j\omega})\} = \frac{1}{2\pi} \int_{-\pi}^{\pi} R(e^{j\theta}) W_b(e^{j(\omega-\theta)}) d\theta , \qquad (7.2.20)$$

where $W_b(e^{j\omega})$ is the Fourier transform of the Bartlett window and is given by (see Problem 7.5)

$$W_b(e^{j\omega}) = \frac{1}{M} \left(\frac{\sin(\frac{\omega M}{2})}{\sin(\frac{\omega}{2})} \right)^2 . \qquad (7.2.21)$$

Hence, the expected value of the periodogram is the convolution (in frequency) of the true spectrum with the Fourier transform of the window. This convolution with the window 'smooths' the spectrum and limits the **resolution** attainable by the estimator. By considering the expected value of the periodogram, the quantities involved are deterministic. We may think of the properties of the window (in particular, the main-lobe width, side-lobe levels and side-lobe roll-off rate, see, for example, [7]) as determining the properties of the spectrum. In particular, the main-lobe width determines the resolution and the side-lobes determine the 'leakage'. In qualitative terms, as M increases the main-lobe of $W_b(e^{j\omega})$ becomes narrower (more 'impulse-like') and the resolution improves. Conversely, as M decreases, the main-lobe becomes broader and the resolution poorer. We illustrate these points in the example below.

d) Sampling in Frequency

As we have already noted, practically the periodogram is computed using a finite length DFT as

$$R_p(k) = \frac{1}{N} \left| \sum_{n=0}^{N-1} x(n) e^{-\frac{j2\pi kn}{N}} \right|^2 \quad ; k=0,1,...,N-1 \ . \quad (7.2.22)$$

That is, we take frequency samples spaced at intervals:

$$\omega_k = \frac{2\pi k}{N} \quad ; k=0,1,...,N-1 \ ,$$

where N is the length of the transform. In the simplest case the transform length is equal to the segment length $N=M$. The density of the samples in frequency may always be increased, however, by zero-extending (zero-padding) the data sequence prior to transforming, and by using $N>M$. It is important to realize that this does *not* improve the frequency resolution, it merely interpolates between samples in frequency. As we have indicated, the resolution is determined by the length of the data segment (by the resolution of the M point Bartlett window), not by the length of the transform. This point is basic to all data windowing in digital signal processing, and is of sufficient importance to warrant an illustrative example:

Example – Consider the (deterministic) signal

$$x(n) = \cos(\omega_1 nT) + \cos(\omega_2 nT) \ , \quad\quad\quad (7.2.23)$$

where $\omega_1=2\pi 95$, $\omega_2=2\pi 100$, and $T=0.001$, for $n=0,1,...,M-1$. The periodogram for this signal,[5] computed using an N point discrete Fourier transform is shown in Figure 7.2.3. Figure 7.2.3a) shows $M=128$, and $N=128$, while Figure 7.2.3b) shows the same value for M but with $N=512$.[6] We see that increasing N by zero-extension has *not* improved resolution (the two peaks are still

---

5. There is nothing inconsistent about applying the periodogram to a deterministic signal, though comments about the variance of the estimate and stationarity of the data are obviously redundant. Note that many random signals contain a deterministic component.

6. As usual, the plots are shown as continuous functions of frequency.

indistinguishable). Figure 7.2.4 shows the effects of increasing M to 512. In this case the twin sinusoid peaks are clearly distinguishable, and the resolution is significantly improved. The relation between Figure 7.2.4 and Figure 7.2.3b can be thought of as the relation between the transform of a 512 point sequence (Figure 7.2.4) and that of the same sequence multiplied by a 128 point rectangular window (Figure 7.2.3b). It is this window which produces the side-lobes that are evident in Figure 7.2.3b).[7]

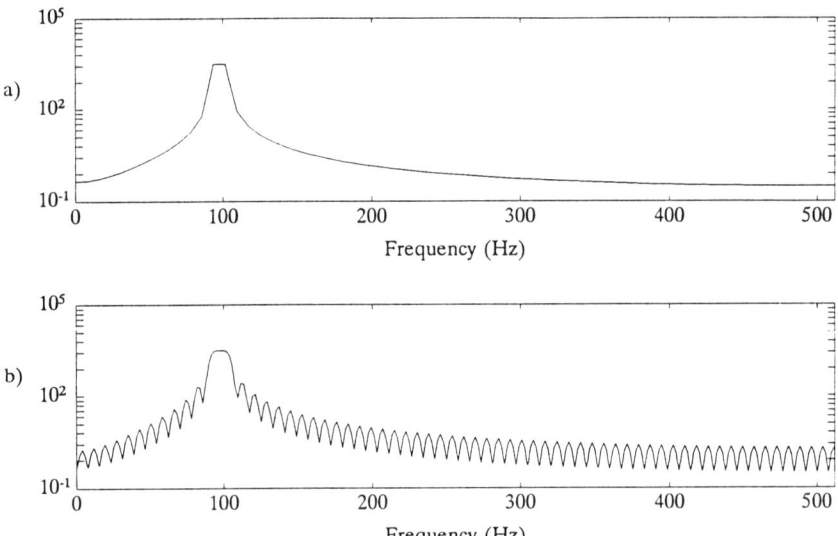

Figure 7.2.3 Periodogram for two closely spaced sinusoids ($\omega_1=2\pi95$, $\omega_2=2\pi100$, $f_s=1kHz$). a) $M=128$, $N=128$, b) $M=128$, $N=512$ by zero-extension.

7.2.2 The Modified Periodogram

The periodogram of equation (7.2.1) is defined as a transform of the raw data sequence $x(n)$, for $n=0,1,...,M-1$. As a generalization of this procedure we may define a **modified periodogram** obtained by applying a data window to the sequence $x(n)$ prior to computing the periodogram. That is, equation (7.2.1) for the raw (simple) periodogram is replaced by

$$R_w(e^{j\omega})=\frac{1}{MC}\,|\sum_{n=0}^{M-1}w_d(n)x(n)e^{-j\omega n}|^2\,, \qquad (7.2.24)$$

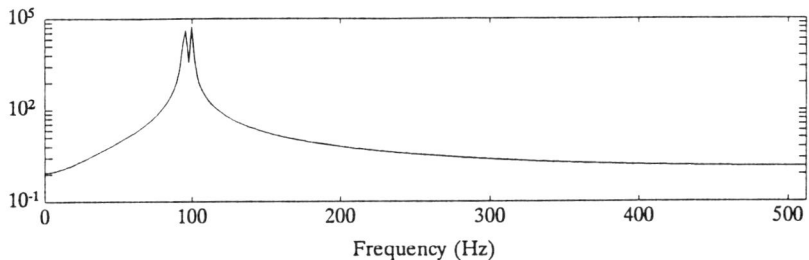

Figure 7.2.4 Periodogram for two closely spaced sinusoids ($\omega_1 = 2\pi 95$, $\omega_2 = 2\pi 100$, $f_s = 1kHz$). $M = 512$, $N = 512$.

where $w_d(n)$ is a data window defined over $n = 0, 1, ..., M-1$ and where

$$C = \frac{1}{M} \sum_{n=0}^{M-1} w_d^2(n) \;, \qquad (7.2.25)$$

is a normalizing constant. Note that the simple periodogram of (7.2.1) can be obtained as a special case of $R_w(e^{j\omega})$ when $w_d(n)$ is rectangular with

$$
\begin{aligned}
w_d(n) &= 1 \quad ; \; n = 0, 1, ..., M-1 \\
&= 0 \quad ; \; otherwise \;.
\end{aligned} \qquad (7.2.26)
$$

Note also that the data window here is quite distinct from the Bartlett window $w_b(n)$ which arises in connection with the simple periodogram. $w_d(n)$ is a data window *explicitly* applied to the data to modify the properties of the resulting spectral estimate. By contrast, $w_b(n)$ simply arises in the *interpretation* of the raw periodogram.

It will be recalled that the simple periodogram is the Fourier transform of the biased estimate $r'(m)$

$$R_p(e^{j\omega}) = \sum_{m=-(M-1)}^{M-1} r'(m) e^{-j\omega m} \;, \qquad (7.2.2)$$

where

$$r'(m) = \frac{1}{M} \sum_{n=0}^{M-|m|-1} x(n) x(n+|m|) \;. \qquad (2.6.18)$$

Similarly, it may easily be shown (see Problem 7.6), that

$$R_w(e^{j\omega})= \sum_{m=-(M-1)}^{M-1} r^{''}(m)e^{-j\omega m} , \qquad (7.2.27)$$

where

$$r^{''}(m)=\frac{1}{MC} \sum_{n=0}^{M-|m|-1} w_d(n)x(n)w_d(n+|m|)x(n+|m|) ,$$

or

$$r^{''}(m)=\frac{1}{MC} \sum_{n=0}^{M-|m|-1} x_1(n)x_1(n+|m|) , \qquad (7.2.28)$$

where $x_1(n)=w_d(n)x(n)$. Hence $r^{''}(m)$ is the biased correlation estimator for the windowed sequence $x_1(n)=w_d(n)x(n)$, and the modified periodogram is the Fourier transform of that correlation estimate.

Properties of the Modified Periodogram

a) Bias

From equation (7.2.28)

$$E\{r^{''}(m)\}=\frac{1}{MC} \sum_{n=0}^{M-|m|-1} w_d(n)w_d(n+|m|)r(m) ,$$

$$=\frac{r(m)}{MC} \sum_{n=0}^{M-|m|-1} w_d(n)w_d(n+|m|)=r(m)g(m) ,$$

where

$$g(m)=\frac{1}{MC} \sum_{n=0}^{M-|m|-1} w_d(n)w_d(n+|m|) ,$$

so that taking the expectation of (7.2.27) and using this result we have

$$E\{R_w(e^{j\omega})\}= \sum_{m=-(M-1)}^{M-1} r(m)g(m)e^{-j\omega m} . \qquad (7.2.39)$$

As with the raw periodogram, the expectation is not equal to the

Fourier transform of $r(m)$, and the result is generally biased. Again, under fairly general restrictions on $w_d(n)$, $R_w(e^{j\omega})$ is asymptotically unbiased [7]. Note that $g(0)=1$, so that for white inputs the modified periodogram is unbiased even for finite segments (as was the raw periodogram).

b) Variance and Consistency

For a general window $w_d(n)$, an expression for the variance of the periodogram is difficult to obtain, even using the approximate methods of the previous section. However, it has been shown [2] that asymptotically, under broad conditions the variance of the periodogram is approximately proportional to the square of the power spectrum:

$$Var\{R_w(e^{j\omega})\} \approx R^2(e^{j\omega}) \ . \tag{7.2.30}$$

This result mirrors that for the raw periodogram and, as in that case, this illustrates that $R_w(e^{j\omega})$ is *not* a consistent estimator for the power spectrum.

7.2.3 Averaging Periodograms − Bartlett's Method [8]

Neither the raw nor modified periodograms described so far are of much practical value because they are not consistent estimators, and therefore they do not produce 'smooth' estimates of the spectrum. We need to reduce the variability in the spectral estimate so as to produce smooth spectra. The variance of the raw spectral estimate can be reduced by averaging raw periodograms. For a sequence $x(n)$ for $n=0,1,...,M-1$, this can be achieved by subdividing the sequence into K consecutive segments of length L (so that $M=KL$). Define

$$x^{(i)}(n)=x(n+iL) \quad ; \ i=0,1,...,K-1 \ , \tag{7.2.31}$$

$$; \ n=0,1,...,L-1 \ .$$

Each of the subsequences $x^{(i)}(n)$ is then used to form a raw periodogram $R_p^{(i)}(e^{j\omega})$

$$R_p^{(i)}(e^{j\omega})=\frac{1}{L}\left|\sum_{n=0}^{L-1}x^{(i)}(n)e^{-j\omega n}\right|^2 \quad ; \ i=0,1,...,K-1 \ . \tag{7.2.32}$$

Finally the K raw spectra are averaged to form

$$R_A(e^{j\omega}) = \frac{1}{K} \sum_{i=1}^{K} R_p^{(i)}(e^{j\omega}) \ . \tag{7.2.33}$$

$R_A(e^{j\omega})$ is referred to as **Bartlett's estimate**. The basic properties of this estimator are:

a) Bias

For a zero-mean stationary input, the mean value of $R_A(e^{j\omega})$ is identical to that for any individual segment:

$$E\{R_A(e^{j\omega})\} = \frac{1}{K} \sum_{i=1}^{K} E\{R_p^{(i)}(e^{j\omega})\} = E\{R_p^{(i)}(e^{j\omega})\} \ . \tag{7.2.34}$$

The second equality follows because the mean values for each segment are identical. From equations (7.2.18) and (7.2.19), for this L point segment $E\{R_p^{(i)}(e^{j\omega})\}$ is given by

$$E\{R_p^{(i)}(e^{j\omega})\} = \sum_{m=-(L-1)}^{L-1} \left[1 - \frac{|m|}{L}\right] r(m) e^{-j\omega m} \ ,$$

so that from (7.2.34)

$$E\{R_A(e^{j\omega})\} = \sum_{m=-(L-1)}^{L-1} \left[1 - \frac{|m|}{L}\right] r(m) e^{-j\omega m} \ . \tag{7.2.35}$$

Comparing this equation with (7.2.18) for the raw M point periodogram we see that we may again think of $E\{R_A(e^{j\omega})\}$ as the transform of a Bartlett windowed correlation sequence. However, the length of the window is reduced from M to L points. Consequently, for a given sequence length the averaged periodogram will have a higher bias compared to the raw periodogram. Similarly, since frequency resolution increases with segment length, the resolution obtained by the averaged periodogram will be worse than that for the corresponding raw periodogram.

b) Variance and Consistency

General expressions for the variance of the averaged periodogram are difficult to obtain. We consider only the simple case of non-overlapping segments taken from a zero-mean white input. For such signals the raw periodograms for the individual segments are statistically independent and identical. It is easy to show that the variance of the sum of K independent identically distributed random variables is $1/K$ times the variance of the individual variables (see Problem 7.8). As we saw in the previous section, the variance of the raw periodogram is largely independent of length. Hence

$$Var\{R_A(e^{j\omega})\} = \frac{1}{K} Var\{R_p^{(i)}(e^{j\omega})\} \approx \frac{1}{K} Var\{R_p(e^{j\omega})\} \ . \quad (7.2.36)$$

Thus, the variance of the averaged periodogram decreases monotonically as the number of averages K increases. Moreover, as $M\to\infty$, then both L and K increase without bound, and as $K\to\infty$, $Var\{R_A(e^{j\omega})\}\to 0$ as does the bias, and hence the averaged periodogram *is* a consistent estimator.

Note that $1/K$ represents an upper limit on the variance reduction that can be obtained by averaging. Such an improvement can only be obtained when the averaging is performed on raw spectra that are uncorrelated from segment to segment, that is when the analysis is performed on non-overlapping segments of a white input. For general inputs, the constituent blocks of data are consecutive and the corresponding raw spectra are correlated. This in turn means that less variance reduction is obtained (see also Problem 7.9).

Overall, we see that breaking a sequence into a number of smaller segments and averaging the periodograms of each reduces the variance, thereby producing a smoother spectral estimate, but that this is traded against increased bias and reduced resolution. For a fixed length sequence, therefore, the choice of segment length must be a trade-off between the conflicting requirements of high resolution and low variance.

Example – Consider the sampled signal:

$$x(n) = \cos(\omega_0 n T) + w(n) \ , \quad (7.2.37)$$

where $w(n)$ is a zero-mean white noise signal with standard deviation σ_w=0.3, ω_0=2π100, and where T=0.001. Let us assume that we have available some 2048 samples of $x(n)$. We restrict attention to non-overlapping segments and assume that computation is performed via a radix-2 FFT algorithm [7]. Figure 7.2.5 shows the averaged periodograms for segment lengths L=2048 (K=1, the raw periodogram), L=1024 (K=2), L=512 (K=4), and L=256 (K=8). As expected, decreasing the segment length and increasing the number of averages decreases the resolution but produces a smoother spectrum.

Finally on this topic, we note two further refinements of the averaged periodogram method:

i) Data Windowing

As with the raw periodogram it is possible to apply a data window to the raw data. For segments of length L we have

$$R_W^{(i)}(e^{j\omega}) = \frac{1}{LC} \left| \sum_{n=0}^{L-1} x^{(i)}(n) w_d(n) e^{-j\omega n} \right|^2 , \qquad (7.2.38)$$

where $w_d(n)$ is a data window taking non-zero values over n=0,1,...,L−1 and C is defined as a normalizing constant with

$$C = \frac{1}{L} \sum_{n=0}^{L-1} w_d^2(n) . \qquad (7.2.39)$$

The estimate is then formed as

$$R_{WA}^{(i)}(e^{j\omega}) = \frac{1}{K} \sum_{1}^{K} R_W^{(i)}(e^{j\omega}) . \qquad (7.2.40)$$

As with the raw periodogram, the choice of $w_d(n)$ depends on the particular objectives with respect to resolution and leakage effects. It can be shown [9], that independent of this choice, the variance reduction obtained by averaging is comparable to that achieved by averaging using raw periodograms.

ii) Overlapping Data Segments

Given a fixed length M point segment of data, it is possible to

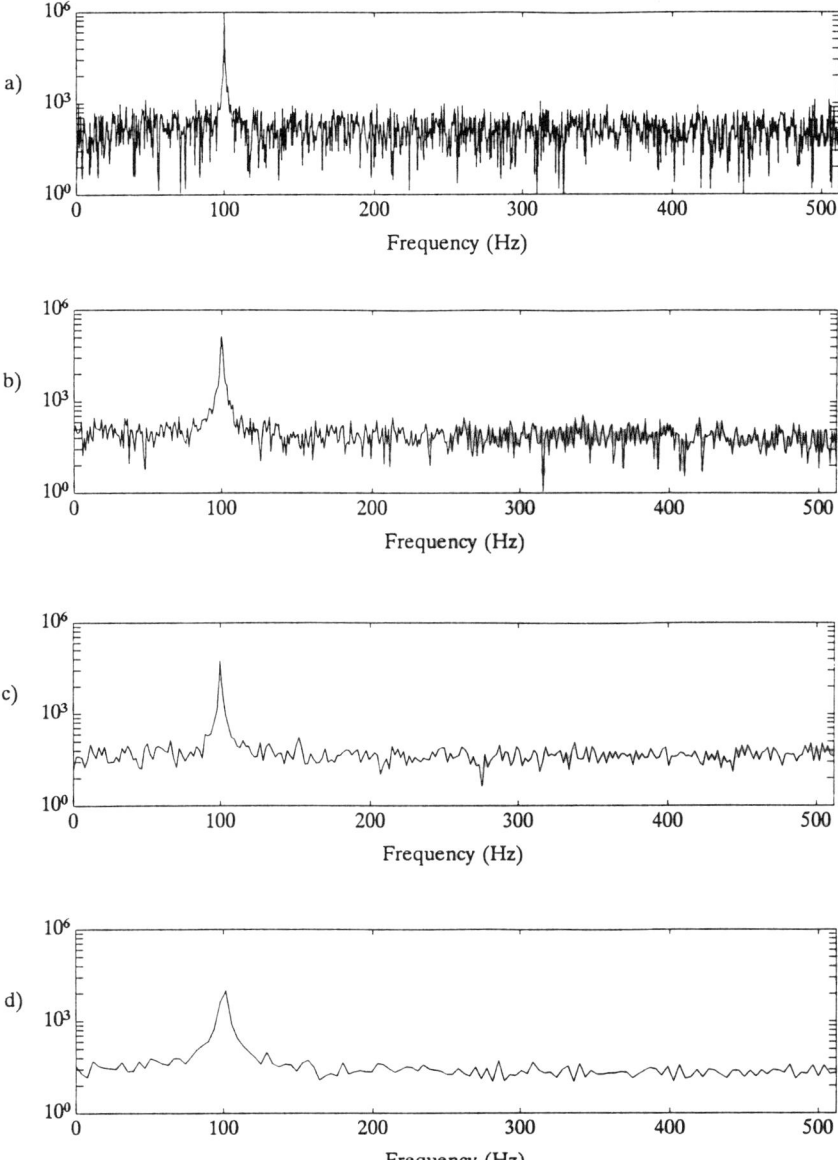

Figure 7.2.5 Averaged periodograms for a sinusoid in white noise. A sinusoid at frequency $(\omega_0 = 2\pi 100)$ is embedded in $N(0, 0.3)$ iid noise. The sample rate is 1 kHz, and a total of 2048 samples are available. No overlapping was employed. a) $L = 2048$, $(K = 1)$, the raw periodogram, b) $L = 1024$, $(K = 2)$, c) $L = 512$, $(K = 4)$, d) $L = 256$, $(K = 8)$.

increase the number of segments available for averaging raw periodograms, while retaining the segment length L, by overlapping successive data windows rather than using strictly consecutive segments. Such overlap is commonly expressed as a percentage, and the greater the percentage overlap, the greater the number of segments from a given data sequence. However, even if the signal $x(n)$ is white, the raw periodograms obtained from these overlapping segments will *not* be independent since they contain some of the same data samples, and the variance will not reduce as $1/K$. In fact, by progressively increasing the percentage overlap, a point is reached where no further variance reduction occurs. The precise location of this cutoff point depends on the nature of the data,[8] but 50% or 75% overlap are commonly used (see also Computer Problem 7.1).

7.3 Blackman-Tukey Spectral Estimation

As we indicated in Section 7.1, one approach to spectral estimation is to first estimate the autocorrelation and then Fourier transform that estimate to obtain an estimate of the power spectrum. In fact, the raw periodogram of the previous section was shown to be equivalent to such an approach in the sense that $R_p(e^{j\omega})$ is equal to the Fourier transform of the biased autocorrelation estimate $r'(m)$. In this section, we also generate spectral estimates by Fourier transforming correlation estimates. The approach is more general than the raw periodogram, however, because we apply a window to the raw correlation estimates prior to transforming. That is, given $x(n)$ for $n=0,1,...,M-1$, we form a spectral estimate with

$$R_{BT}(e^{j\omega}) = \sum_{n=-(M_1-1)}^{M_1-1} w_l(n) r'(n) e^{-j\omega n} , \qquad (7.3.1)$$

where $r'(n)$ is the biased correlation estimator of equation (2.6.18), and where $w_l(n)$ is a symmetric data window of length $2M_1-1$, sometimes referred to as a **lag window** [3]. Equation (7.3.1) is

---

8. For Gaussian data with a Hanning window, Marple [4] reports that 65% overlap gives the best variance reduction.

known variously as the **Blackman-Tukey (BT)** or **weighted covariance** spectral estimate [10] or as a **correlogram**. Note that when $w_l(n)$ is rectangular and $M_1 = M$, the estimate is equivalent to the raw periodogram, though the computational procedure is different. In general, the symmetry requirement is imposed on the data window to ensure that $R_{BT}(e^{j\omega})$ is the transform of a symmetric sequence. This in turn ensures that $R_{BT}(e^{j\omega})$ is real and even. Also, M_1 is chosen to be less than M and the lag window typically tapers away from the center. This places less emphasis on higher lag values. These values are less reliable having been obtained by averaging fewer samples, and therefore having a higher variance (see Section 2.6). As with the data windowing of the modified periodogram, the effect of the multiplication by the lag window can be interpreted as a convolution in the frequency domain.[9] Equation (7.3.1) can be rewritten as

$$R_{BT}(e^{j\omega}) = \frac{1}{2\pi} \int_{-\pi}^{\pi} W_l(e^{j(\omega-\theta)}) R_p(e^{j\theta}) d\theta \ , \tag{7.3.2}$$

where, in addition to converting the multiplication to a convolution, we have used the relation (7.2.2). From (7.3.2) the Blackman-Tukey estimator may be interpreted as the convolution (in frequency) of the raw periodogram with the lag window. We certainly would like $R_{BT}(e^{j\omega})$ to have the non-negative property of the true spectrum. From (7.3.2), we see that a sufficient condition for this is that $W_l(e^{j\omega})$ should be non-negative:

$$W_l(e^{j\omega}) \geq 0 \quad ; \ -\pi < \omega \leq \pi \ . \tag{7.3.3}$$

An equivalent condition is that $w_l(n)$ be positive semi-definite in the sense that

$$\sum_{i=0}^{M_1-1} \sum_{j=0}^{M_1-1} a_i w_l(i-j) a_j \geq 0 \ , \tag{7.3.4}$$

for any a_i (see also the definitions of Section 2.1). The condition ensures that the convolution of the two non-negative transforms produces a non-negative result. As we have observed, this is a

---

9. By taking $w_l(n)$ to be non-zero only over $n = -(M_1-1)$ to M_1-1, the limits in (7.3.1) can be extended to $-\infty$ to $-\infty$.

sufficient condition for the resulting estimate to be non-negative. It is not a necessary condition, however. As a simple counter-example; if $w_l(n)$ is rectangular, then it is not a positive semi-definite sequence, but the resulting Blackman-Tukey estimate is non-negative being equivalent to the periodogram. Note that [3], the Bartlett and Parzen windows have non-negative transforms, but many other popular windows including rectangular, Hamming, Hanning and Kaiser do not [7]. Note also that this requirement for a positive semi-definite sequence explains why the biased correlation estimator $r'(n)$, is used in (7.3.1) rather than the unbiased estimator $\hat{r}(n)$. It will be recalled (see Section 2.6) that while $r'(n)$ is positive semi-definite, the unbiased estimator $\hat{r}(n)$ is not necessarily positive semi-definite. Thus even with a positive definite lag window, the use of this estimator may not produce a non-negative spectrum.

Properties of the Blackman-Tukey Estimator

a) Bias

From (7.3.2) the expected value of $R_{BT}(e^{j\omega})$ is

$$E\{R_{BT}(e^{j\omega})\} = \frac{1}{2\pi} \int\limits_{-\pi}^{\pi} W_l(e^{j(\omega-\theta)}) E\{R_p(e^{j\theta})\} d\theta \ .$$

Since $R_p(e^{j\omega})$ is asymptotically unbiased, then for sufficiently large M we may write

$$E\{R_p(e^{j\omega})\} \approx R(e^{j\omega}) \ ,$$

and hence

$$E\{R_{BT}(e^{j\omega})\} \approx \frac{1}{2\pi} \int\limits_{-\pi}^{\pi} W_l(e^{j(\omega-\theta)}) R(e^{j\theta}) d\theta \ . \qquad (7.3.5)$$

Hence, in general $R_{BT}(e^{j\omega})$ is biased. As with the periodogram, however, as $M_1 \to \infty$, then the spectral characteristics of the window becomes more and more impulse-like:

$$W_l(e^{j\omega}) \to A\delta(\omega) \ , \qquad (7.3.6)$$

and $R_{BT}(e^{j\omega})$ is asymptotically unbiased provided $A=1$. This occurs if

$$\frac{1}{2\pi}\int_{-\pi}^{\pi}W_l(e^{j\omega})\,d\omega=\frac{1}{2\pi}\int_{-\pi}^{\pi}A\,\delta(\omega)\,d\omega=1=w_l(0)\ ,$$

where the latter equality follows from basic properties of the discrete Fourier transform.

b) Variance and Consistency

The Blackman-Tukey estimate has variance

$$Var\{R_{BT}(e^{j\omega})\}=E\{R_{BT}^2(e^{j\omega})\}-E^2\{R_{BT}(e^{j\omega})\}\ .$$

This is generally difficult to evaluate, but for M sufficiently large an approximate relation is given by [11]

$$Var\{R_{BT}(e^{j\omega})\}\approx\frac{R^2(e^{j\omega})}{M}\left[\sum_{m=-(M_1-1)}^{M_1-1}w_l^2(m)\right]\ ,\qquad(7.3.7)$$

where M_1 is the length of the lag window. Note that from (7.3.7), as $M\rightarrow\infty$, $Var\{R_{BT}(e^{j\omega})\}\rightarrow0$, and the Blackman-Tukey estimate is consistent.

Generally, the effect of the convolution of equation (7.3.2) is to smooth the spectral estimate. Note that as the window becomes shorter, the main lobe becomes broader and the smoothing is increased. Against this, as with the periodogram reducing the length of the lag window also decreases resolution.

7.4 Parametric Spectral Estimation

7.4.1 Introduction

We now turn our attention to modern methods of spectrum estimation. We begin with the observation that none of the methods described so far rely on any particular model for the generation of the observed data, and therefore all of these techniques may be classified as non-parametric. In this section, we estimate the spectrum by assuming a model for the structure of the signal, estimating the parameters of the model, and generating a spectral estimate from those parameters. We may think of this process as improving the spectral estimate by utilizing prior knowledge of the data structure via the model. Also, we recognize that in the classical methods described in Sections 7.2 and 7.3, the spectra are derived from data which is either explicitly or implicitly windowed. This assumes that the data or correlation values outside the window are zero. This is clearly unreasonable in most cases and it is this window effect which limits the resolution obtained by the estimator. When a model is used, the model is assumed to generate all data samples, not just those which are actually observed. In this way, no windowing occurs and the resolution limitations may be overcome. This is particularly useful whenever we must rely on a short segment of data to form the estimate, either because this is all that is available, or because the signal is statistically time-varying. Of course, the cost of this resolution gain is that the model must fit the data, or the results will be poor.

We restrict attention to linear processes, and to generation models that are finite order rational transfer functions:

$$H(z) = \frac{b_0 + b_1 z^{-1} + \ldots + b_N z^{-N}}{1 - a_1 z^{-1} - \ldots - a_M z^{-M}} . \qquad (2.4.2)$$

We recall from Section 2.4 that a sequence $x(n)$ generated by supplying a zero-mean iid sequence as input to a system of the form (2.4.2), corresponds to an ARMA(M,N) process. Furthermore, the spectrum of such a process is given by

$$R_{xx}(e^{j\omega}) = |H(e^{j\omega})|^2 \sigma_w^2 ,$$

$$= \left| \frac{b_0 + b_1 e^{-j\omega} + \ldots + b_N e^{-j\omega N}}{1 - a_1 e^{-j\omega} - \ldots - a_M e^{-j\omega M}} \right|^2 \sigma_w^2 , \qquad (2.4.5)$$

where σ_w^2 is the power of $w(n)$. Equation (2.4.2) also includes the AR and MA processes as special cases corresponding to $N=0$ (with $b_0=1$) and $M=0$, respectively. Of the three possibilities, we shall focus primarily on the AR case which produces the simplest spectral estimates, and we shall only briefly sketch out methods for MA and ARMA spectrum estimation. In practice, the AR model is used more often than the MA. This is partly because the poles of the AR model facilitate the modelling of narrow spectral peaks far more efficiently (with fewer coefficients) than do the zeros of the MA. Hence, AR spectrum estimators tend to produce 'high resolution' spectra. Practically, the AR model has a further advantage since it is much easier to estimate the parameters of this model than those of the MA. This is because, as we shall see, the AR parameters can be obtained as the solution of a set of linear equations. This contrasts with both MA and ARMA estimators where the solution of nonlinear equations is required in order to identify the parameters. Note that, if a process is **invertible** (see Problem 2.15), then even if the wrong model is selected, a reasonable approximation to the process structure may be obtained provided a high enough model order is used. Actually, we know that since we are only interested in the power spectrum, we can always find a **spectrally equivalent** model for any linear process (see Section 2.5). Consequently, by choosing a high enough order, we may obtain a good approximation to any of the three forms by using an AR model.

7.4.2 AR Spectral Estimation

a) Overall Structure

Assume that we observe samples $x(0), x(1), ..., x(L-1)$ of a zero-mean stationary sequence, drawn from an M-th order AR model, generated by supplying a zero-mean iid sequence $w(n)$ as input to a system

$$H(z) = \frac{1}{1 - a_1 z^{-1} - ... - a_M z^{-M}} \; . \qquad (2.4.3)$$

We may express the generation of the AR(M) process $x(n)$ via the difference equation relation:

$$x(n) = a_1 x(n-1) + a_2 x(n-2) + ... + a_M x(n-M) + w(n) \; , \qquad (2.4.9)$$

and the spectrum of $x(n)$ takes the form

$$R(e^{j\omega}) = \frac{\sigma_w^2}{|1 - a_1 e^{-j\omega} - ... - a_M e^{-jM\omega}|^2} \; . \qquad (2.4.6)$$

Suppose we choose to model $x(n)$ via an AR(p) model. We form estimates $c_1, c_2, ..., c_p$, of the parameters of the model, and an estimate $\hat{\sigma}_w^2$ of the residual noise power. Finally, we form the spectrum estimate using:

$$R_{AR}(e^{j\omega}) = \frac{\hat{\sigma}_w^2}{|1 - c_1 e^{-j\omega} - \ldots - c_p e^{-j\omega p}|^2} , \qquad (7.4.1)$$

or

$$R_{AR}(e^{j\omega}) = \frac{\hat{\sigma}_w^2}{|\underline{f}^t \underline{s}|^2} , \qquad (7.4.1b)$$

where $\underline{f} = [1, -c_1, \ldots, -c_p]^t$, and where $\underline{s} = [1, e^{-j\omega}, \ldots, e^{-j\omega M}]^t$. $\underline{s}$ is often called a **frequency scanning vector** or **steering vector**.

b) Estimation of the Parameters

If the parameter estimates $c_1, c_2, \ldots, c_p$, correspond exactly to the true parameters $a_1, a_2, \ldots, a_M$, then $R_{AR}(e^{j\omega})$ will exactly equal the true spectrum. Generally, of course, the estimates will not be perfect and $R_{AR}(e^{j\omega})$ will be approximate. We have already discussed the problem of estimating the parameters of the AR process in previous Chapters. As we have seen, there are various ways to approach the problem. In Section 3.3.1 we examined the relationship between linear prediction and the AR model. In Section 6.5, we developed the forward and backwards linear predictors and illustrated their relation to the digital lattice structure. As we shall see, all of these concepts are significant for the study of the AR spectral estimator, each leading to a different estimator for the parameters. In fact, the various estimation possibilities lead to spectra that have significantly differing properties:

i) Yule-Walker Spectral Estimate — Beginning with the linear predictor, recall from Section 3.3.1 that the single-step predictor of order p has the form

$$\hat{x}(n) = \sum_{i=1}^{p} c_i x(n-i) , \qquad (3.3.2)$$

with prediction error

$$e(n) = x(n) - \hat{x}(n) . \qquad (3.3.3)$$

Recall further that the least-squares predictor coefficients obtained by minimizing

$$J = E\{(x(n) - \hat{x}(n))^2\} , \qquad (3.3.4)$$

satisfies the normal equations

$$R\underline{c} = \underline{g} , \qquad (3.3.5)$$

where R is the $(p \times p)$ autocorrelation matrix for the signal $x(n)$, with coefficients $R(i,j) = r(|i-j|)$, where $\underline{g} = [r(1), r(2), ..., r(p)]^t$ and where $\underline{c} = [c_1, c_2, ..., c_p]^t$. If the signal is generated via an AR(M) model as in (2.4.9), and provided $p \geq M$, then the signal is now given by (2.4.9), while the prediction has the form (3.3.2). Combining these we have

$$J = E\{e^2(n)\} = E\{[(a_1 - c_1)x(n-1) + ... + (a_M - c_M)x(n-M) +$$
$$+ (0 - c_{M+1})x(n-M-1) + ... + (0 - c_p)x(n-p) + w(n)]^2\} .$$

As indicated in Section 3.3, the solution which minimizes J is (intuitively) given by

$$\underline{c} = [a_1, a_2, ..., a_M, 0, ..., 0]^t .$$

Hence the AR parameters, and thus the AR spectrum, are exactly identified by the least-squares predictor. As we have already observed, in practice the correlation values $r(m)$ are replaced by estimates $r'(m)$, obtained from the available data. For example, we may directly estimate the correlation coefficients using

$$r'(m) = \frac{1}{L} \sum_{n=0}^{L-m-1} x(n)x(n+m) \quad ; 0 \leq m , \qquad (2.6.18)$$

which corresponds to the biased correlation estimator (see Section 2.6). Using these or similar estimates we form R and $\underline{g}$. The normal equations (3.3.5) may then be solved for the predictor coefficients. Also, for the AR model we may rearrange (2.4.9) as

$$w(n) = x(n) - a_1 x(n-1) - ... - a_M x(n-M) .$$

Then, multiplying both sides by $x(n)$ and taking the expectation gives

$$\sigma_w^2 = r(0) - a_1 r(1) - \ldots - a_M r(M) \ . \tag{7.4.2}$$

Equation (7.4.2) gives a relation between the correlation coefficients, the AR parameters, and the power of the generating white noise. In practice, we estimate σ_w^2 by replacing $a_1, \ldots, a_M$ by $c_1, \ldots, c_p$, and replacing $r(i)$ by $r'(i)$, yielding

$$\hat{\sigma}_w^2 = r'(0) - c_1 r'(1) - \ldots - c_p r'(p) \ . \tag{7.4.3}$$

As we saw in Section 3.3, we may combine (3.3.5) and (7.4.3) to obtain a single set of equations for the prediction error filter (but where in this case the correlation coefficients are replaced by estimates):

$$\begin{bmatrix} r'(0) & r'(1) & . & . & r'(p) \\ . & r'(0) & . & . & . \\ . & r'(1) & . & . & . \\ . & . & . & . & . \\ r'(p) & . & . & . & r'(0) \end{bmatrix} \begin{bmatrix} 1 \\ -c_1 \\ -c_2 \\ . \\ . \\ -c_p \end{bmatrix} = \begin{bmatrix} \hat{\sigma}_w^2 \\ 0 \\ . \\ . \\ 0 \end{bmatrix} \ . \tag{7.4.4}$$

Finally, we solve these equations for $c_1, c_2, \ldots, c_p$ and $\hat{\sigma}_w^2$, and then form the spectral estimate as

$$R_{YW}(e^{j\omega}) = \frac{\hat{\sigma}_w^2}{|1 - c_1 e^{-j\omega} - \ldots - c_p e^{-j\omega p}|^2} = \frac{\hat{\sigma}_w^2}{|\underline{f}^t \underline{s}|^2} \ , \tag{7.4.5}$$

where $\underline{f} = [1, -c_1, \ldots, -c_p]^t$ is the vector of prediction error filter coefficients. Estimation of the AR spectrum via solution of the normal equations produces the so-called **Yule-Walker AR** spectral estimate. The biased estimator for the correlation coefficients given by (2.6.18) is typically used since it produces positive semi-definite estimates. The Yule-Walker algorithm is intuitively simple, and produces good estimates when a reasonable number of samples is available. When only a short segment is available, however, other AR estimation methods can provide higher resolution.

ii) Burg Method [12] – The Burg method for estimating the AR parameters and forming an AR spectral estimate, is based on minimizing the sum of the forward and backwards prediction errors as discussed in Section 6.5. In particular, in the Burg method the coefficients are obtained by choosing reflection coefficients γ_i which minimize

$$J_p = E\{[e_p^f(n)]^2 + [e_p^b(n)]^2\} ,\qquad (7.4.6)$$

where $e_p^f(n)$, $e_p^b(n)$ are the forward and backwards prediction errors corresponding to

$$e_p^f(n) = x(n) - c_1 x(n-1) - c_2 x(n-1) - ... - c_p x(n-p) ,\qquad (6.5.2)$$

and

$$e_p^b(n) = x(n-p) - b_1 x(n-p+1) - b_2 x(n-p+2) - ... - b_p x(n) .\qquad (6.5.5)$$

We recall from Section 6.5 that the Burg predictor coefficients are obtained by computing reflection coefficients:

$$\gamma_i = \frac{2E\{e_{i-1}^f(n)e_{i-1}^b(n-1)\}}{E\{[e_{i-1}^f(n)]^2 + [e_{i-1}^b(n-1)]^2\}} \quad ; i=1,2,...,p ,\qquad (7.4.7)$$

where

$$e_0^f(n) = e_0^b(n) = x(n) .$$

Or, using a 'block estimation' approach, we replace the expectation operator by a data based estimate:

$$\gamma_i = \frac{2\sum\limits_{j=i}^{N-1} e_{i-1}^f(j)e_{i-1}^b(j-1)}{\sum\limits_{j=i}^{N-1}\left[[e_{i-1}^f(j)]^2 + [e_{i-1}^b(j-1)]^2\right]} \quad ; i=1,2,...,p ,\qquad (6.5.34)$$

where

$$e_i^f(n) = e_{i-1}^f(n) - \gamma_i e_{i-1}^b(n-1) ,\qquad (6.5.19)$$

and

$$e_i^b(n) = e_{i-1}^b(n-1) - \gamma_i e_{i-1}^f(n) ,\qquad (6.5.20)$$

are the relations between the prediction errors and the reflection coefficients. These reflection coefficient estimates are subsequently used to obtain predictor coefficients via the reflection coefficient to predictor coefficient recursions of Section 6.5. That is

$$f_j^{(j-1)}=0, \quad f_0^{(j)}=1 \;, \tag{6.5.28}$$

$$f_i^{(j)}=f_i^{(j-1)}-\gamma_j f_{j-i}^{(j-1)} \qquad ; \; i=1,2,...,j$$
$$; \; j=1,2,...,p \;. \tag{6.5.29}$$

Here $\underline{f}$ is the p-th order forward prediction error filter, where $\underline{f}=[f_0^{(p)},f_1^{(p)},...,f_p^{(p)}]^t=[1,-c_1^{(p)},...,-c_p^{(p)}]^t$ and the coefficients $c_i^{(p)}$ are the p-th order forward predictor coefficients. Finally, σ_w^2 is estimated and the Burg spectral estimate is obtained as:

$$R_{BU}(e^{j\omega})=\frac{\hat{\sigma}_w^2}{|1-c_1^{(p)}e^{-j\omega}-...-c_p^{(p)}e^{-j\omega p}|^2}=\frac{\hat{\sigma}_w^2}{|\underline{f}^t\underline{s}|^2} \;. \tag{7.4.8}$$

The Burg method is known to produce very high resolution spectral estimates from short data segments. However, the method is very sensitive to the selected model order. If a high model order is used, spurious spectral peaks tend to arise. This is because in practice errors in the estimation of $r(i)$ produce non-zero values for $c_i^{(p)}$ for $i=M+1,M+2,...,p$, even if the AR(M) model is appropriate. These non-zero parameters represent additional poles in the generating model, corresponding to additional (spurious) peaks in the spectrum.

The Burg method is also associated with a phenomenon known as **line splitting** [13]. This is a phenomenon where a single peak in the spectrum is represented by two distinct peaks in the Burg spectral estimate – the spectral line is 'split'. Note that this may also occur in the Yule-Walker estimate, especially when the biased correlation estimate is used [14]. When applied to the classical problem of estimating the spectrum for a sinusoid embedded in white noise, the estimate of the sinusoidal frequency obtained via the Burg method may also exhibit bias. This bias reportedly varies as a function of the starting phase of the sinusoid within the observed data [15]. Fortunately, both of these effects occur primarily with very short data segments.

Model Order Selection

Having determined that an AR model is appropriate, before the general spectral estimation procedure can be performed, we must select a model order. We naturally prefer to choose M as low as possible consistent with the structure of the data (this is the principle of **parsimony** in time-series modelling mentioned in Section 2.4). In practical terms parsimony is appropriate for computational reasons, and because the quality of the spectral estimate may actually degrade if the model order is chosen too high, with the introduction of artifacts such as false peaks into the result. However, if the model order is too low, one obtains an oversmoothed, low resolution spectrum. Moreover, if the process is actually MA, ARMA, or is simply corrupted by additive noise, then the actual order needed for an AR model would be infinite, and any finite order will result in biased parameter estimates. Practically, it is common to sequentially compute the spectrum for many different orders, and to determine which is the best order according to some suitable criterion. We know that there exists a monotonic relation between prediction order and error energy (See Section 3.3). One might simply monitor $J_{\min} = \hat{\sigma}_w^2$, as p increases until an acceptable residual energy is obtained. However, it is not clear what represents an 'acceptable residual'. A better approach might be to choose the lowest order such that the residual is white, since removal of the AR structure should leave a white error. However, how is the residual spectrum to be estimated?

Generally, it is clear that the question of order selection is non-trivial, and a number of criteria have been proposed. These include:

Final Prediction Error [16]

$$FPE(p) = \hat{\sigma}_{pw}^2 \left[\frac{L + (p+1)}{L - (p+1)} \right], \tag{7.4.9}$$

where L is the data length and $\hat{\sigma}_{pw}^2$ represents the variance of the residual associated with order p. The FPE is the product of two factors: $\hat{\sigma}_{pw}^2$ which decreases as p increases, and the rest of (7.4.9), which increases with p. The order selected is the minimum value of FPE. This criterion is reported to underestimate the true model order [4].

Akaike's Information Criterion [17]

$$AIC(p)=L\ln(\hat{\sigma}^2_{pw})+2p \ . \tag{7.4.10}$$

Again, the two terms act in opposite senses, the first decreasing with increasing p and the second increasing with p. The model order selected is the value that minimizes AIC. Practically, AIC is often considered to overestimate the true order [18].

Criterion Autoregressive Transfer Function (CAT) [19]

$$CAT(p)=\left[\frac{1}{L}\sum_{j=1}^{p}\frac{1}{[L/(L-j)]^{1/2}\hat{\sigma}_{pw}}\right]-\frac{1}{[L/(L-p)]\hat{\sigma}_{pw}} \ . \tag{7.4.11}$$

Once again, the model order p is the value that minimizes CAT.

Overall, the obvious conclusion is that no one formula can provide a satisfactory estimate of model order in all situations. In fact, some research [20], suggests that none of these criteria works well. As so often in signal processing, experience, trial-and-error, and subjective judgement play a crucial role in the final decision about the model order to choose.

AR Spectral Estimation and Maximum Entropy

The AR spectral estimate is often associated with the term **Maximum Entropy**. This association stems from an interpretation of the AR estimate given by Burg [21]: Given the autocorrelation sequence for lags $r(0), r(1), ..., r(p)$, there are an infinite number of ways that one may construct correlation values $r(m)$ for $m>p$, while still retaining positive definiteness for the overall sequence. Burg suggested that the extrapolation be performed in such a manner as to maximize the uncertainty (entropy) of the signal, subject to the given autocorrelation values $r(0), r(1), ..., r(p)$. We note [4] that the Maximum Entropy method and AR spectral estimators are precisely equivalent if the signal is Gaussian and if the exact correlation values are known. However, if the data is not Gaussian, or the correlation values are estimated from the data rather than being exact, then $R_{AR}(e^{j\omega})$ is not a Maximum Entropy spectral estimator.

Relation of the AR Spectral Estimates to the Correlogram

We recall from Section 2.4, that given the AR coefficients $a_1, a_2, ..., a_p$, we may calculate the correlation values $r(m)$, via the relation

$$r(m) = \sum_{k=1}^{p} a_k r(m-k) \qquad ; \, m > 0 \, ,$$

$$= \sum_{k=1}^{p} a_k r(m-k) + \sigma_w^2 \; ; \, m = 0 \, ,$$

$$= r(-m) \; ; \, m < 0 \, . \tag{7.4.12}$$

For the AR spectrum we may write

$$R_{AR}(e^{j\omega}) = \frac{\sigma_w^2}{|1 - c_1^{(p)} e^{-j\omega} - ... - c_p^{(p)} e^{-j\omega p}|^2} \, .$$

The coefficients $c_i^{(p)}$ have similar relation to (7.4.12), with correlation values $r(i)$ replaced by $\hat{r}(i)$. Hence

$$R_{AR}(e^{j\omega}) = \sum_{k=-\infty}^{\infty} \hat{r}(k) e^{-j\omega k} \, , \tag{7.4.13}$$

where, of course, the relation between $\hat{r}(k)$ and $r(k)$ depends on both the accuracy of the estimates of a_i, and on the validity of the AR model. In comparison with the non-parametric correlogram we have for the Blackman-Tukey estimator of Section 7.3 we have

$$R_{BT}(e^{j\omega}) = \sum_{n=-(M_1-1)}^{M_1-1} w_l(n) r'(n) e^{-j\omega n} \, . \tag{7.3.1}$$

Of course, the precise relation between (7.4.13) and (7.3.1) is unclear since it depends on the relation between the correlation sequences used in each. However, the general result is clear enough — in $R_{BW}(e^{j\omega})$ the correlation is assumed to be zero outside some finite range (M_1-1) to (M_1-1). No such windowing occurs with the AR estimator. Since there is no windowing effect, the resolution limitations and leakage effects associated with the window do not occur.

7.4.3 MA Spectral Estimation

Recall from Section 2.4, that for the MA(N) process we have generating model:

$$H(z)=b_0+b_1z^{-1}+...+b_Nz^{-N} , \qquad (2.4.4)$$

producing an MA(N) sequence $x(n)$ as

$$x(n)=b_0w(n)+b_1w(n-1)+...+b_Nw(n-N) , \qquad (2.4.10)$$

where $w(n)$ is a zero-mean iid sequence. For this MA process (see Problem 2.16)

$$r(m)=b_m b_0 \sigma_w^2+...+b_N b_{N-m}\sigma_w^2 \qquad ; 0\leq m\leq N$$

$$=0 \quad ; m>N$$

$$r(m)=r(-m) . \qquad (7.4.14)$$

One approach to estimating the spectrum is to estimate the b_i's from (7.4.14). However, these equations are nonlinear in the parameters and are often difficult to solve. In fact, the utility of this approach is limited since, given estimates $\hat{b}_i$, say, corresponding to estimates $\hat{r}(m)$ of $r(m)$, then the MA power spectral estimate is obtained as the transform

$$R_{MA}(e^{j\omega})= \sum_{m=-N}^{N} \hat{r}(m)e^{-j\omega m} . \qquad (7.4.15)$$

This is identical in form to the non-parametric (classical) estimates described in Sections 7.1-7.3. The difference is that, in the classical methods the data are used to calculate the spectrum directly, or via the correlation estimate. In MA spectrum estimation, the data are used to estimate the MA parameters, and the spectrum is then estimated using

$$R_{MA}(e^{j\omega})=| \hat{b}_0+\hat{b}_1 e^{-j\omega}+...+\hat{b}_N e^{-j\omega N}|^2\hat{\sigma}_w^2 . \qquad (7.4.16)$$

An MA model might be appropriate if the spectrum is characterized by broad spectral peaks and sharp nulls. However, a disadvantage of this approach is that if the MA model is not appropriate, then a very high model order is required to obtain good resolution. Thus the MA spectrum is not considered as a high resolution method, and has received less attention than other parametric methods for spectral estimation.

7.4.4 ARMA Spectral Estimation

ARMA models are potentially more efficient than MA or AR, allowing the possibility to represent both poles and zeros in the observed data, with relatively few parameters. In spite of this, there are fewer algorithms for ARMA spectral estimation than for AR. This is because the *simultaneous* estimation of both MA and AR parameters requires the solution of *nonlinear* equations. Iterative solutions for these equations have been proposed. However, they are not guaranteed to be stable, or to converge to the desired parameter values. A simpler, though sub-optimal approach is to estimate the AR and MA parameters *separately*, then combine the estimates to form the ARMA spectrum. Thus we consider the ARMA(M,N) spectrum

$$R_{xx}(z)=H(z)H(z^{-1})\sigma_w^2 \ ,$$

with

$$H(z)=\frac{B(z)}{A(z)}=\frac{b_0+b_1z^{-1}+\cdots+b_Nz^{-N}}{1-a_1z^{-1}-...-a_Mz^{-M}} \ . \qquad (2.4.2)$$

We may summarize this 'separate approach' as follows: Beginning with the AR parameters, we estimate

$$\hat{A}(z)=1-\sum_{i=1}^{M}\hat{a}_iz^{-i} \ , \qquad (7.4.17)$$

we then inverse filter $x(n)$ as

$$u(n)=x(n)-\hat{a}_1x(n-1)-\hat{a}_2x(n-2)-...-\hat{a}_Mx(n-M) \ . \qquad (7.4.18)$$

If the parameter estimates $\hat{a}_i$ are exact, then $u(n)$ represents the MA component of the signal. We may then use an MA estimation method applied to $u(n)$ to produce correlation estimates $\hat{r}_{uu}(m)$. From this the MA spectrum is

$$R_{MA}(e^{j\omega})=\sum_{m=-N}^{N}\hat{r}_{uu}(m)e^{-j\omega m} \ . \qquad (7.4.15)$$

Finally, the ARMA spectrum is

$$R_{ARMA}(e^{j\omega}) = \frac{R_{MA}(e^{j\omega})}{\left|1 - \sum_{k=1}^{M} \hat{a}_k e^{-j\omega k}\right|^2} . \qquad (7.4.19)$$

The estimation of the AR parameters may be accomplished via the **modified Yule-Walker method**. We can show (see Problem 7.12) that the ARMA parameters are related to the autocorrelation coefficients as

$$r(m) = \sum_{i=1}^{M} a_i r(m-i) + \sigma_w^2 \sum_{i=m}^{N} b_i h(i-m) \quad ; 0 \le m \le N ,$$

$$= \sum_{i=1}^{M} a_i r(m-i) \quad ; m > N ,$$

$$r(m) = r(-m) . \qquad (7.4.20)$$

where $h(i)$ are the impulse response coefficients corresponding to the equivalent causal generating system (see Section 2.4, equation (2.4.1)). For $m = N+1, N+2, ..., N+M$, we may write these relations in matrix form as:

$$\begin{bmatrix} r(N) & r(N-1) & . & r(N-M+1) \\ r(N+1) & r(N) & . & . \\ . & . & . & . \\ r(N+M-1) & . & . & r(N) \end{bmatrix} \begin{bmatrix} a_1 \\ a_2 \\ . \\ a_M \end{bmatrix} = \begin{bmatrix} r(N+1) \\ r(N+2) \\ . \\ r(N+M) \end{bmatrix} .(7.4.21)$$

This set of equations is known as the **Modified Yule-Walker Equations**. We observe that they form a linear set with a non-symmetric Toeplitz coefficient matrix. The key point is that given $r(N), r(N+1), ..., r(N+M)$, equations (7.4.21) allow calculation of the AR parameters *separately* from the MA parameters. Once the AR parameters have been estimated, we may inverse filter using (7.4.18). The filtered sequence $u(n)$ is then used to construct correlation coefficients $\hat{r}_{uu}(m)$. We transform using (7.4.15), and finally we calculate the ARMA spectrum using (7.4.18).

7.5 *Minimum Variance Spectral Estimation*

In this section, we develop another spectral estimator known variously as the **Maximum Likelihood Spectral Estimator** or the **Minimum Variance Spectral Estimator** [22],[23]. This estimator is non-parametric, but achieves significantly higher resolution than the classical methods of Sections 7.1 to 7.3. We obtain the estimator by solving an apparently unrelated constrained minimization problem: Consider a sequence of measurements $x(n)$ for $n=0,1,...,M-1$, consisting of a (complex) sinusoid of frequency ω_0 and phase ϕ, and an additive zero-mean stationary random noise $v(n)$ so that

$$x(n)=Ae^{j(\omega_0 n+\phi)}+v(n) \quad ; \; n=0,1,...,M-1 \; ,$$
$$=A_c s(n)+v(n) \; , \tag{7.5.1}$$

where $A_c=e^{j\phi}$ and where $s(n)=e^{j\omega_0 n}$. In vector form, we may write these observations as

$$\underline{x}=A_c \underline{s}+\underline{v} \; , \tag{7.5.1b}$$

where $\underline{s}=[s(0),s(1),...,s(M-1)]^t$, $\underline{v}=[v(0),v(1),...,v(M-1)]^t$ and $\underline{x}=[x(0),x(1),...,x(M-1)]^t$. Suppose we apply a (complex) linear weighting to the measurements $\underline{x}$ using coefficients $\underline{a}=[a_0,a_1,...,a_{M-1}]^t$ to produce y as

$$y=\underline{a}^t\underline{x} \; . \tag{7.5.2}$$

We intend y to be an estimate of A_c, that is to estimate the complex constant representing the amplitude and phase of the sinusoid in (7.5.1). We may measure the power of y as

$$E\{|y|^2\}=E\{y^*y\}=\underline{a}^H E\{\underline{x}^*\underline{x}^t\}\underline{a} \; ,$$
$$=\underline{a}^H R_{xx}\underline{a} \; , \tag{7.5.3}$$

where $R_{xx}=E\{\underline{x}^*\underline{x}^t\}$. As defined in Section 2.1, R_{xx} is the correlation matrix for the complex signal $x(n)$, and has elements $R_{xx}(i,j)=E\{x^*(i)x(j)\}$.

Suppose we seek to minimize the variance of y while simultaneously producing an unbiased estimate of the complex sinusoid. We may write this as the minimization of:

$$J = E\{|y - \bar{y}|^2\} ,$$ (7.5.4a)

subject to

$$\bar{y} = E\{y\} = A_c .$$ (7.5.4b)

Expanding J we have

$$J = E\{|y|^2\} - |A_c|^2 ,$$

so that the solution which minimizes J will equivalently minimize the power:

$$J = E\{|y|^2\} = \underline{a}^H R_{xx} \underline{a} ,$$

again, subject to the constraint that the output is an unbiased estimate of A_c. The unbiasedness constraint is achieved by imposing the condition

$$\underline{a}^t A_c \underline{s} = A_c ,$$

or

$$\underline{a}^t \underline{s} = 1 .$$ (7.5.5)

Note that with this condition

$$E\{y\} = E\{\underline{a}^t \underline{x}\} = E\{\underline{a}^t A_c \underline{s}\} + E\{\underline{a}^t \underline{v}\} ,$$

or

$$E\{y\} = E\{\underline{a}^t A_c \underline{s}\} = A_c ,$$ (7.5.6)

and the result of the linear weighting is an unbiased estimate of the sinusoid amplitude and frequency. Using Lagrange multipliers we write the criterion J as

$$J = \underline{a}^H R_{xx} \underline{a} - \lambda(\underline{a}^t \underline{s} - 1) ,$$ (7.5.7)

where λ is a (possibly complex) scalar constant. We solve this minimization problem by differentiating[10] J with respect to $\underline{a}$ and equating to zero, this yields

$$\nabla_{\underline{a}} J = 2R_{xx}\underline{a} - \lambda \underline{s}^* = \underline{0} \ . \tag{7.5.8}$$

Hence, rearranging we have

$$\underline{a}_{MV} = \frac{\lambda}{2} R_{xx}^{-1} \underline{s}^* \ . \tag{7.5.9}$$

Since $\underline{a}_{MV}$ must satisfy the constraint (7.5.5) we have

$$\lambda \underline{s}^H R_{xx}^{-1} \underline{s} = 2 \ ,$$

or

$$\lambda = \frac{2}{\underline{s}^H R_{xx}^{-1} \underline{s}} \ ,$$

which is real. Substituting into (7.5.9) we have

$$\underline{a}_{MV} = \frac{R_{xx}^{-1} \underline{s}^*}{\underline{s}^H R_{xx}^{-1} \underline{s}} \ . \tag{7.5.10}$$

The power associated with $\underline{a}_{MV}$ is easily shown to be (see Problem 7.14)

$$E\{|y|^2\} = \frac{1}{\underline{s}^H R_{xx}^{-1} \underline{s}} \ , \tag{7.5.11}$$

and (7.5.11) represents the minimum power, subject to the specified unbiasedness constraint.

So far we have solved a constrained minimization problem, but we have not indicated how this helps solve the spectral estimation problem. We now note that, in reality $\underline{s}$ and thus $E\{|y|^2\}$ are functions of ω_0. Since y is an (unbiased) estimate of the sinusoid amplitude and phase at frequency ω_0, the power $E\{|y|^2\}$, is an estimate of the squared amplitude at that frequency. We indicate this explicitly by writing

---

10. See, for example [24] for details of complex vector differentiation (see also Section 8.3).

$$R_{MV}(e^{j\omega_0})=\frac{1}{\underline{s}^H(\omega_0)R_{xx}^{-1}\underline{s}(\omega_0)} \ , \qquad (7.5.12)$$

and call this the Minimum Variance Spectral Estimator (MVSE). $R_{MV}(e^{j\omega_0})$ represents the minimum power subject to a constraint that the estimator has distortionless response at frequency ω_0. This has the effect of minimizing contributions in the output at ω_0 due to other frequencies. That is, we minimize leakage from other frequencies. Practically, we obtain the spectral estimate by sequentially calculating $R(e^{j\omega_k})$ for $\omega_k=2\pi k/M$, for $k=1,2,...,M$.

Narrowband Filtering Interpretation of the MVSE

The vector $\underline{a}_{MV}$ that minimizes

$$J=E\{\,|\,y-A_c\,|^2\} \quad subject \ to \ E\{y\}=A_c \ , \qquad (7.5.4)$$

is identical to that which minimizes (see Problem 7.15):

$$J_p=E\{\,|\,y_v\,|^2\} \ subject \ to \ E\{y\}=A_c \ ,$$

where

$$y_v=\underline{a}^t\underline{v} \ ,$$

and where y remains as defined by (7.5.2). Once again, the unbiasedness constraint can be written in the form (7.5.5). Thus, analogous to (7.5.7), we have

$$J=\underline{a}^H R_{vv}\underline{a}-\lambda(\underline{a}^t\underline{s}-1) \ . \qquad (7.5.13)$$

Following a similar minimization procedure we conclude that

$$\underline{a}_{MV}=\frac{R_{vv}^{-1}\underline{s}^*}{\underline{s}^H R_{vv}^{-1}\underline{s}} \ , \qquad (7.5.14)$$

which is equivalent to (7.5.10). Using (7.5.2) the output y has the form

$$y=\frac{\underline{s}^H R_{vv}^{-1}\underline{x}}{\underline{s}^H R_{vv}^{-1}\underline{s}} \ . \qquad (7.5.15)$$

An interesting special case arises when $\underline{v}$ is white, for then we have

$$y=\frac{\underline{s}^{H}\underline{x}}{\underline{s}^{H}\underline{s}}=\frac{1}{M}\underline{s}^{H}\underline{x} \ ,$$

or

$$y=\frac{1}{M}\sum_{i=0}^{M-1}x(i)e^{-j\omega_0 i} \ .$$

We may think of this as the output of a filter $h(n)$

$$y=\sum_{i=0}^{M-1}h(n-i)x(i) \ , \tag{7.5.16}$$

at time $n=0$, where

$$h(m)=\frac{1}{M}e^{j\omega_0 m} \quad ; \ m=-(M-1),-(M-2),...,0 \ , \tag{7.5.17a}$$

$$=0 \qquad\qquad ; \ elsewhere \ .$$

y is our estimate of A_c, and is obtained as the output of a bandpass filter $h(n)$, centered on frequency ω_0, with amplitude response (see Problem 7.17):

$$|H(e^{j\omega})|=\frac{1}{M}\left|\frac{\sin[M(\omega-\omega_0)]}{\sin(\omega-\omega_0)}\right| \ . \tag{7.5.17b}$$

The Minimum Variance Spectral Estimate can be considered as the measured output power from a set of such narrowband filters. Generally, the form of the filters varies according to the nature of the inputs. They are centered on ω_k and are defined, for each k, by the appropriate solution to (7.5.12) with ω_0 replaced by ω_k.

The periodogram of Section 7.2 can also be interpreted as the output from a set of bandpass filters. To see this we note that we may express the periodogram (7.2.1) as [3]

$$R_p(e^{j\omega_0})=M\left|\sum_{i=0}^{M-1}h(n-i)x(i)\right|^2 \ ,$$

at time $n=0$, where

$$h(m) = \frac{1}{M} e^{j\omega_0 m} \quad ; \; m = -(M-1), -(M-2), ..., 0 \; ,$$

$$= 0 \qquad\quad ; \; elsewhere \; .$$

Once again $h(m)$ is an M point FIR bandpass filter with response given by (7.5.17b). The periodogram estimate is thus obtained by squaring and scaling the output from an FIR bandpass filter centered on frequency ω_0. For the case of white inputs, the Minimum Variance and Periodogram estimates are identical. For more general inputs, the filters corresponding to the periodogram remain as given by (7.5.17). For the minimum variance estimate, the filters are optimized in the sense that (7.5.12) minimizes energy leakage into each frequency 'bin'. This means, in general, that the filters vary according to both the input type and the particular frequency band. Note, however, that a disadvantage of the minimum variance procedure, as compared to the periodogram, is that the optimization requires prior knowledge of the correlation matrix R_{xx}. In practice, as elsewhere, R_{xx} is replaced by a data based estimate. Note also, that the amplitude scale of the estimates produced by the minimum variance estimator is generally not accurate. For example, if $x(n)$ is white, then from (7.5.12)

$$R_{MV}(e^{j\omega}) = \frac{\sigma_w^2}{M} \; , \tag{7.5.18}$$

so that compared to the true spectrum, a scale factor M has been introduced.

The MV estimator is sometimes referred to as the **Maximum Likelihood (ML) spectral estimator**. This is because for Gaussian noise, the ML estimate of sinusoidal amplitude has the same form as the minimum variance estimate [3]. However, the term Minimum Variance is more general since this describes the estimator (7.5.12), independent of the noise form.

Example – As a numerical example, consider the problem of estimating the spectrum of the AR process:

$$x(n) = \sqrt{2}\, x(n-1) - x(n-2) + v(n) \; , \tag{7.5.19}$$

with $v(n)$ zero-mean, Gaussian iid with $\sigma_v^2 = 1$. For this experiment, assume that a total of $N=512$ observations of $x(n)$ sampled at $f_s=200\,Hz$ are available to form the estimate. Equation (7.5.19) is an AR(2) process. The poles of the generating system (see Section 2.4) are given by the roots of

$$1-\sqrt{2}\,z^{-1}+z^{-2}=0 \;,$$

or

$$z=\frac{1}{\sqrt{2}}\pm j\frac{1}{\sqrt{2}} \;, \tag{7.5.20a}$$

so that

$$|z|=1, \quad arg(z)=\pm\frac{\pi}{4} \;. \tag{7.5.20b}$$

Thus, the process has a sharp resonance at radian frequency $\omega=\pi/4$. For the $200\,Hz$ sample rate assumed, this translates to $25\,Hz$. In this example, the correlation matrix was estimated using the time average

$$\hat{R}=\frac{1}{N}\sum_{i=0}^{N}\underline{x}_i\underline{x}_i^t \;, \tag{7.5.21}$$

where $\underline{x}_i=[x(i),x(i-1),...,x(i-M+1)]^t$, with $x(i)=0 \; ; i<0$. Obviously, this is only one of the possibilities for estimating R. Each estimator will produce a different spectrum. However, for for large N, with $M\ll N$, these differences will be slight. Figure 7.5.1 shows the minimum variance spectrum obtained for $M=16$ and 64. Figure 7.5.2 shows the averaged periodogram obtained for the same data. This was obtained by averaging M point rectangular non-overlapping windows over the $N=512$ data points (see Section 7.2). We see that the resolution of the AR spectrum is superior to the averaged periodogram in each case.

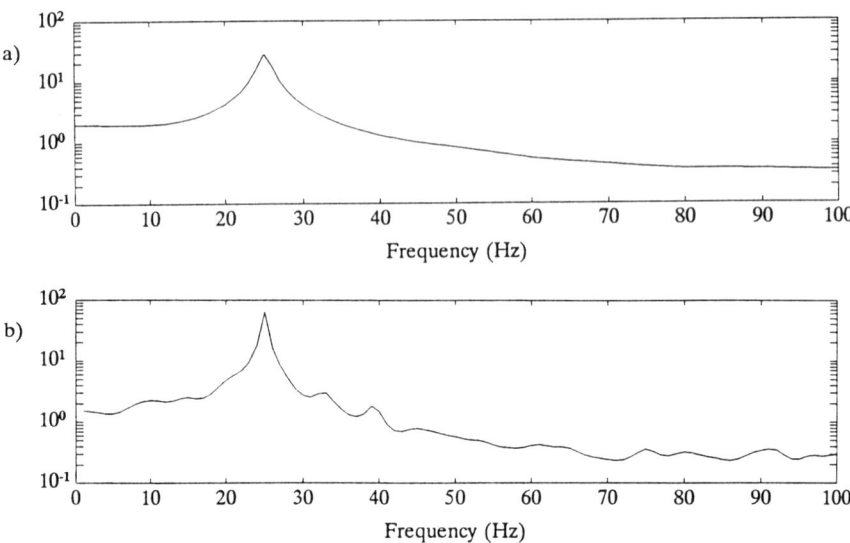

Figure 7.5.1 Minimum Variance spectral estimates obtained from a total of 512 observations of an AR(2) process. a) $M=16$, b) $M=64$.

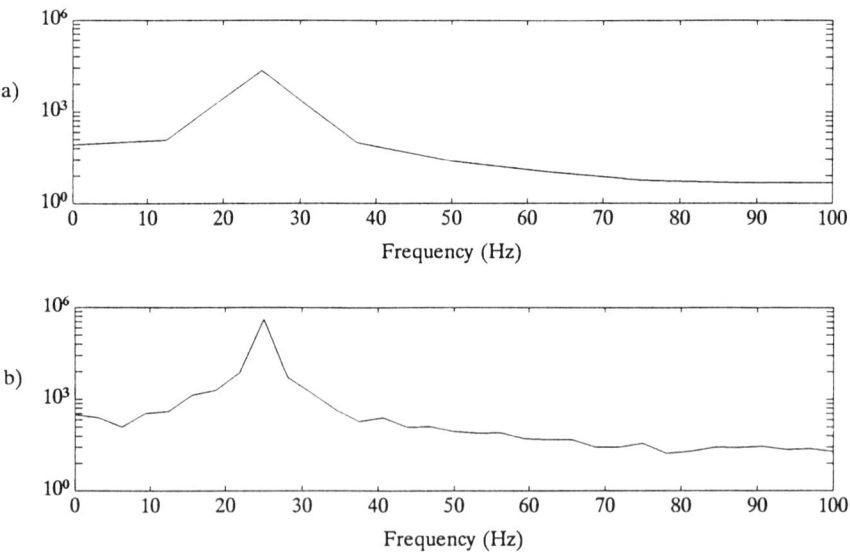

Figure 7.5.2 Averaged periodogram spectral estimates obtained from 512 observations of an AR(2) process. a) $M=16$, b) $M=64$.

Appendix 7A — Alternate Form for the Power Spectrum

Consider a zero-mean stationary sequence $x(n)$. In this appendix we examine the claim that the definition of the power spectrum:

$$R(e^{j\omega}) = \lim_{M \to \infty} \left\{ \frac{1}{2M+1} E\{ | \sum_{n=-M}^{M} x(n)e^{-j\omega n} |^2 \} \right\}, \qquad (7.1.2a)$$

is 'nearly equivalent' to the power spectral definition of equation (7.1.1)

$$R(e^{j\omega}) = \sum_{n=-\infty}^{\infty} r(n)e^{-j\omega n}. \qquad (7.1.1)$$

We begin by expanding (7.1.2a) as

$$R(e^{j\omega}) = \lim_{M \to \infty} \left\{ \frac{1}{2M+1} E\{ \sum_{n=-M}^{M} \sum_{m=-M}^{M} x(n)x(m)e^{-j\omega(n-m)} \} \right\}. \qquad (A1)$$

Since $x(n)$ is zero-mean stationary we then have:

$$R(e^{j\omega}) = \lim_{M \to \infty} \left\{ \frac{1}{2M+1} \sum_{n=-M}^{M} \sum_{m=-M}^{M} r(n-m)e^{-j\omega(n-m)} \right\},$$

or

$$R(e^{j\omega}) = \lim_{M \to \infty} \left\{ \frac{1}{2M+1} \left[r(2M)e^{j\omega 2M} + 2r(2M-1)e^{j\omega(2M-1)} + ... \right. \right.$$

$$\left. \left. ... + 2M + 1 r(0) + 2Mr(1)e^{-j\omega} + ... + r(2M)e^{-j\omega 2M} \right] \right\}, \qquad (A2)$$

or

$$R(e^{j\omega}) = \lim_{M \to \infty} \left\{ \frac{1}{2M+1} \sum_{k=-2M}^{2M} (2M+1-|k|)r(k)e^{-j\omega k} \right\},$$

$$= \lim_{M \to \infty} \left\{ \sum_{k=-2M}^{2M} (1 - \frac{|k|}{2M+1})r(k)e^{-j\omega k} \right\}. \qquad (A3)$$

A sufficient condition for (A3) to reduce to (7.1.1) is that $r(k)$ decays as k increases such that

$$\sum_{k=-\infty}^{\infty} |k| r(k) < \infty . \tag{A4}$$

This condition holds for almost all practically occurring sequences.

Problems

7.1 For the special case of $x(n)$ a zero-mean white noise sequence:

a) Show that the periodogram

$$R_p(e^{j\omega}) = \frac{1}{M} | \sum_{n=0}^{M-1} x(n) e^{-j\omega n} |^2 , \tag{7.2.1}$$

is an unbiased estimate of $R(e^{j\omega})$.

b) Evaluate the bias for the more general modified estimate

$$R_w(e^{j\omega}) = \frac{1}{M} | \sum_{n=0}^{M-1} x(n) w_d(n) e^{-j\omega n} |^2 ,$$

where $w_d(n)$ is an arbitrary data window.

7.2 For the periodogram defined by equation (7.2.1), show that the covariance

$$cov\{R_p(e^{j\omega_1}) R_p(e^{\omega_2})\} = E\{R_p(e^{j\omega_1}) R_p(e^{j\omega_2})\}$$
$$- E\{R_p(e^{j\omega_1})\} E\{R_p(e^{j\omega_2})\} ,$$

for a Gaussian white input with variance σ^2, is given by

$$cov\{R_p(\omega_1)R_p(e^{j\omega_2})\}=\sigma^4\left\{\left[\frac{\sin[(\omega_1+\omega_2)M/2]}{M\sin[(\omega_1+\omega_2)/2]}\right]^2+\right.\qquad(7.2.12)$$

$$\left.+\left[\frac{\sin[(\omega_1-\omega_2)M/2]}{M\sin[(\omega_1-\omega_2)/2]}\right]^2\right\}.$$

7.3 Consider the first order MA process:

$$x(n)=w(n)+aw(n-1),$$

with $\sigma_w^2=1$. If the sample rate is $f_s=6Hz$, evaluate the bias

$$B(k)=E\{R_p(k)\}-E\{R(k)\},$$

where $R_p(k)$ is the $M=4$ point periodogram of $x(n)$ defined by

$$R_p(k)=\frac{1}{M}\left|\sum_{n=0}^{M-1}x(n)e^{-j\frac{2\pi nk}{M}}\right|^2\quad;\ k=0,1,...,3.$$

and $R(k)$ is the true spectrum sampled at $\omega=2\pi k/M$ for $k=0,1,...,3$.

7.4 Repeat question **7.3** if the process $x(n)$ is the first order AR process with generating model

$$x(n)=0.5x(n-1)+w(n)\quad;\sigma_w^2=1,$$

for an arbitrary sample rate f_s, and using an $M=2$ point periodogram.

7.5 Show that the Fourier transform of the Bartlett window

$$w_b(m)=\frac{M-|m|}{M}\quad;|m|<M,$$
$$=0\quad\quad\quad;otherwise,$$

is given by

$$W(e^{j\omega})=\frac{1}{M}\left|\frac{\sin(\frac{\omega M}{2})}{\sin(\frac{\omega}{2})}\right|^2.\qquad(7.2.21)$$

7.6 Consider the modified periodogram

$$R_w(e^{j\omega}) = \frac{1}{MC} \left| \sum_{n=0}^{M-1} w_d(n)x(n)e^{-j\omega n} \right|^2 , \qquad (7.2.24)$$

where $w_d(n)$ is a data window defined over $n=0,1...,M-1$ and where

$$C = \frac{1}{M} \sum_{n=0}^{M-1} w_d^2(n) , \qquad (7.2.25)$$

is a normalizing constant. Show that

$$R_w(e^{j\omega}) = \sum_{n=-(M-1)}^{M-1} r^{''}(m)e^{-j\omega m} , \qquad (7.2.27)$$

where

$$r^{''}(m) = \frac{1}{MC} \sum_{n=0}^{M-|m|-1} w_d(n)x(n)w_d(n+|m|)x(n+|m|) .$$

7.7 Given the modified periodogram

$$R_w(e^{j\omega}) = \frac{1}{MC} \left| \sum_{n=0}^{M-1} w_d(n)x(n)e^{-j\omega n} \right|^2 , \qquad (7.2.24)$$

where

$$C = \frac{1}{M} \sum_{n=0}^{M-1} w_d^2(n) , \qquad (7.2.25)$$

show that

$$E\{R_w(e^{j\omega})\} = \frac{1}{2\pi MC} \int_{-\pi}^{\pi} R(e^{j\theta}) \left| W_d(e^{j(\omega-\theta)}) \right|^2 d\theta .$$

7.8 Suppose that

$$E\{R_p^{(i)}(e^{j\omega})\} = c ,$$

where c is a constant, and that

$$cov\{R_p^{(i)}(e^{j\omega})R_p^{(j)}(e^{j\omega})\} = \sigma^2 \quad ; i=j ,$$
$$= 0 \quad ; i \neq j .$$

Find the variance for the averaged periodogram $R_A(e^{j\omega})$

$$R_A(e^{j\omega}) = \frac{1}{K}\sum_{i=1}^{K} R_p^{(i)}(e^{j\omega}) , \qquad (7.2.33)$$

and compare this to that of the raw periodogram $R_p^{(i)}(e^{j\omega})$.

7.9 Repeat the previous question, where as before

$$E\{R_p^{(i)}(e^{j\omega})\} = c ,$$

but where now

$$cov\{R_p^{(i)}(e^{j\omega})R_p^{(j)}\} = \sigma^2 \qquad ; i=j$$
$$= a\sigma^2 \qquad ; i \neq j ,$$

where $0 < a < 1$.

7.10 For a Gaussian iid input $x(n)$, with the Blackman-Tukey spectral estimator

$$R_{BT}(e^{j\omega}) = \sum_{n=-(M-1)}^{M-1} w_l(n)r'(n)e^{-j\omega n} , \qquad (7.3.1)$$

find an expression for the correlation $E\{R_{BT}(e^{j\omega_0})R_{BT}(e^{j\omega_1})\}$.

7.11 Consider an AR(M) process, $x(n)$, corrupted by an additive zero-mean iid noise $v(n)$, with variance σ_v^2,

$$y(n) = x(n) + v(n) .$$

Show that the resulting spectrum $R_{yy}(e^{j\omega})$ is ARMA (M,M).

7.12 Consider an ARMA(M,N) process. Show that the ARMA parameters are related to the autocorrelation coefficients as

$$r(m) = \sum_{i=1}^{M} a_i r(m-i) + \sigma_w^2 \sum_{i=m}^{N} b_i h(i-m) \quad ; 0 \leq m \leq N ,$$

$$= \sum_{i=1}^{M} a_i r(m-i) \quad ; m > N ,$$

$$r(m)=r(-m) \ . \tag{7.4.20}$$

where $h(i)$ are the impulse response coefficients corresponding to the equivalent causal generating system.

7.13 Consider the complex signal

$$x(i)=A_c e^{j\omega_0(i-1)}+v(i) \quad ; \ i=1,2,...,M \ , \tag{1}$$

where $A_c=e^{j\phi}$, with ϕ a constant phase, and $v(i)$ is a zero-mean stationary noise sequence.

a) With equation (1) written in vector form:

$$\underline{x}=A_c\underline{s}+\underline{v} \ ,$$

find a suitable transformation matrix B such that $\underline{\hat{x}}=B\underline{x}$, where $\underline{\hat{x}}$ is a (complex) constant signal embedded in (complex) noise.

b) If

$$\hat{y}=\underline{a}_1^t\underline{\hat{x}} \ ,$$

find an optimal Minimum Variance Unbiased (MVUB) estimator $\underline{a}_1$ for A_c using the observations $\underline{\hat{x}}$.

c) Find the relation between the output power $E\{|\hat{y}|^2\}$ and the output power for the corresponding MVUB estimator of A_c derived from $\underline{x}$.

7.14 Consider signal measurements

$$x(n)=A_c e^{j\omega_0 n}+v(n) \quad ; \ n=0,1,...,M-1 \ ,$$

where $A_c=e^{j\phi}$. Show that the power associated with the result of the minimum variance weighting:

$$y=\underline{a}_{MV}^t\underline{x} \ ,$$

where

$$\underline{a}_{MV}=\frac{R_{xx}^{-1}\underline{s}^*}{\underline{s}^H R_{xx}^{-1}\underline{s}} \ , \tag{7.5.10}$$

is given by

$$E\{|y|^2\} = \frac{1}{\underline{s}^H R_{xx}^{-1} \underline{s}} \ . \tag{7.5.11}$$

7.15 For the signal model

$$x(i) = A_c e^{j\omega_0 i} + v(i) \quad ; i=1,2,...,M \ ,$$

where $v(i)$ is stationary, zero-mean noise, and A_c is constant, show that the problem:

Minimize $E\{|y - A_c|^2\}$ subject to $E\{y\} = A_c$, where

$$y = \underline{a}^t \underline{x} \ ,$$

is equivalent to the problem:

Minimize $E\{|y_v|^2\}$ subject to $E\{y\} = A_c$, where

$$y_v = \underline{a}^t \underline{v} \ .$$

Hence show that

$$\underline{a}_{MV} = \frac{R_{vv}^{-1} \underline{s}^*}{\underline{s}^H R_{vv}^{-1} \underline{s}} \ ,$$

and find an expression for $Var\{y\}$ in terms of the noise correlation matrix $R_{vv} = E\{\underline{v}^* \underline{v}^t\}$.

7.16 Given observations of the form

$$x(i) = A_c e^{j\omega_0(i-1)} + v(i) \quad ; i=1,2,...,M \ ,$$

where the noise $v(i)$ is zero-mean iid with variance σ_v^2, use the results of the previous question to show that

$$Var\{y\} = \frac{\sigma_v^2}{M} \ .$$

7.17 In the bandpass filter interpretation of spectral estimation, show that an impulse response

$$h(m) = \frac{1}{M} e^{j\omega_0 m} \quad ; \, m = -(M-1), -(M-2), ..., 0 \, ,$$

$$= 0 \qquad\qquad ; \, elsewhere \, ,$$

has amplitude response

$$|H(e^{j\omega})| = \frac{1}{M} \left| \frac{\sin[M(\omega-\omega_0)]}{\sin(\omega-\omega_0)} \right| \, .$$

7.18 a) Show that for a stationary zero-mean input $x(n)$, the least-squares backward prediction errors at stages $i, j \leq p$ are mutually orthogonal with

$$E\{e_i^b(n) e_j^b(n)\} = 0 \qquad\qquad ; \, i \neq j \, ,$$

$$= E\{[e_j^b(n)]^2\} \qquad ; \, i = j \, ,$$

where

$$e_i^b(n) = x(n-i) - b_1^{(i)} x(n-i+1) - ... - b_i^{(i)} x(n) \, ,$$

where $b_i^{(j)}$ is the i-th backwards predictor coefficient for prediction order j. What can you say about the sequence $E\{[e_1^b(n)]^2\}, E\{[e_2^b(n)]^2\}, ..., E\{[e_p^b(n)]^2\}$?

b) In vector notation the equations for the backwards prediction errors can be written as

$$\underline{e}_n^b = L \underline{x}_n \, ,$$

where $\underline{x}_n = [x(n), x(n-1), ..., x(n-p)]^t$,
$\underline{e}_n^b = [e_0^b(n), e_1^b(n), ..., e_p^b(n)]^t$,

and where

$$L = \begin{bmatrix} 1 & 0 & 0 & . & . \\ -b_1^{(1)} & 1 & 0 & . & . \\ -b_2^{(2)} & -b_1^{(2)} & 1 & . & . \\ . & . & & . & . & 0 \\ -b_p^{(p)} & -b_{p-1}^{(p)} & . & . & 1 \end{bmatrix} \, .$$

Show that

$$R_{xx}^{-1} = L^t D^{-1} L \ ,$$

where D is the diagonal matrix

$$D = diag\{\sigma_0^2, \sigma_1^2, ..., \sigma_p^2\} \ ,$$

with $\sigma_i^2 = E\{[e_i^b(n)]^2\}$.

c) For an AR(M) signal $x(n)$, show that

$$\frac{1}{R_{MV}(e^{j\omega})} = \sum_{i=0}^{M-1} \frac{1}{R_{AR}^{(i)}(e^{j\omega})} \ ,$$

where $R_{AR}^{(i)}(e^{j\omega})$ is the i-th order AR spectrum obtained as

$$R_{AR}^{(i)}(e^{j\omega}) = \frac{\sigma_i^2}{|1 - \sum_{k=1}^{i} a_k^{(i)} e^{-j\omega k}|^2} \ ,$$

where $a_i^{(j)}$ is the i-th forward predictor coefficient for order j, and where

$$R_{MV}(e^{j\omega}) = \frac{1}{\underline{s}^H R_{xx}^{-1} \underline{s}} \ ,$$

with $\underline{s} = [1, e^{j\omega}, ..., e^{j(M-1)\omega}]^t$, where σ_i^2 is the prediction error energy associated with the prediction.

Computer Problem 1 – Evaluation of Periodogram Methods

Write a program to calculate power spectra using windowed, averaged periodograms. The program should embody the following parameters:

i) window type:
    a) Rectangular
    b) Hamming
    c) Bartlett.

ii) Window length – Selectable as 2^m ; $m \leq 10$.

iii) Window overlap (%).

iv) Maximum data length $(M \leq 15,000$ samples).

These parameters should be selectable as options at run-time.

a) *i)* Generate 2048 points of Gaussian random noise.

ii) Calculate the 1024 point raw periodogram using half of this data.

iii) Calculate the spectral variance defined by

$$J = \frac{1}{M} \sum_{k=0}^{M-1} \left(R_p(e^{j\omega_k}) - R(e^{j\omega_k}) \right)^2 ,$$

where $\omega_k = 2\pi k / M$.

b) Calculate the averaged periodogram for the following parameters:

i) Rectangular window, 0% overlap, $m=9$.
ii) Rectangular window, 0% Overlap, $m=7$.
iii) Rectangular window, 0% overlap, $m=5$.

Evaluate J in each case and tabulate the results. Does the variance behave as you would expect?

c) Construct a similar experiment to b) above to evaluate the effects of overlap on the variance. What conclusions can be drawn?.

d) Construct a similar experiment to evaluate the effect of window type on the variance J.

Computer Problem 2 – AR Spectral Estimation

Write a program to produce an AR spectral estimate. The parameters should be calculated via the Yule-Walker method (see Section 7.4). For a given set of data $x(n)$ for $n=0,1,...,N-1$, say, estimate the correlation coefficients via

$$r'(m) = \frac{1}{N} \sum_{n=0}^{N-m-1} x(n)x(n+m) \quad ; 0 \leq m ,$$

$$r'(-m) = r'(m) .$$

Using these correlation coefficients, your program should follow the procedure of Section 7.4 to estimate the AR parameters $c_1, c_2, ..., c_p$, and residual σ_w. Consider the AR(2) process

$$x(n) = 1.7x(n-1) - 0.9x(n-2) + w(n) \ ,$$

where $w(n)$ is zero-mean Gaussian iid with variance σ_w^2.

a) Calculate and plot the spectrum $R_{xx}(e^{j\omega})$.

b) Generate 2048 points of data from $x(n)$ (assume $w(n)$ is Gaussian with unit variance). Calculate and plot the averaged raw periodogram for $M=16$, 32 and 128, using all the available data, but with non-overlapping successive windows.

c) Set $p=2$ and calculate the AR spectrum. Plot the result.

d) Calculate and tabulate the spectral variance as defined in Computer Problem 1 above, for $p=2,3,...,12$.

e) Calculate and plot the spectra corresponding to $p=4,8$ and 12.

References

1. E.A.Robinson, "A historical perspective of spectrum estimation," *Proc. IEEE*, vol. 70, pp. 885-906, 1982.

2. G.M.Jenkins and D.G.Watts, *Spectral Analysis and its Applications*, Holden-Day, 1968.

3. S.M.Kay, *Modern Spectral Estimation Theory and Applications*, Prentice-Hall, 1988.

4. S.L.Marple, *Digital Spectral Analysis with Applications*, Prentice-Hall, 1987.

5. *Proc.IEEE*, "Special issue on spectral estimation," Sept. 1982.

6.   S.M.Kay and S.L.Marple, "Spectrum analysis: a modern perspective," *Proc. IEEE*, vol. 69, pp. 1380-1419, 1981.

7.   A.V.Oppenheim and R.W.Schafer, *Discrete Time Signal Processing*, Prentice-Hall, 1989.

8.   M.S.Bartlett, *An Introduction to Stochastic Processes with Special Reference to Methods and Applications*, Cambridge University Press, 1953.

9.   P.D.Welch, "The use of the Fast Fourier Transform for the estimation of power spectra: a method based on time averaging over modified periodograms," *IEEE Trans. Audio, Electroacoust.*, vol. AU-15, pp. 70-73, 1967.

10.   R.B.Blackman and J.W.Tukey, *The Measurement of Power Spectra from the Point of View of Communications Engineering*, Dover, 1958.

11.   J.G.Proakis and D.G.Manolakis, *Introduction to Digital Signal Processing*, Macmillan, 1988.

12.   J.P.Burg, "Maximum entropy spectral analysis," *Proc. 37th Meeting of the Society of Exploration Geopysicists*, 1967. Reprinted in *Modern Spectrum Analysis*, D.G.Childers (Ed), IEEE Press.

13.   P.F.Fougere, E.J. Zawalick and H.R.Radoski, "Spontaneous line splitting in maximum entropy power spectrum analysis," *Phys. Earth Planet Inter.*, vol. 12, pp. 201-207, 1976.

14.   S.M.Kay and S.L.Marple, "Source of and remedy for line splitting in autoregressive spectrum analysis," *Proc. IEEE Int. Conf. Acoust., Speech, Signal Processing*, pp. 151-154, 1979.

15.   D.N.Swingler, "Frequency errors in MEM processing," *IEEE Trans. Acoust., Speech, Signal Processing*, vol. ASSP-28, pp. 257-259, 1980.

16.   H.Akaike, "Power spectrum estimation through autoregressive model fitting," *Ann. Inst. Stat. Meth.*, vol. 21, pp. 407-419, 1969.

17.   H.Akaike, "A new look at statistical model identification," *IEEE Trans. Automat. Contr.*, vol. AC-19, pp. 716-723, 1974.

18.   M.B.Priestley, *Spectral Analysis and Time Series, Vols I and II*, Academic Press, 1981.

19. E.Parzen, "Some recent advances in time series modelling," *IEEE Trans. Automat. Contr.*, vol. AC-19, pp. 723-730, 1974.

20. T.J.Ulrych and R.W.Clayton, "Time series modeling and maximum entropy," *Phys. Earth Planet. Inter.*, vol. 12, pp.188-200, 1976.

21. J.P.Burg, "Maximum entropy spectral analysis," PhD dissertation, Department of Geophysics, Stanford University, 1975.

22. J.Capon, "High resolution frequency-wavenumber spectrum analysis," *Proc. IEEE*, vol. 57, pp. 1408-1418, 1969.

23. R.T.Lacoss, "Data adaptive spectral analysis methods," *Geopysics*, vol. 36, pp. 661-675, 1971.

24. R.A.Horn and C.R.Johnson, *Matrix Analysis*, Cambridge University Press, 1985.

chapter eight

Adaptive Arrays

8.1 Array Signal Processing

8.1.1 Introduction

Array signal processing is concerned with the application of signal processing techniques to the outputs from a spatially distributed array of sensors. The power of array processing lies in the possibility of achieving improvements in measured Signal-to-Noise Ratio (SNR) as compared to that attainable with a single input sensor. The spatial dimension introduced by processing a distributed array of sensors allows directional discrimination between signals and interference. Sensor arrays are used in many applications including radar, sonar, seismology and tomography [1]. The signals measured in these applications are either electromagnetic or acoustic in origin. In all cases the array is designed to receive signals generated by propagating waves [2]. Physically, there are many differences between the various arrays, ranging from the form of energy transported by the wave, to the type of sensors, to the physical dimensions of the array. In spite of these differences, however, many of the processing objectives of both electromagnetic and acoustic problems are essentially the same. As in the time domain, the broad objectives of signal processing applied to such systems are the extraction of significant physical information from measured data. Our primary interest here is in optimal and adaptive arrays, that is arrays which perform optimally in response to unknown and/or time-varying signals. In most cases, the objective of adaptive array processing is to detect or enhance signal inputs, while simultaneously reducing unwanted interference. We shall be primarily concerned with the problem of preserving signals incident to the array from a particular direction, and employing the array as a spatial filter to attenuate interferences from other directions. The array can also be viewed as a mechanism for improving SNR by averaging many sensor measurements together. As with time domain processing our objective in introducing adaptive arrays is to allow flexible response in the absence of prior information and in the presence of temporal, and in this case spatial, fluctuations.

Although we are mainly concerned with optimal and adaptive systems we shall begin by reviewing, briefly, the principles of array processing and beamforming. We note that it is impossible in a single chapter such as this to completely review the subject, and the material contained here is necessarily incomplete. For a more complete description of the various properties of arrays both adaptive and otherwise, the interested reader is referred to the texts by Monzingo and Miller [3], by Haykin [1], Hudson [4] and most recently, by Compton [5]. Additionally, tutorial papers on arrays and adaptive arrays include works by Dudgeon [2], Applebaum [6] and Griffiths [7].

8.1.2 Basic Array Properties

An array comprises a set of spatially distinct sensor elements. The physical form of the sensors varies according to the medium. For example, hydrophones are used in sonar and marine geophysical systems, microphones in air and geophones for land based seismology. Our concern here is not with the form of the sensor, only with the measurement generated by the sensor. For the i-th sensor of a set of N, we denote this measured output signal as $x_i(t)$.[1] We assume that the transfer function between the signal incident at the sensor and the measured output is unity. We also assume that the sensor is **omni-directional** − that is, it responds equally to incident signals from all directions.

The spatial organization of the array sensors can be defined arbitrarily in three dimensions. For the sake of simplicity we shall restrict attention here to linear, equispaced arrays in one dimension (line arrays), with inter-element spacing d. (see Figure 8.1.1). Extension of results for the line array to higher dimensions, and to more general geometries is straightforward, though algebraically more involved. Extension to non-uniform and to random spacing of elements is somewhat more difficult [8]. We concentrate on

---

1. In contrast to most of the work in earlier Chapters we admit the general possibility of $x_i(t)$ complex. This is consistent with most of the array processing literature and also allows considerable simplification of the analysis of array properties.

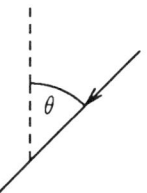

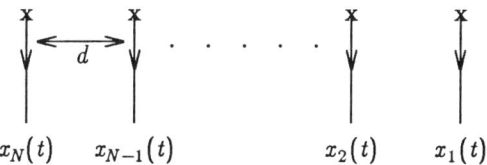

$$x_N(t) \qquad x_{N-1}(t) \qquad\qquad x_2(t) \qquad x_1(t)$$

Figure 8.1.1 A signal is incident at angle θ to an array comprising a line of N equispaced sensors.

linear processing of the received signals. That is, where the output from the array is obtained as a linear combination of the sensor inputs. Thus we may write

$$y(t) = \sum_{i=1}^{N} w_i x_i(t) \ , \tag{8.1.1}$$

where the w_i's are weightings applied to the sensor measurements and where $y(t)$ is the final array output (see Figure 8.1.2).

At this stage we consider only **narrowband arrays**. A narrowband array is one in which the array is designed to operate with signals restricted to a narrow spread of frequencies around a single center frequency ω_0, say. For such an array, each sensor is weighted using a single complex factor

$$w_i = w_{iR} + j w_{iI} \ . \tag{8.1.2}$$

w_{iR} and w_{iI} can also be thought of as in-phase and quadrature components, so that the imaginary component is simply delayed

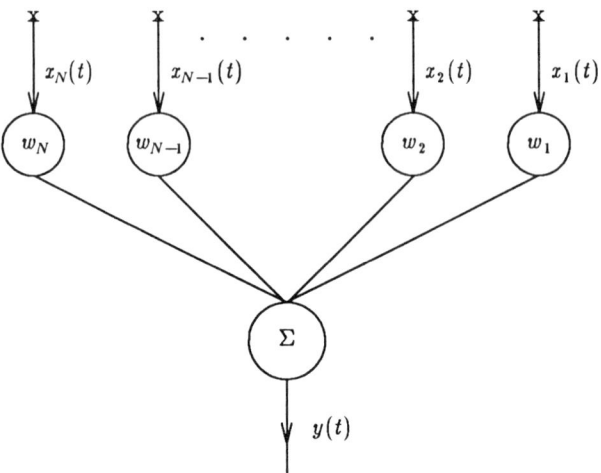

Figure 8.1.2 Linear weighting applied to an array of sensors measurements $x_i(t)$ producing array output $y(t)$.

by 90^0 at the array center frequency.

Signals, Noise and Interference

Consider a signal originating at some point in space with arbitrary coordinates x, y, z. The measured signal at sensor j is related to $s(t)$ via an impulse response function, $h_j(t)$, say, which models the transmission path between the coordinates of the source and those of the jth sensor. Here, for simplicity we have assumed that the transmission path is linear and time-invariant. We have

$$s_j(t) = h_j(t) * s(t) , \qquad (8.1.3)$$

$h_j(t)$ models the transmission of energy from the source to the j-th sensor. If no distortion or attenuation occurs during transmission, then $h_j(t)$ is simply a delay dependent on the distance between source and sensor, and on the propagation velocity. The received signal then has the form

$$s_j(t) = s(t - t_j) . \qquad (8.1.4)$$

The delay t_j is dependent on the angle between the signal source and the sensor. In particular, if the medium is homogeneous, that is if the physical variables influencing the transmission are constant throughout the medium, then the transmission velocity is constant. Energy from a point source then radiates spherically. At distinct time intervals, points which lie on the same spherical surface centered on the source are in phase and constitute a wavefront. If the distance from the source to the array is great compared to the dimension of the array, then the arc of the spherical wavefront may be approximated by a straight line (that is, if the arc is short relative to the radius of the circle to the source). This is referred to as the far-field. This type of propagating energy is referred to as **plane wave propagation**. It should be kept in mind that generally the transmission velocity will only be approximately constant. For example, in sonar measurements the velocity of sound actually depends on water temperature, pressure and salinity (see, for example, [9]). Consequently, sound speed actually varies as a function of numerous factors including water depth, geographical location, weather conditions, even the time of day.

In a similar fashion to the signal $s(t)$, we may define $u(t)$ as an 'undesired' signal or interference. Like $s(t)$, this interference is received at the sensor after propagating from a distant point source. This propagation can also be modelled using an impulse response function similar to $h_j(t)$.

A final component in the received signal is the ubiquitous measurement noise, which we will denote $v(t)$ and which may be generated locally at each sensor, or remotely within the medium, but which in either case is not associated with a particular direction of arrival.

Example − SNR Gain by Summation

A simple example serves to illustrate one of the basic properties of arrays. Suppose a signal $s(t)$ is incident at broadside ($\theta=0^0$ in Figure 8.1.1) to a linear array with unit weighting $w_i=1$ for $i=1,2,...,N$. The signal measurement at each sensor is assumed to be corrupted by a measurement noise $v_i(t)$ which is uncorrelated from sensor to sensor. That is,[2]

---

2. This is the spatial equivalent of white noise, as evidenced by the correlation structure (8.1.5).

$$E\{v_i(t)v_j(t)\}=\sigma_v^2\delta(i-j) \ , \tag{8.1.5}$$

so that we have

$$x_i(t)=s(t-D)+v_i(t) \ . \tag{8.1.6}$$

Now, the input SNR may be defined by

$$SNR_i=\frac{s^2(t)}{\sigma_v^2} \ . \tag{8.1.7}$$

The array output is

$$y(t)=\sum_{i=1}^{N} x_i(t) \ ,$$

$$=Ns(t)+\sum_{i=1}^{N} v_i(t) \ . \tag{8.1.8}$$

The output SNR is thus

$$SNR_o=\frac{N^2 s^2(t)}{E\{(\sum_{i=1}^{N} v_i(t))^2\}} \ ,$$

which, as a consequence of (8.1.5), is given by

$$SNR_o=\frac{N^2 s^2(t)}{N\sigma_v^2}=N\frac{s^2(t)}{\sigma_v^2} \ ,$$

or

$$SNR_o=N\times SNR_i \ . \tag{8.1.9}$$

This simple example illustrates an important feature of the array — the use of N spatially distributed elements allows an increase in SNR by a factor of N. Moreover, the result is in no way dependent on the *temporal* correlation of $v_i(t)$. The result is a limit, however, in that if

$$E\{v_i(t)v_j(t)\}\neq 0 \ ,$$

then the SNR gain obtained by simple summation will be reduced.

Aside from the directional characteristics of the source $s(t)$, each received signal has a set of spectral (frequency domain) characteristics. The simplest form of source is a monochromatic (that is single frequency) plane progressive wave of frequency ω, say

$$s(t)=e^{j(\omega t-kr)} , \qquad (8.1.10)$$

where r is distance in the direction of travel, and k is the **wave number** with

$$k=\frac{2\pi}{\lambda} , \qquad (8.1.11)$$

where λ is the wavelength $\lambda=c/f$, so that

$$k=\frac{2\pi f}{c} ,$$

or

$$k=\frac{\omega}{c} , \qquad (8.1.12)$$

with $\omega=2\pi f$, where c is the propagation velocity for the medium, which is assumed constant. All points along a line drawn perpendicular to the direction of propagation are in phase and form a wavefront (see Figure 8.1.3).

8.1.3 The Elements of Beamforming

The Array Pattern Function

Consider a monochromatic plane wave incident to a linear array of omnidirectional sensors at an angle θ to the vertical (see Figure 8.1.3). Assume that the 'phase-center' for the array is at element number 1,[3] so that we have

$$x_1(t)=e^{j\omega t} . \qquad (8.1.13)$$

---

3. This choice is arbitrary. Assuming the response is in phase at some other sensor simply produces a constant phase shift across the array and does not affect the pattern function.

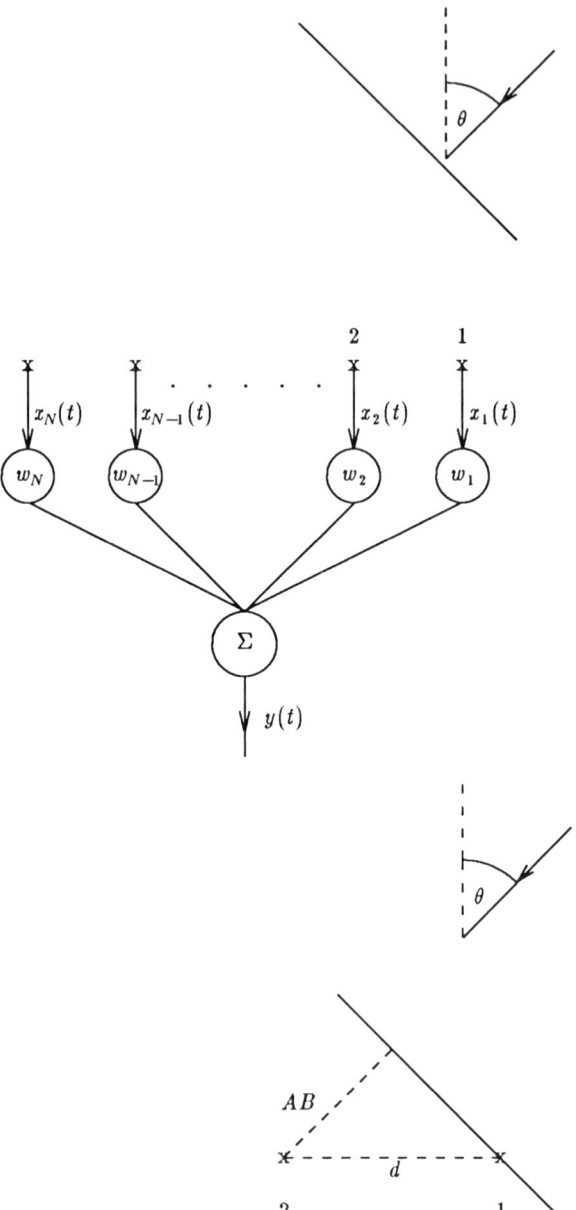

Figure 8.1.3 Plane wave source incident to a line array from an angle θ. A wavefront must travel distance AB after arriving at sensor 1 before being received at sensor 2.

That is, we assume that the wave is 'in phase' at sensor number 1. At sensor 2, the relative phase is determined by the distance AB in the diagram, which by simple geometry is

$$AB = d\sin\theta . \tag{8.1.14}$$

Hence the signal received at sensor 2 is

$$x_2(t) = e^{j[\omega t - kr]} ,$$

$$= e^{j[\omega t - kd\sin\theta]} .$$

Similarly, for the third sensor we have

$$x_3(t) = e^{j[\omega t - 2kd\sin\theta]} ,$$

and in general for the i-th sensor

$$x_i(t) = e^{j[\omega t - k(i-1)d\sin\theta]} . \tag{8.1.15}$$

Summing the responses, assuming unit weighting $w_i = 1$, produces the beamformer output $y(t)$ as

$$y(t) = \sum_{m=0}^{N-1} e^{j[\omega t - mkd\sin\theta]} ,$$

$$= e^{j\omega t} \sum_{m=0}^{N-1} e^{-jkmd\sin\theta} ,$$

$$= \left(\frac{1 - e^{-jkNd\sin\theta}}{1 - e^{-jkd\sin\theta}} \right) e^{j\omega t} ,$$

$$y(t) = A(\theta) e^{j\omega t} . \tag{8.1.16}$$

Hence, the array output is the product of the input to a single sensor (sensor 1), and a factor A. This factor is dependent on the angle of the source θ, and on k, d and N, though we usually suppress this and write simply $A(\theta)$. The magnitude of A is given by:

$$|A(\theta)| = \left| \frac{1 - e^{-jNkd\sin\theta}}{1 - e^{-jkd\sin\theta}} \right| ,$$

$$= \left| \frac{\sin(\frac{Nkd}{2}\sin\theta)}{\sin(\frac{kd}{2}\sin\theta)} \right| . \tag{8.1.17}$$

The directional characteristic can also be normalized by division by the peak response value

$$|G(\theta)| = \frac{|A(\theta)|}{N} . \tag{8.1.18}$$

$G(\theta)$ is called the **pattern function** for the array [9]. By fixing N, k and d, this function may be plotted on a linear scale versus θ, or may be given in the form of a polar plot. A typical example of both forms is given in Figures 8.1.4a) and b).

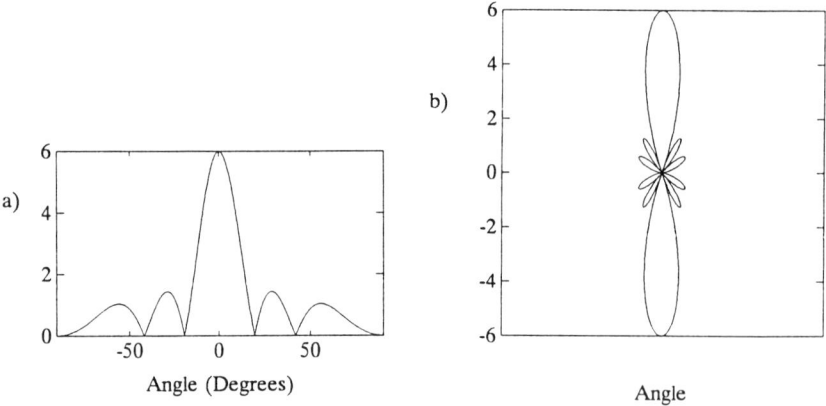

Figure 8.1.4 Array amplitude response as a function of plane wave angle. a) Rectangular plot, b) Polar plot.

Both plots show the manner in which the magnitude of an incident signal is selectively attenuated according to the angle of incidence. The array is thus a form of spatial filter. At 0^0 (broadside), the signal is passed with unit gain. At other angles of incidence the signal is attenuated. The response is characterized by a **main-lobe** and a number of **side-lobes**. The terms are similar to those used in time domain windowing and the comparison need not end there since we may characterize the beam pattern in terms of the 3dB bandwidth of the main-lobe, maximum side-lobe level and asymptotic roll-off of the side-lobes. Note that the directional response of the array changes if any of the parameters k, d or N changes. We can more effectively examine the effects of these parameter variations by parameterizing the pattern function using:

$$u = \frac{\sin\theta}{\lambda} , \qquad\qquad (8.1.19)$$

substituting $k = \frac{2\pi}{\lambda}$ we may write (8.1.17) as

$$|G(u)| = \frac{1}{N} \left| \frac{\sin(N\pi d u)}{\sin(\pi d u)} \right| . \qquad\qquad (8.1.20)$$

We see from this equation that the pattern repeats after an interval $1/d$, and that zeros occur at $1/Nd$, implying that the main-lobe width is $2/Nd$, from zero to zero. Hence the main-lobe becomes narrower, relative to the entire pattern, and thus the array more selective, as N, (or more precisely as the effective array length Nd) increases. We may plot the array pattern function for any value of u. However, because

$$u = \frac{\sin\theta}{\lambda} ,$$

then in reality u is limited to the range

$$-\frac{1}{\lambda} \leq u \leq \frac{1}{\lambda} , \qquad\qquad (8.1.21)$$

because θ traverses $-\pi/2$ to $\pi/2$ over this range. Hence, (8.1.21) corresponds to the range of u for the real part of the array response. The real array response is thus restricted to that part of the response for which the limits in (8.1.21) apply. In particular, since the pattern repeats after a span of $1/d$ for u, the real part of the response depends on the range of (8.1.21) versus

$$-\frac{1}{2d} \leq u \leq \frac{1}{2d} . \qquad\qquad (8.1.22)$$

Or, comparing (8.1.21) and (8.1.22), on $2d$ versus λ.

Figures 8.1.5, 8.1.6a)-d) show examples of the real response for four cases. We see that when $d<\lambda/2$, the main-lobe occupies a wide range of angles. By contrast, when $d>\lambda/2$ we obtain a much narrower main-lobe but the side-lobes are increased because at $u=\lambda/2$, we are further along the pattern than $1/2d$.

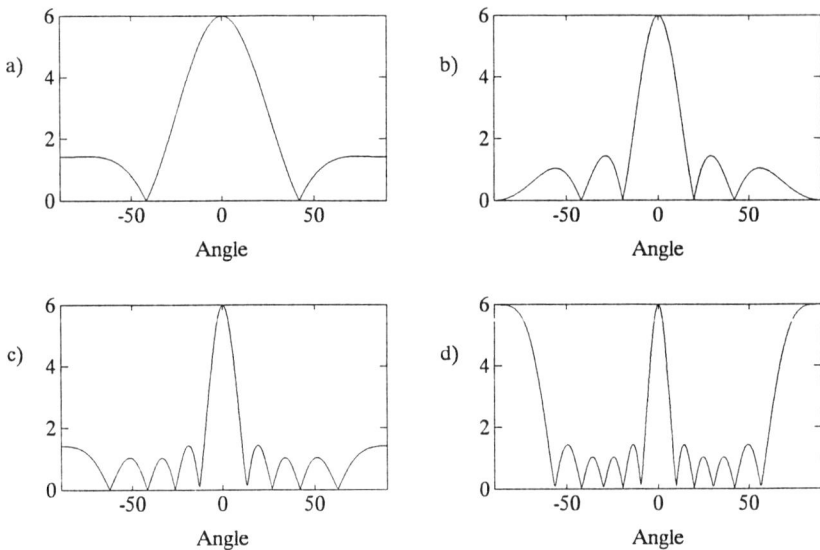

Figure 8.1.5 Array amplitude response for four element spacings: a) $d<\lambda/2$, b) $d=\lambda/2$, c) $\lambda/2<d<\lambda$, and d) $d=\lambda$.

An extreme example occurs when $d=\lambda$ when the side-lobes are of equal magnitude to the main-lobe at $\theta=\pm\pi/2$, that is endfire. In this case the array provides no directional discrimination against signals incident at such angles. These increased side-lobes are referred to as **grating lobes** and represent the effects of spatial aliasing. Just as in the time domain, to avoid spatial aliasing it is necessary to sample in space at least twice per cycle. Thus to avoid aliasing we should choose

$$d\leq\lambda/2 \ . \tag{8.1.23}$$

The array is considered optimal when $d=\lambda/2$ because at that spacing the main-lobe is at its narrowest before the side-lobes start to grow. We note that since

$$\lambda=\frac{c}{f} \ ,$$

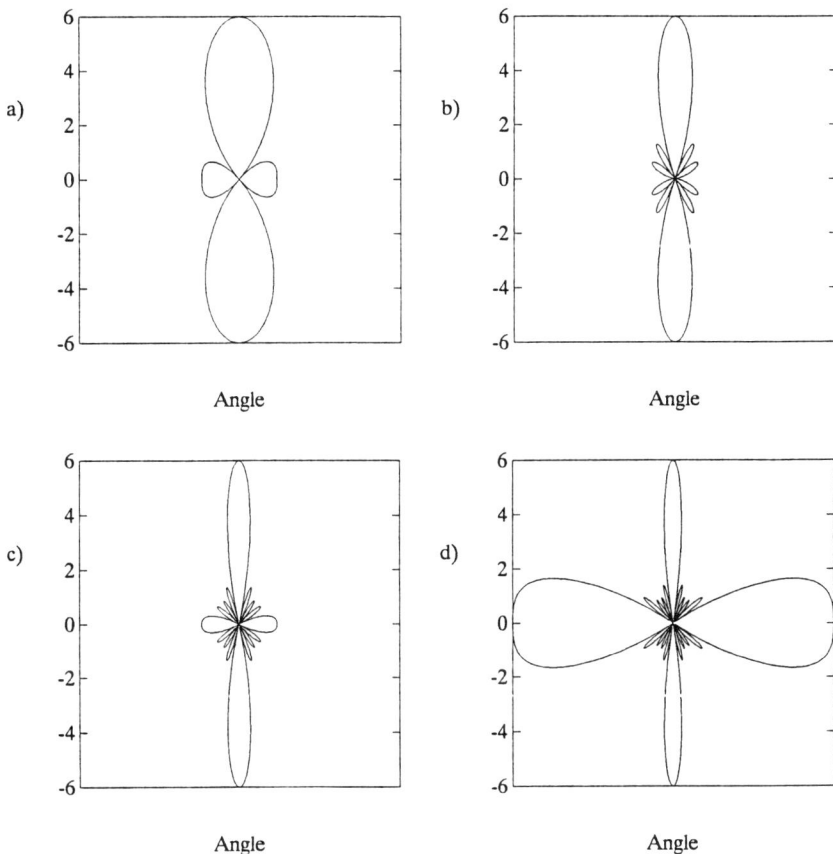

Figure 8.1.6 Polar form plot for array amplitude response for four element spacings: a) $d<\lambda/2$, b) $d=\lambda/2$, c) $\lambda/2<d<\lambda$, and d) $d=\lambda$.

then the corresponding frequency is f_0, given by

$$d=\frac{\lambda}{2}=\frac{c}{2f_0} \ ,$$

or

$$f_0=\frac{c}{2d} \ . \tag{8.1.24}$$

Hence the array is 'optimal' with respect to frequency f_0 which may be considered as an **upper frequency operating limit** for the array. This is because if f_0 increases, λ_0 decreases, the ratio d/λ increases beyond $d=\lambda/2$, and grating lobes occur. When

$d/\lambda<1/2$ no grating lobes are generated but the array is considered inefficient since it delivers less spatial discrimination than it is capable of.

The monochromatic plane wave (8.1.10) has a wavelength of λ in the direction of propagation. Along the axis of the array the wave also exhibits sinusoidal variation with position. In this case the wavelength is given by

$$\frac{\lambda}{|\sin\theta|} \ . \tag{8.1.25}$$

The minimum occurs when $\theta=\pm\pi/2$ (endfire) when the wavelength is equal to λ. The maximum occurs when $\theta=0$, or π (broadside) when all points along the axis are in phase and the wavelength is infinite.

Beam Steering

As we have seen, simple summation produces an array response which has a main-lobe centered at broadside. However, we may redirect, or steer, the array to produce a main-lobe at some other angle θ_0, say. In some early arrays this was achieved by physically rotating the array (see, for example, [9]). A more practical method is to delay the outputs from the individual sensors electrically. The delays used are linearly related between the sensors and thus simulate the effects of physically rotating the array (see Figure 8.1.7). Suppose we have a monochromatic plane wave, incident to the array at angle θ. From Figure 8.1.7, at the i-th element the delayed output is

$$x_i\big(t-\Delta(i-1)\big)=e^{j[\omega(t-\Delta(i-1))-k(i-1)d\sin\theta]} \quad ; \ i=1,2,...,N$$

To steer the array to angle θ_0, the delay at sensor i must cancel the phase at that angle, that is

$$e^{j[\omega(t-\Delta(i-1))-k(i-1)d\sin\theta_0]}=e^{j\omega t} \ .$$

Hence we require

$$\omega\Delta=-kd\sin\theta_0 \ . \tag{8.1.26}$$

Using $k=\dfrac{\omega}{c}$ we have

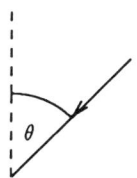

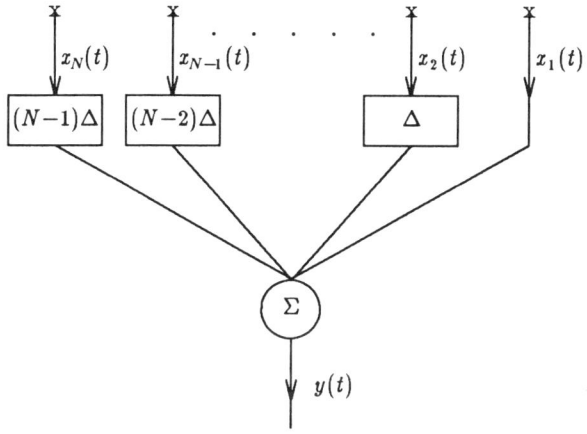

Figure 8.1.7 The use of delays to steer plane wave signals in a linear array.

$$\Delta = -\frac{d}{c}\sin\theta_0 \ ,$$

or

$$\theta_0 = arcsin\left[\frac{-c\Delta}{d}\right] \ . \tag{8.1.27}$$

Thus, the incorporation of delay $\Delta(i-1)$ at the i-th sensor steers the main-lobe to an angle θ_0 given by (8.1.27).

Using different sets of delays to produce multiple outputs, beams may be formed in multiple directions simultaneously. This allows the gain of the array to be brought to bear at as many different angles as desired. This **delay and sum** computation is the classical approach to beamforming. Note that for digital implementations, it is generally more efficient and reduces hardware complexity to replace this time domain processing with equivalent frequency domain processing using the FFT algorithm [1].

8.2 Array Shading

As we have seen, increasing the number of elements in an array increases the spatial resolution that is attainable and also decreases the side-lobe levels. On the other hand, the size and complexity of the array is increased. In practice, therefore, the array size is chosen as a compromise between these conflicting requirements. Given that N is fixed by such practical considerations, we may still modify the array characteristics by weighting or **shading** the measurements. In general terms, the objectives of such weighting are to produce an array pattern function $G(\theta)$ that is as close to a pure impulse in the desired direction as possible. The issues involved in such shading are entirely analogous to those in time domain windowing. In fact, the arrays considered thus far have *de facto* used unit weighting:

$$w_i = 1 \quad ; \ i = 1, 2, ..., N \ .$$

This can be considered equivalent to applying a rectangular window in the spatial domain. More generally, summing the weighted sensor measurements, the response to a monochromatic plane wave incident at angle θ is

$$y(t) = \sum_{n=0}^{N-1} w_{n+1} e^{j(\omega t - kn d \sin\theta)} \ , \tag{8.2.1}$$

or

$$y(t) = \left(\sum_{n=0}^{N-1} w_{n+1} e^{-j\frac{2\pi nd}{\lambda}\sin\theta} \right) e^{j\omega t} \ .$$

As before we let

$$u = \frac{\sin\theta}{\lambda} \ , \tag{8.1.19}$$

to obtain

$$y(t) = G(u)e^{j\omega t} , \qquad (8.2.2)$$

where the array pattern function $G(u)$ is then given by

$$G(u) = \left(\sum_{n=0}^{N-1} w_{n+1} e^{-j2\pi n u d} \right) . \qquad (8.2.3)$$

$G(u)$ is the Fourier transform of the finite sampled sequence $w_1, w_2, w_3, \ldots, w_N$ with respect to the frequency variable u, and with an assumed sample interval d. The modulus of the array pattern function is the modulus of that Fourier transform. Thus, $w_1, w_2, \ldots, w_N$ is a window, and $G(u)$ is the Fourier transform of that window. We have already seen the transform of one example – the rectangular window, as depicted in Figure 8.1.4.

While shading schemes can be used to optimize certain general characteristics of the array they do nothing to solve the problem of interferences arriving at the array from particular directions. In fact, we may steer nulls in particular directions while controlling the array response in the desired direction. For an array with N elements the set of N complex weights gives a total of N specific conditions which can be imposed. For example, it is possible to null $(N-1)$ interferences from distinct directions while simultaneously preserving a desired signal (from another direction). In order to demonstrate this utility we will first give a more general form for the signals and interferences to be processed.

A More General Signal Description

We define a narrowband signal as one for which

$$I(t) = m(t)e^{j\omega t} , \qquad (8.2.4)$$

where $m(t)$ is an amplitude modulation which is considered sufficiently slowly varying that it can be taken as constant across the array. A **steering vector** $\underline{s}$, associated with angle θ is defined as the $(N \times 1)$ vector whose elements are the relative phase delays associated with a plane wave incident to a linear equispaced array from an angle θ. That is

$$\underline{s}(\theta)=[1,e^{-jkd\sin\theta},e^{-j2kd\sin\theta},...,e^{-j(N-1)kd\sin\theta}]^t \ . \qquad (8.2.5)$$

We may also define a vector of measurements $\underline{x}(t)$ as

$$\underline{x}(t)=[x_1(t),x_2(t),...,x_N(t)]^t \ .$$

For a narrowband signal $I(t)$ incident to the array at angle θ we may write

$$\underline{x}(t)=I(t)\underline{s}(\theta)=m(t)\underline{s}(\theta)e^{j\omega t} \ . \qquad (8.2.6)$$

For M separate sources we may write

$$\underline{x}(t)=I_1(t)\underline{s}_1+I_2(t)\underline{s}_2+...+I_M(t)\underline{s}_M \ ,$$

$$=\sum_{i=1}^{M} I_i(t)\underline{s}_i \ , \qquad (8.2.7)$$

where $\underline{s}_i=\underline{s}(\theta_i)$ and the explicit dependence of $\underline{s}$ on θ has been suppressed to simplify the notation. If the sources are centered at sensor 1, we may write[4]

$$\underline{x}(t)=\left[\sum_{i=1}^{M} m_i(t)\underline{s}_i\right]e^{j\omega t} \ . \qquad (8.2.8)$$

Next we define Q, a matrix of dimension $(N\times M)$ whose columns consist of the steering vectors associated with the M sources:

$$Q=[\underline{s}_1,\underline{s}_2,...,\underline{s}_M] \ ,$$

and a vector

$$\underline{m}=[m_1(t),m_2(t),...,m_M(t)]^t \ .$$

From (8.2.8) we may write

$$\underline{x}(t)=Q\underline{m}e^{j\omega t} \ . \qquad (8.2.9)$$

---

4. Note that we may write $\underline{x}(t)$ in this form even if the sources are not phase aligned at one sensor — all that is required is a constant phase shift applied to each steering vector. This refinement is ignored in the interests of simplicity.

Defining a vector of complex weighting coefficients as $\underline{w}=[w_1,w_2,...,w_N]^t$, the array output is then given by

$$y(t)=\underline{w}^t\underline{x}(t)=\underline{w}^t Q\underline{m}e^{j\omega t} . \qquad (8.2.10)$$

We now return to the question of directing nulls in the array response. In terms of the steering vector description, the gain g say, applied to a signal from direction θ is

$$g=\underline{s}^t(\theta)\underline{w} . \qquad (8.2.11)$$

For M sources we have M distinct gains $g_1,g_2,...,g_M$ with

$$g_1=\underline{s}^t_1\underline{w} , \quad g_2=\underline{s}^t_2\underline{w} ,..., g_M=\underline{s}^t_M\underline{w} ,$$

or in combination

$$g=Q^t\underline{w} , \qquad (8.2.12)$$

where $g=[g_1,g_2,...,g_M]^t$. Equations (8.2.12) form the basis for the desired control of the array response. In particular, if g_i is the gain associated with direction θ_i we may form a null in that direction by finding the set of weights $\underline{w}$ for which

$$g_i=\underline{s}^t_i\underline{w}=0 . \qquad (8.2.13)$$

More generally, through equations (8.2.12) we may specify the gain in each of M directions. For example, if we consider the first source from direction θ_1 to be signal, and the remaining $(M-1)$ as interferences then we should set:

$$g_1=\underline{s}^t_1\underline{w}=1 ,$$

$$g_i=\underline{s}^t_i\underline{w}=0 \quad ; i\neq1 ,$$

or

$$Q^t\underline{w}=[1,0,...,0]^t . \qquad (8.2.14)$$

The solution of these equations gives the required weight vector $\underline{w}$. If the number of sources M equals the number of sensors N, then Q is square and the equations can generally be solved for $\underline{w}$. If $M<N$ we may add extra constraints and still solve the set. If the

number of sources is greater than the number of sensors no exact solution is usually possible, though a least-squares solution may be.

Example – Consider a 3 element array with complex weights w_1, w_2, w_3. Assume that a desired signal is incident from angle $\theta=0^0$ and that interferences are incident from $\theta=30^0$, and $\theta=90^0$. We will find the weight values required to suppress the interferences while preserving the desired signal. It is assumed that the signal and interferences are monochromatic plane waves with the inter-element spacing d set so that $d/\lambda=1/2$. We start by setting a gain vector using (8.2.14) as

$$\underline{g}=\begin{bmatrix} \underline{s}_1^t \\ \underline{s}_2^t \\ \underline{s}_3^t \end{bmatrix} \underline{w}=\begin{bmatrix} 1 \\ 0 \\ 0 \end{bmatrix} ,$$

where, using the definition for $\underline{s}(\theta)$ of (8.2.5) we have

$$\begin{bmatrix} 1 & 1 & 1 \\ 1 & e^{-j\pi\sin(\pi/6)} & e^{-j2\pi\sin(\pi/6)} \\ 1 & e^{-j\pi\sin(\pi/2)} & e^{-j2\pi\sin(\pi/2)} \end{bmatrix} \underline{w}=\begin{bmatrix} 1 \\ 0 \\ 0 \end{bmatrix} ,$$

$$\begin{bmatrix} 1 & 1 & 1 \\ 1 & -j & -1 \\ 1 & -1 & 1 \end{bmatrix} \underline{w}=\begin{bmatrix} 1 \\ 0 \\ 0 \end{bmatrix} .$$

Using Cramer's rule [10] we may solve for $\underline{w}$ as

$$\det(Q)=-4 ,$$

and

$$w_1=\frac{\begin{vmatrix} 1 & 1 & 1 \\ 0 & -j & -1 \\ 0 & -1 & 1 \end{vmatrix}}{-4}=1/4+j/4 ,$$

$$w_2=\frac{\begin{vmatrix} 1 & 1 & 0 \\ 1 & 0 & -1 \\ 1 & 0 & 1 \end{vmatrix}}{-4}=1/2 ,$$

$$w_3 = \frac{\begin{vmatrix} 1 & 1 & 1 \\ 1 & -j & 0 \\ 1 & -1 & 0 \end{vmatrix}}{-4} = 1/4 - j/4 \ .$$

The accuracy of this solution is easily verified by substitution of the weight values into (8.2.14). Figure 8.2.1 shows a plot of the array amplitude response for this configuration. As can be seen, the plotted response does indeed achieve the required cancellation in the specified directions.

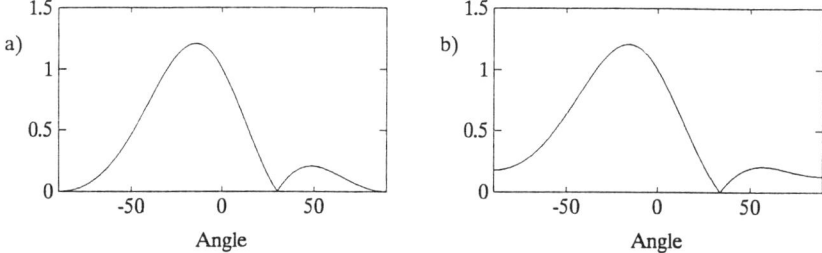

Figure 8.2.1 Array response for an $N=3$ element array with $w_1 = 1/4 + j/4$, $w_2 = 1/2$, $w_3 = 1/4 - j/4$. a) $d/\lambda = 0.5$, b) $d/\lambda = 0.45$.

Solutions of higher order problems are easily obtained numerically from equations (8.2.12). For example, Figure 8.2.2 shows an example with $N=6$. The system was designed to steer nulls at angles $\theta = 20$, 30, 53, −6, and −45^0 while preserving a signal from angle $\theta = 0^0$. Examination of the figure shows that these target objectives have been met.

This solution is not as powerful as it may appear, however, since the control of the array is achieved only for a limited number of signals and interferences, all occurring at the same frequency. Additionally, the design requires prior knowledge of the angle of arrival and the frequency of the source (through d/λ). For the (3×3) example above if $d/\lambda \rightarrow 0.45$, say, the array response calculated using the same weights employed for the response of Figure 8.2.1a) is changed to that of Figure 8.2.1b). (The calculation of the required set of weights is left as an exercise for the reader).

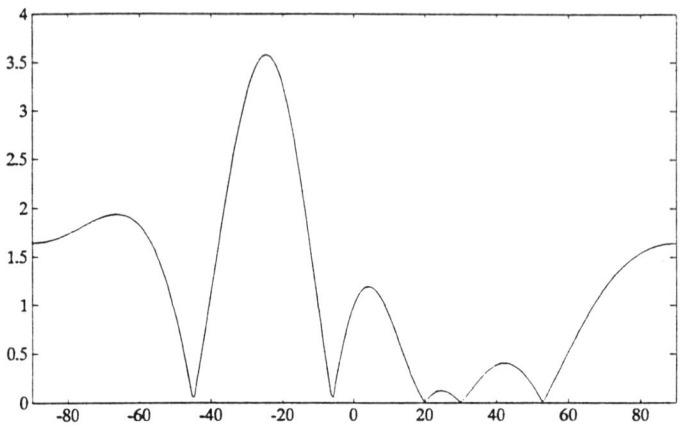

Figure 8.2.2 Array response for an $N=6$ element array. The weights were calculated to produce nulls at 20^0, 30^0, 53^0, -6^0 and -45^0.

This approach requires an unrealistically high level of *a priori* information for most practical problems. In general, it is far better to accept an approximate solution which is *robust* to some model imperfections. The subject of the design of such array processing strategies is discussed in the following sections.

8.3 Optimal Arrays

8.3.1 General

In Section 8.1 we examined a broadside plane wave signal corrupted by spatially uncorrelated noise as input to an array of N sensors. We found that applying equal weights to the received signals, and summing the results gives an output SNR that is improved over the input by a factor of N. In fact, as we shall see, under such conditions this result is optimal in the sense of producing the greatest gain in SNR attainable by any set of weights. However, when the array is subjected to directional interferences, or when the measurement noise is correlated between sensors, then non-uniform weighting can produce superior results. The question is, how should the weights be chosen to optimize array performance for any particular signal and noise conditions?

As usual, there are many ways we can define optimality. For array processing, the four most widely used criteria are [3]:

i) Maximization of output Signal-to-Noise plus Interference Ratio (SNIR)
ii) Minimization of Mean-Squared Error
iii) Maximum Likelihood
iv) Minimum Noise Variance

Here, we focus primarily on the first two. It turns out, however, that for typical input signal conditions, all four criteria produce an optimal set of weights that is identical up to a scale factor. Such a scale factor does not impact the resulting SNR gain.

Whichever criterion is employed, the optimization is performed with respect to complex signals similar to those defined in Section 8.1. In reality, the signals encountered, the processing and the final results are all real. There are two ways to handle this: One possibility is to perform the optimization with respect to the complex signals and simply recognize that the final output should be the real part of the result. That is

$$array\ output = \mathrm{Re}\{y(t)\} = \mathrm{Re}\{\underline{w}^t \underline{x}(t)\} \ . \qquad (8.3.1)$$

Alternatively, it is possible to redefine the complex quantities $\underline{w}$ and $\underline{x}$ in terms of real numbers. For example, if $\underline{w}$ and $\underline{x}$ are vectors with two elements, we may write

$$y(t) = \underline{w}^t \underline{x}(t) \ ,$$

$$= [w_1, \ w_2] \begin{bmatrix} x_1 \\ x_2 \end{bmatrix} \ ,$$

$$= [w_{1R} + jw_{1I}, \ w_{2R} + jw_{2I}] \begin{bmatrix} x_{1R} + jx_{1I} \\ x_{2R} + jx_{2I} \end{bmatrix} \ .$$

The actual output is given by

$$\mathrm{Re}\{\underline{w}^t \underline{x}(t)\} = (w_{1R} x_{1R} - w_{1I} x_{1I}) + (w_{2R} x_{2R} - w_{2I} x_{2I}) \ . \qquad (8.3.2a)$$

This result may also be written as

$$y(t) = \mathrm{Re}\{\underline{w}^t \underline{x}(t)\} = [w_{1R}, \ -w_{1I}, \ w_{2R}, \ -w_{2I}] \begin{bmatrix} x_{1R} \\ x_{1I} \\ x_{2R} \\ x_{2I} \end{bmatrix} \ . \qquad (8.3.2b)$$

If we redefine $\underline{w}$ and $\underline{x}$ as (4×1) vectors with real elements:

$$\underline{w}=[w_{1R},\ -w_{1I},\ w_{2R},\ -w_{2I}]^t\ ,\qquad \underline{x}=[x_{1R},\ x_{1I},\ x_{2R},\ x_{2I}]^t\ ,$$

then the output is redefined in terms of purely real quantities as

$$y(t)=\underline{w}^t\underline{x}(t)\ . \tag{8.2.10}$$

Similarly, for an N element array, we may redefine N element complex vectors in terms of $2N$ element real vectors. In practice, the complex form (8.3.1) is more widely used, although both forms are encountered in the literature and some familiarity with both is desirable. Accordingly we shall examine signal and noise models for arrays in terms of both real and complex formulations.

8.3.2 Signal and Noise Models

a) Real Signal Formulation

As we have seen, the output from an N element array can be written as

$$y(t)=\underline{w}^t\underline{x}(t)\ , \tag{8.2.10}$$

where $\underline{x}(t)$ and $\underline{w}$ are $2N$ element vectors of received signals and weights, respectively, defined by

$$\underline{x}=[x_{1R},\ x_{1I},...,\ x_{NI}]^t\ ,\qquad \underline{w}=[w_{1R},\ -w_{1I},...,-w_{NI}]^t\ .$$

For a random signal $x(t)$, the mean output power is

$$E\{y^2(t)\}=\underline{w}^t E\{\underline{x}(t)\underline{x}^t(t)\}\underline{w}=\underline{w}^t R\underline{w}\ , \tag{8.3.3}$$

where R is the $(2N\times2N)$ correlation matrix defined by

$$R=E\{\underline{x}(t)\underline{x}^t(t)\}=E\{\begin{bmatrix} x_{1R}^2 & x_{1R}x_{1I} & . & . & x_{1R}x_{NI}\\ x_{1R}x_{1I} & x_{1I}^2 & . & . & .\\ . & . & . & . & .\\ . & . & . & . & .\\ x_{1R}x_{NI} & x_{1I}x_{NI} & . & . & x_{NI}^2 \end{bmatrix}\}\ . \tag{8.3.4}$$

Note that in contrast to the definition used in previous chapters, R is a spatial rather than temporal correlation matrix. As always, the expectation operation is defined with respect to the ensemble, rather than with respect to time or space.

The array output power $E\{y^2(t)\}$ can be divided into separate components corresponding to the signal, narrowband directional interferences, and noise:

i) Signal Component

Assuming a narrowband signal consisting of a single source with steering vector s and amplitude modulation $m(t)$ (see equations (8.2.4) and (8.2.5)), the output power due to the signal is

$$E\{y_s^2(t)\}=\pi_s \underline{w}^t \underline{s}\ \underline{s}^t \underline{w} , \qquad (8.3.5)$$

where $\pi_s = E\{m^2(t)\}$ is the signal power, and where from (8.3.2b) and (8.2.5) the real form for $\underline{s}$ is

$$\underline{s}=[1,0,cos(u),-sin(u),cos(2u),...,-sin((N-1)u)]^t$$

where $u=kd\sin(\theta)$. Note that, in this model the amplitude modulation is random but the source direction is fixed.

ii) Interference Component

Directional interferences can be modelled in a similar fashion to the signal. That is, each interference is characterized by a steering vector and an amplitude modulation. M distinct interference sources can be written

$$i(t)=\sum_{k=1}^{M} m_k(t)\underline{s}_k , \qquad (8.3.6)$$

where $m_k(t)$ and $\underline{s}_k$ are amplitude modulations and steering vectors, respectively, for the M interferences $k=1,2,...,M$. Assuming statistically independent amplitude modulations for each source, the correlation of the interferences is given by

$$R_I=\sum_{k=1}^{M} \pi_k \underline{s}_k \underline{s}_k^t , \qquad (8.3.7)$$

where the π_k are the powers associated with the M interference components.

iii) Noise Component

Uncorrelated Noise – Is usually assumed to arise locally at the sensors (measurement noise). As the name suggests, uncorrelated noise has zero covariance between sensors and is characterized by a spatial correlation matrix of the form

$$R_v = \sigma_v^2 I \ , \qquad\qquad (8.3.8)$$

where noise $v(t)$ has power σ_v^2.

Isotropic Noise – This is a common model for 'ambient' noise. The model consists of a large number of statistically independent plane waves, originating on the inside of a sphere of large radius, each with a random orientation [9].

More than one noise form can occur in a measurement. For example, in sonar applications the measurements are often corrupted by both uncorrelated and isotropic noise [9]. Assuming the distinct noise and interference sources are all mutually independent, then the total noise plus interference correlation matrix for such a problem is

$$R_T = R_v + R_I + \sum_{k=1}^{M} \pi_k \underline{s}_k \underline{s}_k^t \ , \qquad\qquad (8.3.9)$$

where R_I is the isotropic noise correlation matrix. The output power due to the noise plus interference components is then

$$E\{y_T^2(t)\} = \underline{w}^t R_T \underline{w} = \underline{w}^t \left(R_v + R_I + \sum_{k=1}^{M} \pi_k \underline{s}_k \underline{s}_k^t \right) \underline{w} \ . \qquad (8.3.10)$$

b) Complex Signal Formulation

In complex form, the array output is given by (8.3.1) where the coefficient vector $\underline{w}$ and the vector of inputs $\underline{x}(t)$ are N component vectors with complex elements. The array output power is

$$E\{|y(t)|^2\} = \underline{w}^H E\{\underline{x}^*(t)\underline{x}^t(t)\}\underline{w} = \underline{w}^H R \underline{w} \ , \qquad (8.3.11)$$

where $*$ denotes complex conjugate, and H denotes conjugate transpose. This expression is similar to that obtained for the real signal formulation but in this case the correlation matrix is the complex form $R = E\{\underline{x}^*(t)\underline{x}^t(t)\}$ (see Section 2.1). R has elements

$$R(i,j) = E\{x_i^*(t)x_j(t)\} . \tag{8.3.12}$$

The signal component of the output power is

$$E\{|y_s(t)|^2\} = \pi_s \underline{w}^H \underline{s}^* \underline{s}^t \underline{w} , \tag{8.3.13}$$

where $\quad \underline{s} = [1, e^{-jkd\sin(\theta)}, e^{-j2kd\sin(\theta)}, ..., e^{-j(N-1)kd\sin(\theta)}]^t$.

This expression is analogous to that for the real formulation, but with the complex form for the weights and steering vector. Similarly, we may write the noise plus interference correlation matrix as

$$R_T = (R_v + R_I + \sum_{k=1}^{M} \pi_k \underline{s}_k^* \underline{s}_k^t) , \tag{8.3.14}$$

where R_v and R_I are the uncorrelated and isotropic noise correlation matrices, respectively. Finally, the noise output power is

$$E\{|y_T(t)|^2\} = \underline{w}^H R_T \underline{w} . \tag{8.3.15}$$

8.3.3 Optimal Array Design Methods

a) Maximize the Signal-to-Noise plus Interference Ratio (SNIR)

The output Signal-to-Noise plus Interference Ratio (SNIR) for the array is defined by

$$SNIR = \frac{signal\ output\ power}{noise\ plus\ interference\ output\ power} . \tag{8.3.16}$$

For signal models of the form described in Section 8.3.2 above, with an array with complex weights $\underline{w}$, using equations (8.3.13) and (8.3.15) we have

$$SNIR = \frac{\pi_s \underline{w}^H \underline{s}^* \underline{s}^t \underline{w}}{\underline{w}^H R_T \underline{w}} . \tag{8.3.17}$$

To maximize *SNIR* we may employ Lagrange multipliers. The optimization proceeds by fixing (constraining) the denominator and maximizing the numerator. Defining a complex Lagrange multiplier λ, we may construct a functional J embodying the desired constraint as

$$J = \pi_s \underline{w}^H \underline{s}^* \underline{s}^t \underline{w} - \lambda [\underline{w}^H R_T \underline{w} - k] , \tag{8.3.18}$$

where k is a complex constant. For an $(N \times N)$ Hermitian matrix C, the derivative of the quadratic form $\underline{x}^H C \underline{x}$, for any $(N \times 1)$ vector $\underline{x}$ is [10]

$$\frac{\partial}{\partial \underline{x}} [\underline{x}^H C \underline{x}] = 2 C \underline{x} . \tag{8.3.19}$$

Using this result we may differentiate (8.3.18) with respect to $\underline{w}$, and equate to zero giving

$$\frac{\partial J}{\partial \underline{w}} = 2(\pi_s \underline{s}^* \underline{s}^t \underline{w} - \lambda R_T \underline{w}) = \underline{0} ,$$

or

$$\pi_s R_T^{-1} (\underline{s}^* \underline{s}^t) \underline{w} - \lambda \underline{w} = \underline{0} ,$$

which may be written

$$(\pi_s A - \lambda) \underline{w} = \underline{0} , \tag{8.3.20}$$

where $A = R_T^{-1} (\underline{s}^* \underline{s}^t)$. The purpose of arranging the optimality condition in this fashion is to demonstrate that the maximization is now in the form of an eigenvalue problem; the eigenvalues correspond to values of SNIR, and the eigenvectors $\underline{w}$ to the corresponding array weights.[5] There are, in general, $\overline{N}$ distinct eigenvalues of the matrix A, and the largest corresponds to the maximum SNIR, $SNIR_{\max}$, say. The eigenvector corresponding to this eigenvalue is the required optimum solution $\underline{w}_{sn}$. From

---

5. To clarify this result, premultiply (8.3.20) by $\underline{w}^H R_T$. This illustrates the equivalence of (8.3.17) and (8.3.20).

equation (8.3.20) we may write

$$\pi_s A \underline{w}_{sn} = \lambda_{\max} \underline{w}_{sn} \; , \tag{8.3.21}$$

but

$$\lambda_{\max} = SNIR_{\max} = \frac{\pi_s \underline{w}_{sn}^H \underline{s}^* \underline{s}^t \underline{w}_{sn}}{\underline{w}_{sn}^H R_T \underline{w}_{sn}} \; . \tag{8.3.22}$$

Substituting into (8.3.20) gives

$$\pi_s A \underline{w}_{sn} = \left(\frac{\pi_s \underline{w}_{sn}^H \underline{s}^* \underline{s}^t \underline{w}_{sn}}{\underline{w}_{sn}^H R_T \underline{w}_{sn}} \right) \underline{w}_{sn} \; .$$

Premultiplying by R_T we have

$$\pi_s \underline{s}^* (\underline{s}^t \underline{w}_{sn}) = \left(\frac{\pi_s \underline{w}_{sn}^H \underline{s}^* (\underline{s}^t \underline{w}_{sn})}{\underline{w}_{sn}^H R_T \underline{w}_{sn}} \right) R_T \underline{w}_{sn} \; .$$

Dividing out the scalar factors on both sides of this equation gives

$$\underline{s}^* = \left(\frac{\underline{w}_{sn}^H \underline{s}^*}{\underline{w}_{sn}^H R_T \underline{w}_{sn}} \right) R_T \underline{w}_{sn} \; , \tag{8.3.23}$$

or

$$\underline{s}^* = c R_T \underline{w}_{sn} \; ,$$

which gives

$$\underline{w}_{sn} = (1/c) R_T^{-1} \underline{s}^* \; , \tag{8.3.24}$$

where c is the scalar value:

$$c = \frac{\underline{w}_{sn}^H \underline{s}^*}{\underline{w}_{sn}^H R_T \underline{w}_{sn}} \; . \tag{8.3.25}$$

The form of the solution may seem unsatisfactory with $\underline{w}_{sn}$ occurring on both sides of (8.3.24). We note, however, that c is a scalar and hence has no impact on the SNR of the result.

Example – As a simple example, consider the case of a unit amplitude narrowband signal incident at broadside to an array of N elements. Assume further, that the measurements at the array sensors are corrupted by uncorrelated noise with variance σ_v^2. We have

$$R_T = \sigma_v^2 I \ , \tag{8.3.8}$$

and
$$\underline{s} = [1,1,...,1]^t \ ,$$

so that

$$\underline{w}_{sn} = (1/c)R_{\ T}^{-1}\underline{s}^* = \frac{1}{c\sigma_v^2}[1,1,...,1]^t \ . \tag{8.3.26}$$

We see that $\underline{w}_{sn}$ consists of equal weightings applied to each sensor output. This is both intuitively appealing, and confirms our earlier assertion that this is, indeed, the optimal weighting in this case. Note that if the variance of the noise was different at each sensor, the optimum solution would consist of weighting each element in inverse proportion to the noise variance for that element.

b) Least-Squares Minimization

An alternative optimization strategy is to choose the array weights so as to minimize a mean-squared error criterion. In this approach, as with the analogous time domain optimization, a desired signal $d(t)$, (often referred to as a **pilot signal** in array applications) is required at the array output (see Figure 8.3.1). The provision of such a desired signal is, of course, the key to this approach and we will discuss this further below. The minimization proceeds as usual with

$$e(t) = d(t) - y(t) = d(t) - \underline{w}^t\underline{x}(t) \ . \tag{8.3.27}$$

For the complex signal formulation we have

$$J = E\{|e^2(t)|^2\} = E\{e^*(t)e(t)\} \ ,$$
$$= E\{|d(t)|^2\} + \underline{w}^H E\{\underline{x}^*(t)\underline{x}^t(t)\}\underline{w} - E\{d^*(t)\underline{w}^t\underline{x}(t)\} -$$
$$- E\{d(t)\underline{w}^H\underline{x}^*(t)\} \ ,$$

$$J = E\{|d(t)|^2\} + \underline{w}^H R\underline{w} - E\{2\mathrm{Re}\{\underline{w}^H d(t)\underline{x}^*(t)\}\} \ . \tag{8.3.28}$$

The solution is now obtained from

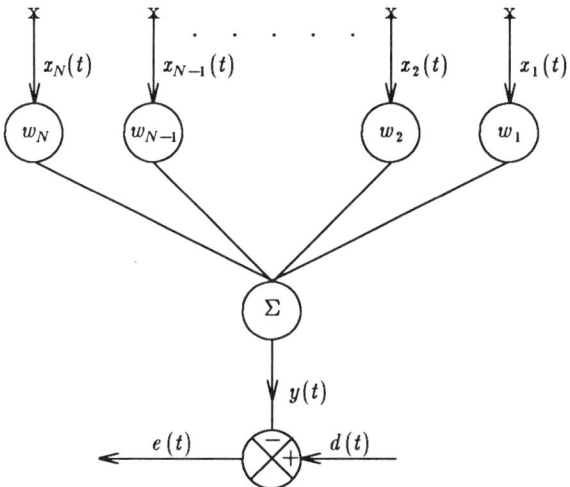

Figure 8.3.1 Least Mean Squares weighting of an array output.

$$\frac{\partial J}{\partial \underline{w}} = \underline{0} \ . \tag{8.3.29}$$

Let us examine the derivative of (8.3.28) term by term. The first term has no dependence on $\underline{w}$ and is thus zero. The second term is a quadratic form. Using (8.3.19), we have

$$\frac{\partial}{\partial \underline{w}}\{\underline{w}^H R \underline{w}\} = 2 R \underline{w} \ .$$

The derivative of the last term of (8.3.28) is

$$\frac{\partial}{\partial \underline{w}}\left[E\{2\mathrm{Re}\{\underline{w}^H d(t)\underline{x}^*(t)\}\} \right] \ .$$

Now,

$$\underline{w}^H \underline{x}^*(t) = [\underline{w}_R - j\underline{w}_I]^t [\underline{x}_R - j\underline{x}_I] = [\underline{w}_R^t \underline{x}_R - \underline{w}_I^t \underline{x}_I - j\underline{w}_I^t \underline{x}_R - j\underline{w}_R^t \underline{x}_I] \ ,$$

where subscripts R and I denote real and imaginary components, respectively, and where in those quantities dependence on t is suppressed to simplify notation. From this equation

$$2\text{Re}\{\underline{w}^H d(t)\underline{x}^*(t)\}=2([\underline{w}_R^t\underline{x}_R-\underline{w}_I^t\underline{x}_I]d_R+[\underline{w}_I^t\underline{x}_R+\underline{w}_R^t\underline{x}_I]d_I) \ . \quad (8.3.30)$$

Now differentiating this expression with respect to real and imaginary components gives

$$\frac{\partial[2\text{Re}\{\underline{w}^H d(t)\underline{x}^*(t)\}]}{\partial\underline{w}_R}=2(\underline{x}_R d_R+\underline{x}_I d_I) \ , \quad (8.3.31\text{a})$$

$$\frac{\partial[2\text{Re}\{\underline{w}^H d(t)\underline{x}^*(t)\}]}{\partial\underline{w}_I}=2(-\underline{x}_I d_R+\underline{x}_R d_I) \ . \quad (8.3.31\text{b})$$

The complete derivative is

$$\frac{\partial}{\partial\underline{w}}=\frac{\partial}{\partial\underline{w}_R}+j\frac{\partial}{\partial\underline{w}_I}=2(\underline{x}_R d_R+\underline{x}_I d_I)-j2(\underline{x}_I d_R-\underline{x}_R d_I) \ ,$$

$$=2\underline{x}^*(t)[d_R+jd_I]=2d(t)\underline{x}^*(t) \ .$$

Finally

$$E\{d(t)\underline{x}^*(t)\}=\underline{g} \ , \quad (8.3.32)$$

and hence combining the derivatives for each of the terms of (8.3.28), we have

$$\frac{\partial J}{\partial\underline{w}}=2R\underline{w}-2\underline{g}=\underline{0} \ , \quad (8.3.33)$$

where $R=E\{\underline{x}^*(t)\underline{x}^t(t)\}$ and $\underline{g}=E\{d(t)\underline{x}^*(t)\}$. Hence

$$\underline{w}_{ls}=R^{-1}\underline{g} \ . \quad (8.3.34)$$

In particular, consider the problem of a where $s(t)$ is a narrowband signal with center frequency ω_0

$$s(t)=m(t)e^{j\omega_0 t} \ , \quad (8.3.35)$$

$$x(t)=s(t)+v(t) \ ,$$

where $v(t)$ is noise, and where $s(t)$ is assumed incident from an angle θ. The vector of received signals is then

$$\underline{x}(t)=d(t)\underline{s}+\underline{v}(t)\ ,\tag{8.3.36}$$

where $\quad s=\left[1,e^{-jkd\sin\theta},\ldots,e^{-jkd(N-1)\sin\theta}\right]^{t}.\quad$ The desired signal $d(t)=s(t)$. From (8.3.35) and (8.3.36),

$$\underline{g}=E\{d(t)\underline{x}^{*}(t)\}=\pi\underline{s}^{*}$$

where π is the power of the signal $d(t)$. Hence, for a single narrowband signal

$$\underline{w}_{ls}=\pi R^{-1}\underline{s}^{*}\ .\tag{8.3.37}$$

Example – As an illustration of the least-squares solution for the array weights, consider the following scenario: An array of four elements is subject to three narrowband plane wave interferences both of unit power incident at angles 50, -35, and -20^{0}, with center frequency such that the array spacing satisfies $d/\lambda=0.5$. The array is also subject to a source of stationary, spatially uncorrelated noise at each sensor, with variance σ^{2}. The desired signal is assumed to be a unit power, monochromatic plane wave incident from broadside, at the same frequency as the interferences. Figure 8.3.2 shows the array pattern for the optimal (complex) weights w_{1},w_{2},w_{3},w_{4} as obtained from equation (8.3.37). The plot shows values of σ^{2} equal to 0, 0.1, 1 and 10. For $\sigma^{2}=0$, the solution is exact because the four elements can completely null out the three monochromatic interferences. The plots for $\sigma^{2}=0.1$, 1.0 and $\sigma^{2}=10$ show the transition from this exact solution to the opposite extreme; when σ becomes large the optimal weights tend to equality, as is illustrated by the previous example.

Equivalence of the Optimal Solutions

Consider the problem of a single narrowband signal $s(t)$ incident to an array at an angle θ. We have seen that the optimal solution corresponding to maximizing the SNIR is

$$\underline{w}_{sn}=(1/c)R_{T}^{-1}\underline{s}^{*}\ ,\tag{8.3.24}$$

where

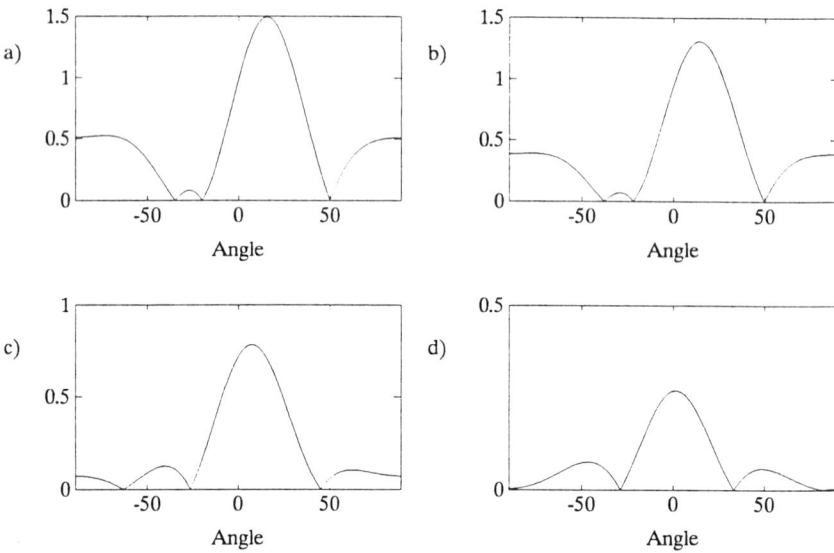

Figure 8.3.2 Array response for an $N=4$ element array with weights calculated according to equation (8.3.37). Inputs consisted of a plane wave signal incident from broadside and interferences from angles 50, -35, and -20^0. Additive uncorrelated noise with variance σ^2 is also present. a) $\sigma^2=0$, b) $\sigma^2=0.1$, c) $\sigma^2=1$, d) $\sigma^2=10$.

$$c=\frac{\underline{w}_{sn}^H \underline{s}^*}{\underline{w}_{sn}^H R_T \underline{w}_{sn}} \; , \tag{8.3.25}$$

while that for the least-squares solution is

$$\underline{w}_{ls}=\pi R^{-1}\underline{s}^* \; , \tag{8.3.37}$$

where

$$R=E\{\underline{x}^*(t)\underline{x}^t(t)\}=\pi \underline{s}^*\underline{s}^t+R_T \; . \tag{8.3.38}$$

We may apply the well-known **matrix inversion lemma** [10]. If A is a non-singular Hermitian matrix, and $\underline{u}$ is a complex vector, then

$$(A+\underline{u}^*\underline{u}^t)^{-1}=A^{-1}-\frac{A^{-1}\underline{u}^*\underline{u}^t A^{-1}}{1+\underline{u}^t A^{-1}\underline{u}^*} \; . \tag{8.3.39}$$

Applying this result to (8.3.37), (8.3.38) gives

$$\underline{w}_{ls} = [\pi \underline{s}^* \underline{s}^t + R_T]^{-1} \pi \underline{s}^* ,$$

$$= \left[\pi R_T^{-1} - \frac{\pi^2 R_T^{-1} \underline{s}^* \underline{s}^t R_T^{-1}}{1 + \pi \underline{s}^t R_T^{-1} \underline{s}^*} \right] \underline{s}^* ,$$

$$= \left[\frac{\pi}{1 + \pi \underline{s}^t R_T^{-1} \underline{s}^*} \right] R_T^{-1} \underline{s}^* , \qquad (8.3.40)$$

but this can be written

$$\underline{w}_{ls} = c_1 R_T^{-1} \underline{s}^* , \qquad (8.3.41)$$

where c_1 is a scalar gain

$$c_1 = \left[\frac{\pi}{1 + \pi \underline{s}^t R_T^{-1} \underline{s}^*} \right] . \qquad (8.3.42)$$

Comparing (8.3.24) and (8.3.41) shows that the optimal forms corresponding to the minimum mean-squared error approach and the maximum SNIR approach are identical to within a scale factor. In fact it can be shown [3] that for similar signal conditions, the optimal solutions obtained from the method of Maximum Likelihood (ML) (assuming Gaussian noise measurements) and from Minimum Variance estimation are also identical to within a scale factor. We conclude that the familiar least-squares solution is therefore a powerful and general optimization strategy for this problem.

Provision of a pilot signal

As we have seen, the least-squares solution requires the provision of a desired or pilot signal. In the ideal case, the desired response is equal to the signal component of the input. In practice, this ideal signal will not be available. However, the system will be useful if the pilot has similar directional and spectral properties to the signal. The pilot should simulate a received signal in the desired look direction and have the correct, or approximately correct carrier frequency. To force the system to have the required properties at several frequencies, the pilot can be designed as a sum of narrowband signals. Similarly, if it is desired to have good response in several directions the pilot can be chosen appropriately. In many cases, these properties are known *a priori* and the solution (8.3.37) can be generated without an explicit pilot signal. In other cases, and for the adaptive systems described in the next Section, the pilot could be physically generated at the source (if the transmitter is cooperative) or it could be generated by a local source, or internally within the system.

8.3.4 Broadband and Narrowband Arrays

So far, we have considered only arrays with output generated by applying a single complex weight to each element. More generally, the form of the weighting depends on whether the array is narrowband or **broadband**. As we have seen, a narrowband array is one in which the array is designed to operate with signals defined in a narrow range around some center frequency. In this case each sensor is weighted using a single complex factor

$$w_i = w_{iR} + jw_{iI} \ .$$

w_{iR} and w_{iI} can also be thought of as in-phase and in quadrature components, the imaginary component being obtained by a delay equivalent to a 90^0 phase shift at the center frequency. As we have seen, this narrowband form is useful when we are designing arrays to deal with monochromatic plane waves as both signal and interference. We have seen how such arrays may be designed to exactly satisfy a finite number of gains in particular directions at a particular frequency. We have also seen, however, that the pattern of the array changes in response to changes in frequency. On the other hand, broadband noise signals cannot be effectively processed using such narrowand arrays. In fact, a signal which occupies a bandwidth, $B = \omega_u - \omega_l$, say, and arrives from a direction θ, generates inter-element phase differences of

$$\Phi_u = \frac{\omega_u d}{c} \sin\theta \ , \tag{8.3.43}$$

$$\Phi_l = \frac{\omega_l d}{c} \sin\theta \ , \tag{8.3.44}$$

corresponding to the upper and lower frequencies in the signal. Actually, a signal defined with such frequency constraints has the capacity to generate every possible phase shift between Φ_l and Φ_u, corresponding to the continuum of frequencies between ω_l and ω_u. A signal with finite (non-zero) bandwidth can be thought of within the array as equivalent to an infinite number of signals arriving at different angles. The single set of N complex elements of the narrowband array can only produce the required pattern corresponding to one of these. A much greater degree of control is provided by the broadband array. In a broadband array, the single complex weighting of the narrowband array is replaced by finite impulse response filter (of length L, say), implemented as a tapped delay line, operating on the output of each sensor (see Section 8.4, Figure 8.4.4). In this case, the array is a filter in space and time – the N distinct sensors give the spatial dimension, and the tapped delay line produces a temporal response. The use of delay in the weighting allows control of the array response for signals of various frequencies (or equivalently, of signals derived from sources with finite angular extent).

In practice, the outputs from each sensor are actually sampled at equal intervals $T=1/f_s$, where f_s is the sampling frequency. Employing the usual notation we suppress the explicit dependence on T and denote the sampled sensor outputs as $x_i(n)$ for $i=1,2,...,N$. Accordingly, the control of the array response is not complete because the sampling produces only a finite set of delays and thus the number of degrees of freedom remain finite. If the sampling interval is sufficiently small, however, and the number of coefficients is large, then the array can approximately control the response over the input signal bandwidth.

One possible way to implement a broadband system is to block transform the received data at each sensor, and then process each frequency separately as a narrowband system. We may think of collecting several or many such systems together to formulate a broadband system.

8.4 Adaptive Systems

8.4.1 Minimum Mean-Squared Error Adaptation for Arrays

As with time domain processing problems, we can develop adaptive versions of optimal arrays. An adaptive processor attempts to produce an array response that is optimal, or approximately optimal in an unknown or statistically changing environment. In fact, there are usually multiple processors which may be developed from any particular optimality criterion. The simplest approach, as with time domain processing, is to use the Minimum Mean-Squared Error (MMSE) criterion, and develop a gradient based LMS algorithm [11]. Given inputs $d(n)$ and $x(n)$, we define an error

$$e(n) = d(n) - y(n) = d(n) - \underline{w}_n^t \underline{x}_n , \qquad (8.4.1)$$

where now the the subscript n indicates the time-dependence of the array weights due to the adaptive nature of the processor. The least-squares formulation attempts to minimize

$$J = E\{e^2(n)\} , \qquad (8.4.2.a)$$

for the real signal formulation, or equivalently

$$J = E\{|e(n)|^2\} , \qquad (8.4.2b)$$

for the complex signal formulation. The LMS update is a steepest descent iteration obtained by replacing the expected squared error by the instantaneous squared error. For the real signal formulation,[6] we have (see Section 4.1)

$$\underline{w}_{n+1} = \underline{w}_n + \frac{\alpha}{2}\nabla\hat{J} , \qquad (8.4.3)$$

where

$$\hat{J} = e^2(n) , \qquad (8.4.4)$$

---

6. For ease of comparison with adaptive filter results of previous Chapters we employ the real signal formulation in this Section.

the adaptation step-size is α, and the 2 appears for convenience. The resulting update is given by

$$\underline{w}_{n+1} = \underline{w}_n + \alpha e(n) \underline{x}_n \ . \tag{8.4.5}$$

This is, of course, equivalent to the time domain LMS update of equation (4.1.14), except that the time domain filter response $\underline{f}$ is replaced by the vector of weights $\underline{w}$. Because of the similar forms for the update equations, the properties of this algorithm follows closely those of the time domain filter. In particular, for inputs $x(n)$ and $d(n)$ both zero-mean stationary, and applying the usual independence assumption (see Section 4.2.2) we may conclude that

$$\lim_{n \to \infty} \{E\{\underline{w}_n\}\} \to \underline{w}_{ls} \ , \tag{8.4.6}$$

where $\underline{w}_{ls} = R^{-1}g$ is the least-squares optimal solution of equation (8.3.34) (for the real signal formulation), provided

$$0 < \alpha < 2/\lambda_{\max} \ , \tag{8.4.7}$$

where $\lambda_{\max}$ is the largest eigenvalue of the spatial correlation matrix of $x(n)$. Similarly, the convergence of the various modes of the adaptive filter is governed by the disparity between the largest and smallest eigenvalues of the correlation matrix R. All of which mirrors the results of Section 4.2 for the time domain LMS algorithm.

One and Two Mode Adaptive Beamformers

The implementation of an adaptive beamformer using the LMS principle is not as straightforward as for the corresponding time domain implementation. This is because the desired signal $d(n)$ contains no explicit spatial information. Unlike the situation with the time domain processor, to reflect the signal direction(s) the desired signal must also be present at the signal input. For example, suppose the desired signal is a monochromatic plane wave incident to the array at some angle θ. The array can be forced to produce a desired response by presenting the pilot at the input with delays δ_i for $i = 1, 2, ..., N$, equivalent to those associated with a signal from the desired 'look-direction'. The pilot is also supplied as desired output from the adaptive processor. The idea is to force the array to pass a signal from the specified direction while minimizing the effects of other inputs. This is illustrated in Figure 8.4.1.

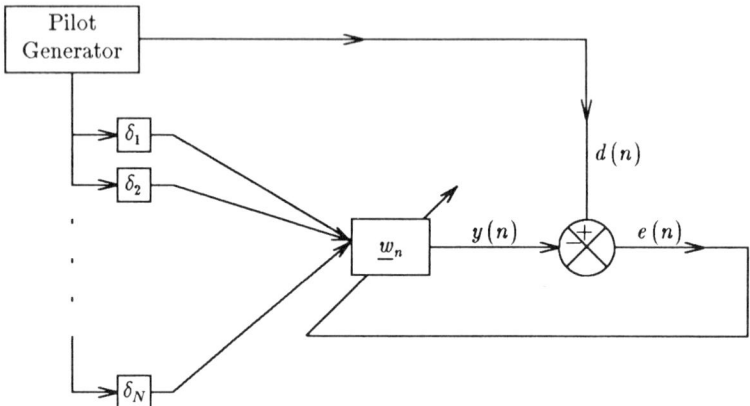

Figure 8.4.1 Least Mean Squares adaptive beamformer.

It is clear from the diagram that when the array adapts to produce a unit gain in the desired direction, the error will be zero. A practical system must also operate on the received signals, of course. Two methods were proposed by Widrow *et al.* [12] dubbed **one and two mode adaptive schemes**, respectively. In the two mode scheme (depicted in Figure 8.4.2) the filter alternates between employing the desired (pilot) signal (mode 1), and adapting the weights to the actual input data with the desired signal equal to zero (mode 2). Thus, in the two mode processor, for mode 1 we have a desired signal $d(n)$ and an input $x(n)=d(n)\underline{s}$, where $d(n)$ and $\underline{s}$ are the desired signal and steering vector for the desired signal, respectively. From equation (8.4.5)

$$\underline{w}_{n+1}=\underline{w}_n+\alpha[d(n)-d(n)\underline{s}^t\underline{w}_n]d(n)\underline{s} . \tag{8.4.8}$$

After some pre-determined number of iterations n_0, say, the algorithm switches to mode 2, with $d(n)=0$, giving update

$$\underline{w}_{n+1}=\underline{w}_n-\alpha[\underline{x}_n\underline{x}_n^t\underline{w}_n] . \tag{8.4.9}$$

After a further n_1 iterations, say, the algorithm reverts to mode 1. The transition from mode 1 to mode 2 involves a set of switches as

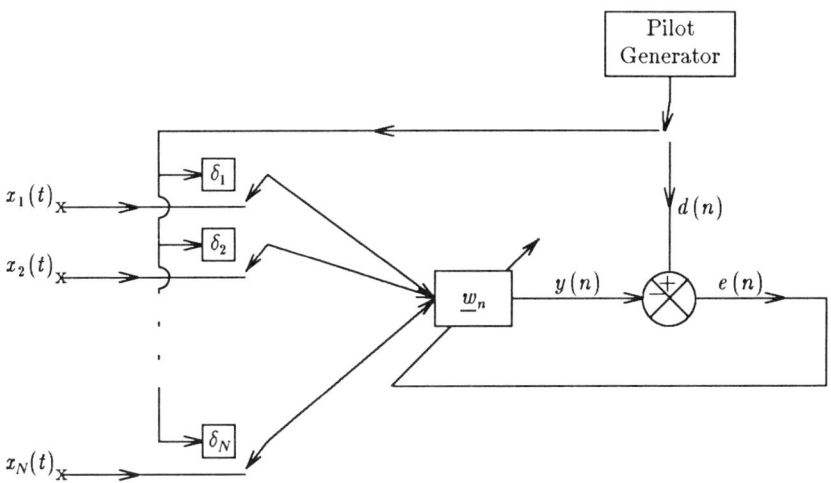

Figure 8.4.2 Two-mode Least Mean Squares adaptive beamformer.

indicated in the diagram. In mode 1 the algorithm attempts to provide a beam in the given look direction(s) as specified by the pilot signal generator. In mode 2, the algorithm attempts to minimize the total output power (the desired signal is set to zero). By changing from mode 1 to mode 2 at a sufficiently high rate, and thus allowing only small changes in the weight vector in any one mode, the algorithm attempts to maintain the beam in the desired direction *and* minimize the output noise power. In an approximate sense this process minimizes the total power of all signals received subject to a 'soft' constraint on the gain and phase of the beam in the look direction(s) of the desired signal.

There are two principal objections to this two mode scheme. Firstly, adaptation is reduced because of the mode switching. Secondly, signals arriving during mode 1 are lost. An alternative, single mode scheme is depicted in Figure 8.4.3. This method eliminates both objections by employing a master processor to perform the adaptation, and a slave processor to process the input data. The master processor utilizes both the pilot and input signals and has an update

$$\underline{w}_{n+1} = \underline{w}_n + \alpha[d(n) - \underline{x}_n^{'t}\underline{w}_n]\underline{x}_n^{'} , \qquad (8.4.10)$$

where $\underline{x}_n^{'} = d(n)\underline{s} + \underline{x}_n$. As its name suggests the slave processor

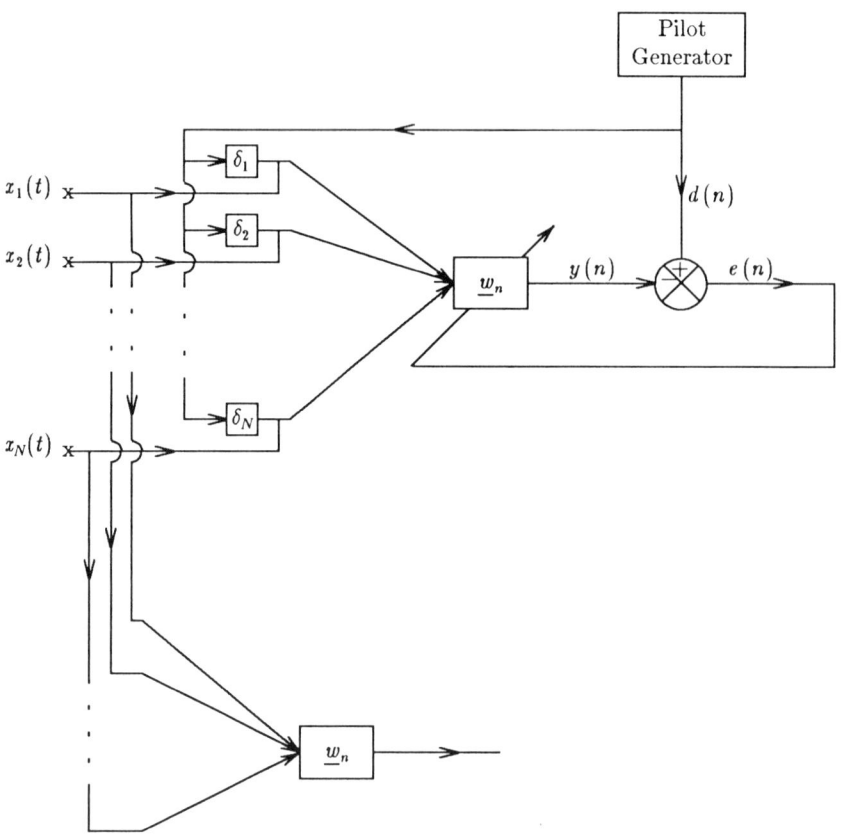

Figure 8.4.3 One-mode adaptive Least Mean Squares beamformer. Master processor (upper component) is adapted using inputs and pilot. Slave processor (lower component) applies weights from master processor to inputs only.

performs no independent adaptation, it merely filters the input data using the weights slaved from the master processor.

In spite of these apparent advantages, we can easily show that the one mode algorithm converges to a solution which is biased from the least-squares optimum of equation (8.3.34). Rearranging equation (8.4.10), we may write

$$\underline{w}_{n+1} = (I - \alpha \underline{x}'_n \underline{x}'^t_n)\underline{w}_n + \alpha d(n)\underline{x}'_n \ . \tag{8.4.11}$$

Now, taking the expectation, and using the independence assumption gives

$$E\{\underline{w}_{n+1}\}=(I-\alpha R^{'})E\{\underline{w}_n\}+\alpha \underline{g}^{'} , \qquad (8.4.12)$$

where $R^{'}=E\{\underline{x}_n\underline{x}_n^{'t}\}$ and $\underline{g}^{'}=E\{d(n)\underline{x}_n^{'}\}$. Equation (8.4.12) is equivalent to equation (4.2.2) of Section 4.2.2, and following similar analysis we may conclude that

$$\lim_{n\to\infty} \{E\{\underline{w}_n\}\}=R^{'-1}\underline{g}^{'}=\underline{w}^{*} , \qquad (8.4.13)$$

subject to

$$0<\alpha<2/\lambda_{\text{max}} , \qquad (8.4.14)$$

where λ_{max} is the largest eigenvalue of the matrix $R^{'}$. Now

$$R^{'}=R+\pi_d\underline{s}\ \underline{s}^{t} , \qquad (8.4.15)$$

where $\pi_d=E\{d^2(n)\}$, and

$$\underline{g}^{'}=\underline{g}+E\{d(n)\underline{s}\} . \qquad (8.4.16)$$

Hence

$$\underline{w}^{*}=[R+\pi_d\underline{s}\ \underline{s}^{t}]^{-1}(\underline{g}+E\{d(n)\underline{s}\}) , \qquad (8.4.17)$$

which is not, in general, equal to the real form of $\underline{w}_{ls}$ given by equation (8.3.34), and hence $\underline{w}^{*}$ is biased.[7]

Griffiths Adaptive Beamformer

Griffiths [13] developed an algorithm which eliminates the bias of the one and two mode adaptive schemes, and also removes the need to provide a pilot signal for the array input. The method is based on a variation of the LMS algorithm in which the desired signal is not explicitly provided at the filter output but is replaced by *a priori* knowledge of the cross-correlation between input and desired output. That is, the LMS update has the form:

---

7. For the two mode system the result is more complicated, but the final result is again that the steady-state mean filter coefficient solution is biased from $\underline{w}_{ls}$ (see reference [13] for details).

$$\underline{w}_{n+1} = \underline{w}_n + \alpha e(n)\underline{x}_n \ ,$$

$$= \underline{w}_n + \alpha d(n)\underline{x}_n - \alpha y(n)\underline{x}_n \ . \qquad (8.4.5)$$

Now, if $d(n)\underline{x}_n$ is replaced by its expected value g, or by an estimate of that quantity, we have the algorithm proposed by Griffiths:

$$\underline{w}_{n+1} = \underline{w}_n + \alpha g - \alpha y(n)\underline{x}_n \ , \qquad (8.4.18)$$

which does not require the provision of an explicit desired signal. The processor does require prior knowledge of the correlation vector, but since the signal and noise are assumed uncorrelated, g is just equal to the cross-correlation between the desired signal and the signal components of the input. Thus g is simply a scaled steering vector (or steering vectors). Moreover, knowledge of the direction(s) of the desired signal is sufficient to determine g up to a scale factor. It is easy to show (see Problem 8.14) that given g, and subject to the usual independence assumptions, this algorithm converges exponentially in the mean to $\underline{w}_{ls}$, provided

$$0 < \alpha < 2/\lambda_{\max} \ , \qquad (8.4.19)$$

where $\lambda_{\max}$ is the largest eigenvalue of R.[8]

8.4.2 Constrained Power Minimization

An alternative, and potentially more powerful approach to beamforming is to construct a **constrained power minimization**. The approach is to design the weight vector so that the total output power is minimized, but subject to a constraint that the gain in some specified 'look-direction' is fixed. In terms of the more general broadband beamformer, we can regard the beamformer as operating on the input in two dimensions – in time and in space. The constrained power minimizer fixes the spatial response of the filter while retaining freedom over the temporal response. Consider the system shown in Figure 8.4.4. If a plane wave is incident at broadside to the array, then the received signal at corresponding

---

8. Note that (8.4.18) is also known as the discrete form of the **Applebaum array** [5].

taps in each filter (that is, along vertical lines in Figure 8.4.4) will be in phase and therefore identical.

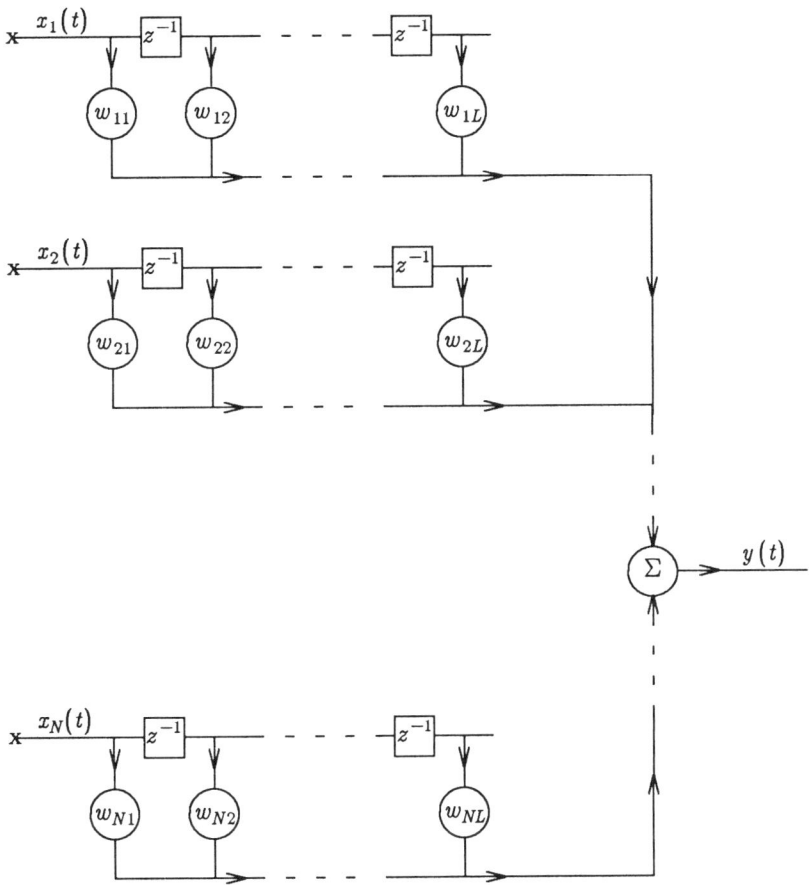

Figure 8.4.4 A broadband array processor. Each input is processed by a separate tapped delay line.

The point is that as far as the broadside signal is concerned, the array is equivalent to a single tapped delay line in which each weight, u_k, say, is equal to the sum of the weights in the corresponding column. That is,

$$u_k = \sum_{i=1}^{N} w_{ik} \, . \tag{8.4.20}$$

Consequently, it is these summation weights that must be constrained to give the desired response. In general, of course, signals incident from other directions will not arrive in phase at the corresponding taps, and will thus be treated differently.[9]

The simplest situation occurs when the desired constraint is that a plane wave incident at broadside is passed by the array with unit gain and no delay, while outputs due to other inputs are minimized. Referring to Figure 8.4.4 again, it is clear that for this case the constraint simply translates to

$$\sum_{i=1}^{N} w_{i1} = u_1 = 1 \; ,$$
$$\sum_{i=1}^{N} w_{ij} = u_j = 0 \qquad ; j > 1 \; . \tag{8.4.21}$$

A more general constraint can be achieved by applying (8.4.20) for $k = 1, 2, ..., L$

$$\sum_{i=1}^{N} w_{i1} = u_1, \; \sum_{i=1}^{N} w_{i2} = u_2, ..., \sum_{i=1}^{N} w_{iL} = u_L \; .$$

Define an augmented LN element weight vector

$$\underline{w} = [w_{11}, w_{21}, ..., w_{N1}, w_{12}, ..., w_{NL}]^t \; , \tag{8.4.22}$$

and a vector of constants

$$\underline{c}_j = [0, 0, ..., 0, 0, ...0, 1, 1, ..., 1, 0, 0, ..., 0]^t \; . \tag{8.4.23}$$

$\underline{c}_j$ is an $(LN \times 1)$ vector with the j-th group of N elements equal to unity, the rest are zeros. Now, we can collect these vectors into an $(LN \times L)$ matrix C as

$$C = [\underline{c}_1, \underline{c}_2, ..., \underline{c}_L] \; , \tag{8.4.24}$$

and the look direction constraint is

---

9. Note that if it is desired to steer the array in one particular direction, this can be achieved by pre-steering the array by inserting suitable delays prior to the constrained processor.

$$\underline{u}=[u_1,u_2,...,u_L]^t \ . \tag{8.4.25}$$

The constraint of equation (8.4.20) can then be written as

$$C^t\underline{w}=\underline{u} \ . \tag{8.4.26}$$

The overall problem is then to minimize the total output power

$$E\{y^2(n)\}=E\{\underline{w}^t\underline{x}_n\underline{x}_n^t\underline{w}\}=\underline{w}^tR\underline{w} \ , \tag{8.4.27}$$

subject to the constraint (8.4.26) where, for this broadband case, the vector $\underline{x}_n$ is an augmented to an NL element data vector:

$$\underline{x}_n=[x_1(n),...,x_N(n),x_1(n-1),...,x_{N-1}(n-L+1)]^t,$$

where the notation $x_i(n-j+1)$ denotes the sample from element i corresponding to time index $n-j+1$. The minimization is performed using Lagrange multipliers. We define a functional:

$$J=\frac{1}{2}\underline{w}^tR\underline{w}+\underline{\lambda}^t(C^t\underline{w}-\underline{u}) \ , \tag{8.4.28}$$

where $\underline{\lambda}$ is the $(L\times1)$ vector of Lagrange multipliers and where the $1/2$ is introduced for convenience. As usual, the minimization is achieved by differentiating J with respect to $\underline{w}$ and equating to zero. This yields

$$\nabla J=2(R\underline{w}+C\underline{\lambda})=0 \ , \tag{8.4.29}$$

or, rearranging

$$\underline{w}_{opt}=-R^{-1}C\underline{\lambda} \ . \tag{8.4.30}$$

Since $\underline{w}_{opt}$ must satisfy the constraint equation (8.4.26), we have

$$-C^tR^{-1}C\underline{\lambda}=\underline{u} \ ,$$

or

$$\underline{\lambda}=-[C^tR^{-1}C]^{-1}\underline{u} \ ,$$

and therefore from (8.4.30)

$$\underline{w}_{opt}=R^{-1}C[C^tR^{-1}C]^{-1}\underline{u} \ . \tag{8.4.31}$$

Adaptive Implementation: Frost's Algorithm [14]

We begin by choosing an initial set of weights $\underline{w}_0$ that satisfy the constraint equation (8.4.26). A suitable choice is

$$\underline{w}_0 = C(C^t C)^{-1} \underline{u} \ . \tag{8.4.32}$$

A gradient based update is given by

$$\underline{w}_{n+1} = \underline{w}_n - \frac{\alpha}{2} \nabla J_n \ , \tag{8.4.33}$$

where J_n is equal to the functional (8.4.28) with $\underline{w}$ replaced by $\underline{w}_n$. Hence, from (8.4.29)

$$\underline{w}_{n+1} = \underline{w}_n - \alpha [R\underline{w}_n + C\underline{\lambda}_n] \ . \tag{8.4.34}$$

Assuming for a moment that R were available, $\underline{\lambda}_n$ would be chosen by requiring that $\underline{w}_{n+1}$ satisfies the constraint equation (8.4.26), so that

$$\underline{u} = C^t \underline{w}_{n+1} = C^t \underline{w}_n - \alpha C^t R\underline{w}_n - \alpha C^t C\underline{\lambda}_n \ .$$

Solving for $\underline{\lambda}_n$ and substituting into (8.4.34) yields

$$\underline{w}_{n+1} = P[\underline{w}_n - \alpha R\underline{w}_n] + C(C^t C)^{-1} \underline{u} \ , \tag{8.4.35}$$

where

$$P = I - C(C^t C)^{-1} C^t \ . \tag{8.4.36}$$

In practice, R is unavailable and is replaced by the instantaneous estimate $\hat{R} = \underline{x}_n \underline{x}_n^t$. This yields the Frost algorithm [14]

$$\underline{w}_{n+1} = P[\underline{w}_n - \alpha y(n)\underline{x}_n] + \underline{w}_0 \ , \tag{8.4.37}$$

$$P = I - C(C^t C)^{-1} C^t \ , \tag{8.4.36}$$

$$\underline{w}_0 = C(C^t C)^{-1} \underline{u} \ . \tag{8.4.32}$$

In the simplest case of the distortionless response with a linear-phase constraint, the optimal processor is also known as the **Minimum Variance Distortionless Look** estimator, or as the

Maximum Likelihood Distortionless Estimator.

The update (8.4.37) has a different form to that of the LMS due to the presence of the constraint. The computation is not so onerous as may be first imagined however, since C and $\underline{u}$ are both fixed and C has a very simple structure. The update satisfies

$$C^t \underline{w}_{n+1} = \underline{u} \ ,$$

at each iteration, as can easily be verified by multiplying both sides of equation (8.4.37) by C^t. For zero-mean stationary inputs, and with the usual independence assumptions, Frost [14] has demonstrated convergence in the mean to $\underline{w}_{opt}$ as given by (8.4.31).

Griffiths-Jim Adaptive Beamformer [15]

Griffiths and Jim [15] developed an alternate implementation of a constrained power minimization beamformer which they called a **generalized side-lobe canceller**. This structure is particularly powerful and flexible, including the Frost beamformer as a special case, but allowing constrained processing with the unconstrained adaptive algorithm. The scheme is depicted in Figure 8.4.5. Referring to the diagram, the delays $\Delta_1, \Delta_2, ..., \Delta_N$, are calculated to steer the array main-beam in the direction of the desired signal. The upper component of the system in the diagram is a conventional beamformer using fixed weights $w_{c1}, w_{c2}, ..., w_{cN}$. These produce an output

$$y_1(n) = \underline{w}_c^t \chi_n \ , \tag{8.4.38}$$

where $\chi_n = [\chi_1(n), \chi_2(n), ..., \chi_N(n)]^t$ are the (steered) inputs and where $\overline{w}_c = [w_{c1} w_{c2}, ..., w_{cN}]^t$. This signal is in turn filtered using a tapped delay line of constraint weights taking the form:

$$y_2(n) = \sum_{m=0}^{L-1} h(m) y_1(n-m) \ , \tag{8.4.39}$$

where $h(m)$ are constraint weights, and L is the number of taps in the filter. The lower section of the processor of Figure 8.4.5 is the side-lobe canceller. The same steered inputs $\underline{x}$ are transformed and then applied to a broadband adaptive beamformer. The transformation is represented via a matrix T as

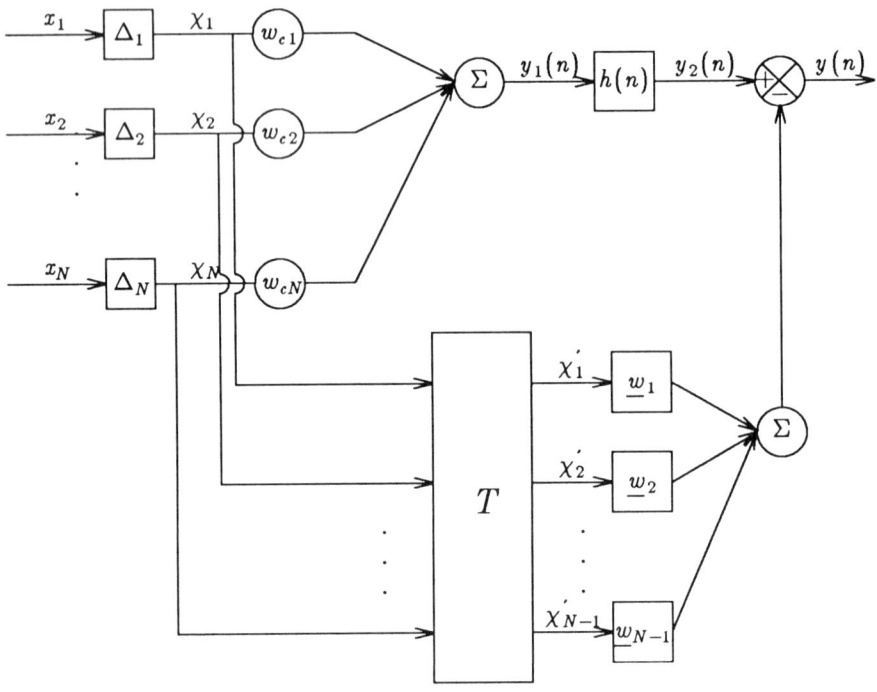

Figure 8.4.5 Griffiths-Jim adaptive beamformer.

$$\underline{\chi}'_n = T \underline{\chi}_n \ . \tag{8.4.40}$$

The purpose of the transformation is to eliminate the signal component $s(n)$ from the inputs to the adaptive beamformer. This is achieved by designing T such that the elements $T(i,j)$ satisfy:

$$\sum_{j=1}^{N} T(i,j) = 0 \ . \tag{8.4.41}$$

That is, the sum of each row of T is equal to zero. This constraint ensures that components arriving in the desired signal direction do not occur in the inputs to the adaptive beamformer. It can easily be verified that the condition (8.4.41) achieves this objective since any signal from the look direction will be in phase after steering delays. Thus, for such a signal

$$\underline{\chi}_n = [s(n), s(n), ..., s(n)]^t ,$$

and hence for such an input

$$\underline{\chi}'_n = T\underline{\chi}_n = \underline{0} .$$

This latter equality follows as a consequence of the constraint (8.4.41). Because of this signal blocking capacity, T is referred to as a **blocking operator**. As indicated in the diagram, T has at most $(N-1)$ rows, so that $\underline{\chi}'_n$ has at most $(N-1)$ elements. The outputs from the blocking operation provide the inputs to a broadband adaptive beamformer with weights $\underline{w}_1, \underline{w}_2, ..., \underline{w}_N$, where $\underline{w}_i = [w_{1i}, w_{2i}, ..., w_{(N-1)i}]^t$. The overall system output $y(n)$ is obtained by subtracting the adaptive beamformer output $y_b(n)$, from $y_2(n)$ of equation (8.4.39):

$$y(n) = y_2(n) - y_b(n) . \tag{8.4.42}$$

It is this system output which is used as the error sequence to drive the adaptive beamformer. Note that the constraint is produced by the blocking operation so that adaptation can take place using the unconstrained algorithm. The update for the adaptive beamformer is thus

$$\underline{W}_{n+1} = \underline{W}_n + \alpha y(n)\underline{\varsigma}_n , \tag{8.4.43}$$

where $\underline{W}_n = [\underline{w}_1^t, \underline{w}_2^t, ..., \underline{w}_L^t]$ is the update vector for index n, and where

$$\underline{\varsigma}_n = [\chi'_1(n), ..., \chi'_{N-1}(n), \chi'_1(n-1), ..., \chi'_{N-1}(n-L+1)]^t.$$

We may be confident that signals from the desired direction will not be cancelled by this processor, because they are not present in the input to the adaptive beamformer due to the effect of the blocking matrix. Note that this structure has parallels with adaptive noise cancellation (see Section 5.1) − blocking the signal component of the input creates a 'reference' signal that consists of the noise only. This reference is filtered, and the result is subtracted from the primary (the beamformed input).

The transformation matrix T may take many forms. We can interpret the rows of T as fixed weightings (beamformers) applied to the inputs. In this view, the constraint (8.4.41) ensures that a null is directed in the signal direction. One simple and intuitively appealing operator is the $[(N-1) \times N]$ matrix:

$$T = \begin{bmatrix} 1 & -1 & 0 & 0 \\ 0 & 1 & -1 & 0 \\ . & 0 & . & . \\ . & . & . & 0 \\ 0 & 0 & 1 & -1 \end{bmatrix}. \qquad (8.4.44)$$

Computationally, this matrix is attractive since its application involves only addition and subtraction operations. This matrix certainly satisfies the constraint equation (8.4.41), and actually corresponds to pairwise subtraction of adjacent steered inputs. Each pairwise subtraction acts as a two-element beamforming operation with a null in the look direction. Consequently, the look direction component of the input is removed and the system satisfies the requirements for a noise cancelling system reference, being entirely noise components. Note that it can be shown that the Griffiths-Jim algorithm is strictly equivalent to Frost's algorithm if, in addition to the condition (8.4.41), the rows of T are mutually orthogonal, and if the fixed beamformer uses uniform weighting $w_{ci} = 1/N$ for $i = 1, 2, ..., N$. The great advantage of the Griffiths-Jim algorithm under these conditions, is that the constrained processor is implemented using the far simpler unconstrained LMS update.

Problems

8.1. Show that the modulus of the steered array response for an N element linear equispaced array with inter-element spacing d, to a monochromatic plane wave incident from angle θ_0 to broadside can be written:

$$|A(\theta)| = \left| \frac{\sin[\frac{\pi N d}{\lambda}(\sin\theta - \sin\theta_0)]}{\sin[\frac{\pi d}{\lambda}(\sin\theta - \sin\theta_0)]} \right|.$$

8.2 a) Derive an expression for the array pattern function for a 7 element linear equispaced array with element spacing d. Sketch the result for:

i) $d=\lambda/2$

ii) $d=\lambda/4$.

b) Give the required delay values and sketch the result if the main beam is rotated to $\pi/7$ radians.

c) Find the pattern function if the array is changed to the four element array shown:

$$\text{x} \quad \text{x} \quad . \quad . \quad \text{x} \quad \text{x}$$
$$d=\overleftrightarrow{\lambda/4}$$

(Note: 'x' refers to an array element, '.' indicates a missing element.)

8.3 A signal is incident to a two element array at an angle 45^0. An interference is incident from $\theta_1=135^0$. Write down the set of equations whose solution gives the required weights. Why can you not solve the equations? Give a physical interpretation to your answer.

8.4 a) Derive the overall array pattern for the array shown (assume equispaced sensors)

$$\text{x} \quad \text{x} \quad \text{x} \quad \text{x} \quad . \quad . \quad . \quad . \quad . \quad \text{x} \quad \text{x} \quad \text{x} \quad \text{x}$$
$$d=\overleftrightarrow{\lambda/2}$$

(Note: 'x' refers to an array element, '.' indicates a missing element.)

b) Give the beam pattern for either of the two sub-arrays.

c) Give the beam pattern assuming each sub-array is replaced by a single omni-directional sensor.

d) Show that the product of the results in **b)** and **c)** is equal to the overall array pattern.

8.5 **a)** Derive an expression for the array pattern function $A(\theta)$ for a three element linear equispaced array with inter-element spacing $d=\lambda/2$. Sketch the result as a function of θ.

   b) If complex weights w_1, w_2, w_3 are applied to the output of the array, design a set of weights to steer the main-lobe of the beam to 30^0 from broadside, and sketch the result.

   c) If a monochromatic plane wave signal is incident to the array at an angle 30^0 to the vertical, design a set of weights which preserve the incident signal, while nulling interfering plane waves from angles 60^0 and 90^0.

   d) If, in addition to the signals and interferences described in **c)**, spatially uncorrelated zero-mean broadband random noise with variance σ_v^2 is present at the sensors, set up (but do not solve) the equations for the least-squares optimal weight set.

8.6 A two element line array is to be steered to an angle 30^0 from broadside as shown in the diagram:

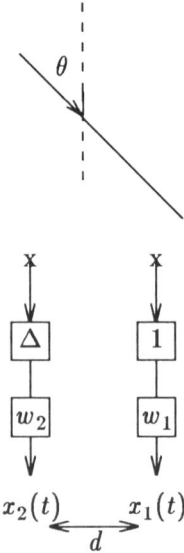

   a) Assuming $d=\lambda/2$, calculate the delay Δ required to produce the desired steering.

b) Subsequent to the steering, complex weights are applied to the signals. Calculate the weights to null an interference at $\theta=60^0$, while preserving a 30^0 signal. (Again, assume $d=\lambda/2$.)

8.7 A monochromatic plane wave interference is incident to a two element array from an angle $\theta=30^0$. Give a general solution for weights w_1, w_2 which will ensure the interference is nulled. (Assume $d=\lambda/2$.)

8.8 Consider the 3 element array shown in the diagram

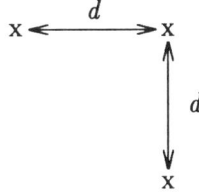

a) Derive an expression for the magnitude of the array response to a plane wave incident at angle θ to the vertical. (Assume $d=\lambda/2$.)

b) Assume the outputs from the array elements are weighted using complex weights w_1, w_2, w_3. Set up and solve equations to determine weights that will preserve a signal incident at angle $\theta=0$ while nulling signals from $\theta=\pi/4$ and $\theta=\pi/2$.

c) Describe how you would approach this problem if, in addition to interferences at $\theta=\pi/2$ and $\theta=\pi/4$, the element outputs were corrupted by zero-mean stationary broadband spatially uncorrelated noise.

8.9 For each of the following examples, find the set of weights w_1, w_2, w_3 that will null interferences θ_1, θ_2 while preserving a broadside signal:

i) $\theta_1=\pi/4$, $\theta_2=\pi/2$.

ii) $\theta_1=\pi/3$, $\theta_2=\pi/2$.

8.10 For the array shown, set up and solve equations defining weights w_1, w_2, w_3 that will preserve a broadside plane wave while nulling interferences (of the same wavelength) at angles 30^0 and 90^0.

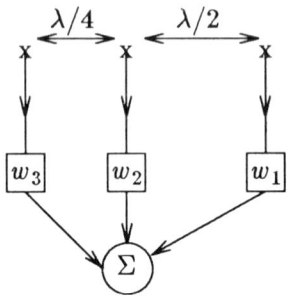

8.11 A signal is incident to an N element array at an angle 30^0. The signal is corrupted by spatial uncorrelated noise with variance σ_v^2. Find the optimal Maximum SNIR weights for the array.

8.12 A narrowband signal is incident to an array at an angle 60^0. The signal is corrupted by broadband spatially uncorrelated noise with variance

$$E\{v_i^2(n)\}=\sigma_i^2 \quad ; i=1,2,...,N \ ,$$

where subscript i refers to sensor index. Find the optimal maximum SNIR weights for the array.

8.13 For an N element linear array, show that using the real formulation for the signals, the set of array weights that minimizes the mean-squared error between the array output and a desired (pilot) signal is

$$\underline{w}_{ls}=R^{-1}\underline{g} \ ,$$

where

$$\underline{g}=E\{d(t)\underline{x}(t)\} \ , \quad R=E\{\underline{x}(t)\underline{x}^t(t)\} \ .$$

8.14 Consider the Griffiths beamformer with update

$$\underline{w}_{n+1}=\underline{w}_n+\alpha\underline{g}-\alpha y(n)\underline{x}_n \ .$$

Show that for $x(n)$ and $d(n)$ zero-mean, jointly stationary then

$$\lim_{n\to\infty} \{E\{\underline{w}_n\}\} \to R^{-1}\underline{g} \ .$$

8.15 For the Frost beamformer with N elements (linear equispaced), with each element connected to a tapped delay line with L taps, set up a constraint equation

$$C^t\underline{w}=\underline{u} \ ,$$

that will amplify an input broadside plane wave by a factor of 3 and will delay the output by 4 samples. If the input consists of white noise which is also uncorrelated from sensor to sensor, find an explicit form for the optimum weight vector.

8.16 Given an N element equispaced linear array with weights w_i for $i=1,2,...,N$, use the Lagrange multiplier method to show that for a broadside signal corrupted by noise $v(n)$ with correlation matrix R_{vv}, the MVDR beamformer is

$$\underline{w}_{MV}=AR_{vv}^{-1}\underline{1} \ ,$$

where $A=1/\underline{1}^t R_{vv}^{-1}\underline{1}$ is a scalar multiplier with $\underline{1}=[1,1,...,1]^t$. Show further that the output variance associated with $\underline{w}_{MV}$ is

$$Var\{output\}=\frac{1}{\underline{1}^t R_{vv}^{-1}\underline{1}} \ .$$

8.17 A monochromatic plane wave signal s is incident to a two sensor linear array with inter-element spacing $d=\lambda/2$, at an unknown angle θ. The signal is corrupted by omni-directional measurement noise at each sensor. The measurements are

$$x_i=s_i+v_i \quad ; i=1,2 \ ,$$

where the noise v_i is real, zero-mean and independent from sensor to sensor. If the spatial correlation matrix $R_{xx}=E\{\underline{x}^*\underline{x}^t\}$ has the form

$$R_{xx}=\begin{bmatrix} 2 & -j/2 \\ j/2 & 2 \end{bmatrix} \ ,$$

find the unknown signal direction and the noise power.

8.18 a) For the Frost adaptive beamformer, find a set of constraints $C^t w = u$, equivalent to filtering a broadside plane wave by a transfer function $H(z) = 1 + az^{-1}$. Give explicit forms for C and $\underline{u}$.

    b) If the input to the array consists entirely of broadband spatially uncorrelated noise, give an explicit form for the optimal weight vector.

8.19 A two element Griffiths-Jim beamformer operates as shown in the diagram:

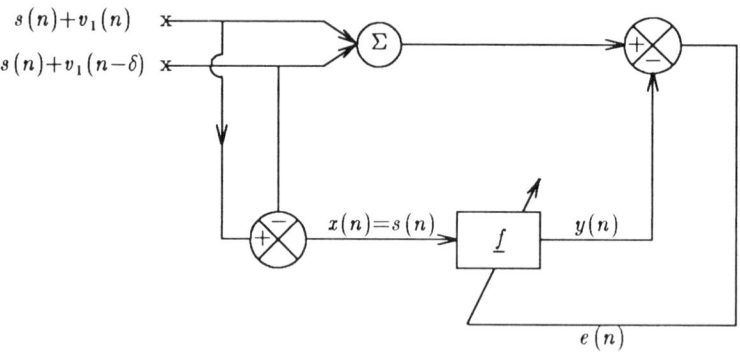

where $s(n) = A\cos\omega_0 nT$, $v(n) = B\cos\omega_1 nT$. If the filter has two coefficients

$$\underline{f} = [f_0, f_1]^t ,$$

find the fixed least-squares solution for the filter $\underline{f}$ and the output y. What happens when $\omega_1 = \omega_0$?

Computer Problem 1

Part 1 — Write a routine to calculate and plot the amplitude response of an array to a unit amplitude narrowband signal at frequency ω_0, with weights w_1, w_2,..., w_N. The inputs to this routine should be the number of weights N, the weight values, frequency ω_0, and the sample interval between plotted points $\delta\omega$. Plot the responses with $\delta\omega = \pi/100$ for $N = 5$ and $N = 8$

with $\omega_i = 1$. Compare your results with analytic predictions obtained using the results of Section 8.1. Using the example of $N=8$, plot the response when element number 6 is absent. Is there a significant impact on the response?

Part 2 − Implement complex form updates for the one and two mode adaptive beamformers. The input signals should be:

    i) Desired signal direction θ.
    ii) Number of elements N.
    iii) Inter-element spacing d.
    iv) Center frequency ω_0.
    v) Sampled input signal $x(n)$.

For the two-mode beamformer, an additional input is provided by the switching interval n_1 (samples). Examine the behavior of your system by plotting the array response using the routine developed in **Part 1** above, every 100 updates. Compare the performance of the two systems when the inputs are:

    i) A monochromatic unit amplitude plane wave signal incident at broadside and Gaussian unit variance white noise, independent from sensor to sensor.

    ii) As *i)* with an additional narrowband interference of amplitude 5, incident at 40^0 to broadside.

For the two-mode algorithm try $n_1=100$. In both cases use $N=13$, $d=\lambda/2$, and perform an ad-hoc search for the 'best' value for α for each algorithm. What can you conclude about the relative merits of the two procedures?

References

1.   *Array Signal Processing*, S.Haykin (Ed), Prentice-Hall, 1985.

2.   D.E.Dudgeon, "Fundamentals of digital array processing,"
      Proc. IEEE, vol. 65, pp. 898-904, 1977.

3.   R.A.Monzingo and T.W.Miller, *Introduction to Adaptive
      Arrays*, Wiley, 1980.

4.   J.E.Hudson, *Adaptive Array Principles*, Peter Peregrinus,
      1981. Reprinted 1989.

5.   R.T.Compton, Jr., *Adaptive Antennas: Concepts and
      Performance*, Prentice-Hall, 1988.

6.   S.P.Applebaum, "Adaptive arrays," *IEEE Trans. Antennas,
      Propag.*, vol. AP-24, pp. 585-598, 1976.

7.   J.W.R.Griffiths, "Adaptive array processing: A tutorial,"
      IEE Proc. Parts F and H, vol. 130, pp. 3-10, 1983.

8.   B.D.Steinberg, *Principles of Aperture and Array System
      Design*, Wiley, 1976.

9.   W.S.Burdic, *Underwater Acoustic System Analysis*,
      Prentice-Hall, 1984.

10.   R.A.Horn and C.R.Johnson, *Matrix Analysis*, Cambridge
      University Press, 1985.

11.   B.Widrow and S.D.Stearns, *Adaptive Signal Processing*,
      Prentice-Hall, 1985.

12.   B.Widrow, P.E.Mantey, L.J.Griffiths and B.B.Goode,
      "Adaptive antenna systems," *Proc. IEEE*, vol. 55, pp. 2143-
      2159, 1967.

13.   L.J.Griffiths, "A simple adaptive algorithm for real-time
      processing in antenna arrays," *Proc. IEEE*, vol. 57, pp.
      1696-1704, 1969.

14.   O.L.Frost III, "An algorithm for linearly constrained
      adaptive array processing," *Proc. IEEE*, vol. 60, pp. 926-935,
      1972.

15.   L.J.Griffiths and C.W.Jim, "An alternative approach to
      linearly constrained adaptive beamforming," *IEEE Trans.
      Antennas, Propag.*, vol. AP-30, pp. 27-34, 1982.

Index

adaptation constant, 164, 341

Adaptive Differential Pulse Code Modulation (ADPCM), 251, 282, 358

adaptive equalization, 282

adaptive line enhancement, 217

adaptive noise cancellation, 205

    for tow-ship noise, 210

    in rooms, 210

    without external reference, 213

Akaike's Information Criterion (AIC), 438

all-pass system, 43, 123

analysis/synthesis, 32, 111

Applebaum array, 508

array processing, 465

array

    adaptive, 501

    broadband, 500, 509

    equivalence of optimal solutions, 497

    Frost's algorithm, 512

    generalized side-lobe canceller, 513

    grating lobes, 476

    Griffiths adaptive beamformer, 507

    Griffiths-Jim algorithm, 513

    constrained power minimization, 508

    least-squares design, 494

    linear, 466

    main-beam and side-lobes, 474

    maximum SNIR design, 491

    narrowband, 467, 500

    optimal designs, 487

    pattern function, 471

    phase center, 471

    shading, 480

    upper frequency operating limit, 477

autocorrelation (see correlation)

autocorrelation and autocovariance formulations, 93

autoregressive (AR) sequence, 30

AR spectral estimates, 431

model order, 437

    and maximum entropy, 438

    relation to correlogram, 439

autoregressive moving average (ARMA) sequence, 30

ARMA spectral estimates, 441

Averaged LMS (ALMS), 287

averaging, 10

Bartlett window, 415

Bartlett's method (for spectral estimation), 421

Bayes rule, 73

beamforming, 471

    delay and sum, 479

beam steering, 478

bit rate, 251

bivariate distribution, 9

Blackman-Tukey spectral estimate, 426, 439

Block LMS (BLMS) algorithm, 296, 356

blocking operator, 515

Burg Method, 435

Cauchy distribution, 19

Cauchy's residue theorem, 57

Cholesky factorization, 98, 132

circular convolution, 355, 373

clipped LMS, 285

coherence, 216

    magnitude squared, 238, 364

    smoothed transform, 364

complex LMS algorithm, 294

complex random signal, 21

consistency, 51

consistent (in mean-square), 69

correlation, 12, 15, 21

    estimate, 54, 97, 104, 110, 115, 433

    lag, 16

    of AR sequence, 38

    sample, 27

correlation matrix, 13, 19

    diagonalization, 169, 192

    eigenvalues and eigenvectors, 162, 169